AF803243

Hunter-Gatherers of the Congo Basin

Hunter-Gatherers of the Congo Basin

Cultures, Histories, and Biology of African Pygmies

Barry S. Hewlett, editor

Routledge
Taylor & Francis Group
New York London

First published 2014 by Transaction Publishers

Published 2017 by Routledge
711 Third Avenue, New York, NY 10017, USA
2 Park Square, Milton Park, Abingdon, Oxon OX14 4RN

First issued in hardback 2017

Routledge is an imprint of the Taylor & Francis Group, an informa business

Copyright © 2014 by Taylor & Francis.

All rights reserved. No part of this book may be reprinted or reproduced or utilised in any form or by any electronic, mechanical, or other means, now known or hereafter invented, including photocopying and recording, or in any information storage or retrieval system, without permission in writing from the publishers.

Notice:
Product or corporate names may be trademarks or registered trademarks, and are used only for identification and explanation without intent to infringe.

Library of Congress Catalog Number: 2013035960

Library of Congress Cataloging-in-Publication Data

Hunter-gatherers of the Congo Basin : cultures, histories and biology of African Pygmies / Barry S. Hewlett, editor.
pages cm
1. Pygmies--Congo River Watershed--History. 2. Pygmies--Congo River Watershed--Social life and customs. 3. Hunting and gathering societies--Congo River Watershed. 4. Ethnology--Congo River Watershed. 5. Human biology--Congo River Watershed. 6. Congo River Watershed--Ethnic relations. 7. Congo River Watershed--Social life and customs. I. Hewlett, Barry S., 1950- editor of compilation, author.
DT650.P94H66 2014
305.896067--dc23

2013035960

ISBN 13: 978-1-4128-5361-3 (pbk)
ISBN 13: 978-1-1385-2556-6 (hbk)

Contents

List of Tables

List of Figures

Acknowledgments

The idea and inspiration for this book emerged at an International Congo Basin Hunter-Gatherer Conference held in Montpellier, France, in September 2010, co-organized by myself, Edmond Dounias, and Kiyoshi Takeuchi. I would like to sincerely thank my co-organizers for the immense amount of time and effort they devoted to making it a vibrant, stimulating, and fun conference. The chapters in the book are not papers from the conference. Leading authorities on topics or theories dealing with Congo Basin hunter-gatherers at the conference were instead asked to write summary overviews and critiques of their areas of specialization.

It was an enormous pleasure to work with all of the authors of the various chapters. The backgrounds, theories, and views of Congo Basin hunter-gatherers vary dramatically between authors, but all were very friendly, responsive, and open to comments and suggestions. It was an honor to work with such an outstanding and professional group of scholars.

The book would not have been possible without the editorial assistance of WSU graduate student Scott Calvert. He carefully read and edited chapters from authors with English as a second language and made insightful and useful comments all along the way. I want to thank Jennifer Nippins at Transaction for her patience in submitting the book and to Hannah Jones for all of her enthusiastic support in the production process.

Of course, none of this book would have been possible without the hospitality, graciousness, and tolerance of the forest peoples from the Congo Basin. They allowed researchers into their homes to conduct (often strange) research on a wide range of topics. We all want to gratefully acknowledge all of this assistance and hope that at least some of the research will be useful to them and to development agencies trying to assist them in the future.

Foreword

Luigi Luca Cavalli-Sforza

It was for me a tremendous pleasure to be asked to write the foreword to this book, *Hunter-Gatherers of the Congo Basin*, generated by my old friend and collaborator, Barry Hewlett. This book represents the work of many old friends of both of us, each of whom contributed research on aspects of different hunter-gatherers and is mostly based on direct field experience with the populations who are the subject of this book.

It may come as a surprise that in order to understand the origin of our species it is necessary to examine African populations, principally taking interest in the direct descendants of their most ancient ancestors. In fact, even if we cannot give precise dates of origin, it is clear that our species originated relatively recently from African ancestry and that a few direct descendants of these early ancestors are still living in basic subsistence-level conditions in the areas where their evolutionary process most likely started.

The changes underlying modern societies have taken place in relatively little time. The rapid development of new living conditions was made possible by the invention of food production, which took only a few thousand years to spread through the world, bringing about a fast and strong cultural and social evolution that caused a substantial modification of the environment in which we live.

My own research with African Pygmies and the travels to visit many other aboriginal peoples have convinced me that we can learn immensely from them, far beyond genetic knowledge. Their careful management of the environment, the deep acquaintance with plants and animals around them, their peaceful style of life, unhindered by statutory laws but based on mutual respect and recognition of equality, are in many ways their cultural legacy to us, the contemporary descendants of the farmer-breeders who replaced hunter-gatherers, built hierarchic societies, and remodeled the face of the planet.

These surviving, transformed groups of hunter-gatherers embody the wisdom of our own antiquity. The changes we have inflicted on our environment will force us to be more careful in avoiding further alterations of the ecosystem in order to continue to live reasonably well, and hopefully increasingly better, in the surroundings where we live at present. It is of course possible, or even likely, that in the course of time we will, or will have to, look for other planets where to expand.

Aka woman with spear, large knife, and baby. (Courtesy of B. Hewlett)

Aka men with hunting nets. (Courtesy of B. Hewlett)

Introduction

The African Pygmies of the Congo Basin are the largest and most diverse group of active hunter-gatherers that remain in the world. At least fifteen different ethnolinguistic groups exist in the Congo Basin, with a total population of 250,000 to 350,000 individuals (see figure 1.1 and table 1.1 in chapter 1 for names, locations, and populations of major ethnic groups). Because it has the largest group of hunter-gatherers in the world, research in this region can provide detailed and rich insights into the commonalities and diversity of a way of life that has characterized most of human history. Pygmies may be able to increase our understanding of the past, but they are modern humans facing the modern human problems of globalization and exploitation. This volume addresses regional and global hunter-gatherer research questions as well as the contemporary social issues confronting foragers in the Congo Basin.

This is the first text to provide a general overview of both the cultural and biological perspectives of African Pygmies. Existing monographs cover one or a few groups and are limited to specialized topics—Bahuchet (1985) focused on Aka ethnoecology, Bailey (1991) described Efé men's subsistence, Hewlett (1991) detailed Aka parenting, and Turnbull (1965) provided vivid descriptions of Mbuti social organization. Cavalli-Sforza (1986) edited the first contemporary book on several African Pygmy groups, but it was limited to biological topics such as genetics, health, and growth. *Hunter-Gatherers of the Congo Basin* is distinct in that it provides a comprehensive overview of current research on Congo Basin hunter-gatherers—from music, social organization, and childcare to health, genetics, and history.

This book is timely because knowledge and understanding of Congo Basin hunter-gatherers' ways of life have accumulated and increased dramatically since the 1970s. The volume has three cohorts of researchers: a senior cohort of Bahuchet, Hewlett, and Ichikawa, who started their research in the early 1970s; an intermediate cohort of Froment, Takeuchi, Moïse, Fürniss, and Sato, who began their research in the 1980s; and a young cohort of Lupo, Verdu, Rupp, and Lewis, who first conducted research in the 1990s. The authors have from at least ten to more than forty years of field experience in central Africa, have made repeated trips to the region, and have supervised several PhD dissertations on Congo Basin hunter-gatherers. Knowledge has increased substantially over the years, but it is widely dispersed in book chapters and journals in different languages. French, Japanese, American, and British researchers have conducted the majority of the research on Congo Basin hunter-gatherers in the last forty years, and each national research group has its own academic traditions, history, and publications. This volume assembles and summarizes the last forty years of research of prominent scholars from around the world.

Hunter-Gatherers of the Congo Basin addresses both topical and theoretical issues. Leading academic authorities provide overviews on the relatively characteristic features of Congo Basin hunter-gatherer life, such as their yodeled polyphonic music (chapters 1 and 7), egalitarianism (chapter 8), multiple child caregiving (chapter 9), and complex relations with neighboring farming groups (chapters 1, 3, 4, and 11). Other chapters address regional and global theoretical questions, such as, why are Pygmies short? (chapters 2 and 5); can tropical forest hunter-gatherers live without the carbohydrates they receive from neighboring farmers? (the "wild yam hypothesis" in chapters 3 and 6); and how do hunter-gatherer children learn to share so extensively? (chapter 9).

Finally, several chapters provide useful research for development agencies trying to understand and design policies to serve the "first peoples" of the Congo Basin. European and the US conservation groups have substantially increased the number of tropical forest parks and reserves in the Congo Basin, and this has led to displacement or dramatically impacted the lifeways of forest peoples. The "bushmeat trade" and transmission of "emerging" tropical rainforest infectious diseases, such as Ebola, are prominent issues in both conservation and public health agencies, and policies on both topics profoundly impact Congo Basin hunter-gatherers. Humanitarian groups, such as UNICEF, have "discovered" the needs of the Pygmies and have started a range of interventions to decrease their marginalization and exploitation. Missionary activities in formal education and health care have expanded in several rural areas occupied by forest foragers because impoverished central African countries are unable to provide public services in these areas or Pygmies are not allowed access to these services because of discrimination or marginalization. Many of the intervention programs target the hunter-gatherers.

The conservationists, humanitarian groups, national governments, and missionaries all intervene in the lives of Pygmies, often with little or no research guiding or informing their policies. All of the chapters should be of interest to Congo Basin development agencies, but the chapters on cultural diversity (chapter 1), political history (chapter 4), health (chapter 5), forager-farmer relations (chapter 11), tropical forest conservation (chapter 12), and representation (chapter 10) should be particularly instructive in establishing public policies.

Terminology

Pygmy

Historically, the groups discussed in this book have been referred to as "Pygmies." The word originated in Europe and emphasizes stature. Homer first used the word "Pygmy" in the Iliad, coming from the Greek word *pugon* (cubit or about 30 cm), to refer to mythical short people who fought cranes. The term was reintroduced by Georg Schweinfurth in 1873 and subsequently adopted by European and American explorers in the Congo Basin in the 1800s to describe the short-statured rainforest hunter-gatherers they encountered in their exploratory travels (see chapters 1, 2, and 5 for more details of the history of Pygmy studies and the use of the term).

Unfortunately, although numerous alternative terms have emerged, none have been agreed upon by academics or the people themselves to replace it. Researchers actively debate whether to use the term *Pygmy* in their publications. French biological and social anthropologists recently decided to use the term *Pygmy* for a collection of papers in *Journal des Africanistes* (2012) because they felt the term was globally recognized, that not all Pygmies were hunter-gatherers, and not all of them lived in the rainforest. Some British and central African social anthropologists, politically active nongovernmental organizations (NGOs), and some development agencies also regularly use the term *Pygmy* because it is widely recognized by the public, is often used to refer to the first peoples in Congo Basin countries, and strengthens the peoples' political position as they seek land rights.

Pygmy is a popular term and easily recognized by a broad group of people in Europe, Japan, the United States, and Africa, and that is why it is in the subtitle of this book. International and local NGOs that use the term *Pygmy* in their titles or literature include the following: Pygmy Survival Alliance, Forest Peoples' Programme, Survival International, Rainforest Foundation, Réseau Recherches Actions Concertées Pygmées, Centre d'Accompagnement des Autochtones Pygmées et Minoritaires Vulnérables, and the Association for the Development of Pygmy Peoples of Gabon. Congo Basin conservation groups, such as World Wildlife Fund and Wildlife Conservation Society, and international human rights groups working in the region, such as UNICEF and Integrated Regional Information Networks (IRIN), also regularly use the term *Pygmy* in their literature.

Some academics and Central African government officers feel *Pygmy* is derogatory or does not adequately represent the people. Republic of Congo (ROC) has outlawed the use of the term because it is viewed as pejorative. Some Cameroonian officials indicate the term is used as an insult and instead use the term "four Bs" to refer to the four forager groups in Cameroon (Baka, Bakola, Bagyeli, and Bedzan) (Robillard and Bahuchet 2012). Pygmy is used somewhat less frequently by US and Japanese anthropologists for a wide variety of reasons; some feel it is a colonial and Western term that emphasizes racial (i.e., physical) features, some feel it has a pejorative meaning in the local area where they conduct research, while others prefer to use other ways to characterize the people, such as their mode of subsistence or thought.

Hunter-Gatherers, Forest Foragers, and Autochtones

Some researchers prefer to use the terms *hunter-gatherers, foragers,* or *forest foragers* to characterize the people. Some researchers are critical of these terms because they are too general, Western-biased, and do not represent all Pygmy groups and are therefore misleading representations. For instance, not all Pygmies live in the forest, such as the Bedzan of Cameroon and the Twa of Rwanda and Burundi, and not all Pygmies are hunter-gatherers or foragers (mobile hunter-gatherers are called "foragers" while sedentary hunter-gatherers are called "collectors") all year. But it is also true that not all Pygmies are very short, such as the Bongo in Gabon and the Twa of the Democratic Republic of Congo. Some are critical of the forager and hunter-gatherer representations because they emphasize mode of production, which is viewed as a Western bias toward economic relationships.

The various critiques remind us to be cautious when using terms to generalize about such a large population and that pronounced diversity exists both between and within ethnic groups. However, it is my position that the terms *forest foragers* and *hunter-gatherers* are not necessarily misleading. No term will represent 100 percent of the ethnic groups considered, so I prefer instead to use an 80 percent criterion: if 80 percent or more of the ethnic groups have a particular feature, it seems reasonable to use the trait to generalize about the people. Of course, it then becomes a question of which features researchers, the local people, or others want to use to describe the people, but here we limit the discussion to the terms that have been used in the literature by field researchers. Chapters 2 and 5 show that the peoples described in this book are almost always shorter than their neighbors; therefore, stature could be used as a defining characteristic. If one examines Bahuchet's list of ethnic groups in table 1.1, in chapter 1, over 80 percent of them live in the rainforest, so it seems reasonable to generalize that most are tropical rainforest groups. The Twa of Rwanda and Burundi no longer live in forests, but their traditional way of life was associated with forests before the trees were destroyed to farm and raise cattle.

The terms *foragers* and *hunter-gatherers* are problematic because many of these people farm at least part of the time. But while most have taken up some form of agriculture, most people view themselves as forest hunting and gathering specialists and often spend several months of the year foraging, both near and far away from villages. Barnard (2002) also points out that the terms *foragers* and *hunter-gatherers* can be used to refer to a mode of subsistence or a mode of thought. Many San peoples in Southern Africa, where Barnard conducts his research, are no longer full-time hunter-gatherers and live in sedentary villages. They continue to hunt and gather from the villages, but they maintain forager modes of thought, such as values and social norms that emphasize egalitarian social-political relations, individual autonomy, immediate return ways of thinking, and extensive sharing and giving. Many Congo Basin foragers have started to farm, some farming as often as their neighbors (e.g., Koya in Gabon (Soengas 2012)) and some growing cash crops such as cacao (e.g., Baka in Cameroon (Oishi 2012)), but even in these contexts, forager modes of thought persist and are distinct from those of their neighbors (e.g., Baka in Cameroon (Yasuoka 2012)). Egalitarian social relations, autonomy, and other features of the forager ethos continue to be transmitted to children in these acculturated and sedentarized contexts (Hirasawa 2005; Nobutaka 2013). Although precise measures do not exist, it is likely that more than 80 percent of Congo Basin foragers continue to hunt and gather at least several months of the year, view themselves as forest hunter-gatherer specialists, or have forager modes of thought. The term *forager-farmer* may be a more precise descriptive term, but at this point I do not to introduce a new term.

Finally, some authors (chapters 4 and 12), Congo Basin government officials (e.g., in ROC, DRC, and Gabon), and human rights groups previously listed refer to Congo Basin forest foragers as "autochtones" or "indigenous." Unlike South America, Australia, and other parts of the world, most African ethnic groups meet the World Bank's criteria to be considered indigenous. All black Africans are indigenous to the continent; Rwandan government officials take this position and consequently do not recognize the Twa as autochtones. But in the African context, the African Commission of Human and People's Rights (ACHPR and IWGIA 2006) assert that

> almost all African states host a rich variety of different ethnic groups.... Basically all of these groups are indigenous to Africa. However, some are in a structurally subordinate position to the dominating groups and the state leading to marginalization and discrimination. It is this situation, which the indigenous concept in its modern analytic form, and the international legal framework attached to it, addresses.

Pygmies are often marginalized, discriminated against, and placed in structurally subordinate positions by non-Pygmies and are therefore considered "indigenous" by international standards (see chapters 11 and 12). Some Congo Basin countries may not use the term *indigenous* in their narratives about Pygmies, but they consistently recognize Pygmies on the radio, TV, and at national celebrations as the "first peoples" of the country.

Personally, I try to limit the use of the term *Pygmies*, but not because it connotes short stature or is inherently derogatory. When I first traveled to Africa in 1971, Ethiopians and Ugandans said I should try to see the Pygmies.

At first I thought they said this because they thought I wanted to see exotic short people, but instead they emphasized their unique music, exceptional dancing abilities, and knowledge of the forest. Africans seldom mentioned their stature. Over time, I came to realize that non-Pygmy peoples in the Congo Basin have their own words to refer to their forager neighbors (e.g., *bambinga* in the Aka area) and that their words have both positive and negative connotations. On the positive side, Pygmies are known for their supernatural abilities (e.g., to turn invisible, transform into an elephant, or cure supernatural ailments); singing and dancing abilities; herbal and forest knowledge; and high fertility (their infants are viewed as resilient). On the negative side, foragers are viewed as primitive, dirty, lazy, and as thieves (see chapter 11 for more details on how non-Pygmies view forest foragers). Farmers in the area where I work call their children *bambinga* when their children disobey, and Boyette (2013) shows that farmer children use the term to insult each other. I try to limit my use of the term *Pygmy* because local Bantu, Oubanguian, or Sudanic words that refer to forest foragers and the French and English terms *Pygmée* and *Pygmy* can have negative and pejorative connotations. Another issue with the term is that it gives the public the incorrect impression that a single culture or ethnic group exists.

Terms for Forest Neighbors

Issues also exist with terms that refer to ethnic groups that live next to, interact with, and often have complex socioeconomic relationships with Pygmies. Researchers use "non-Pygmies," "Bantus," "farmers," "villagers," "neighbors," or "the others" to generalize about these groups. The terms are problematic because they gloss over the extensive diversity between the 150 different non-forager ethnolinguistic groups that exist in the Congo Basin (Joris and Bahuchet 1994). Some groups are primarily fisherfolk, such as the Kwélé of Cameroon or the Monzombo of the Central African Republic (CAR), and seldom farm. Some researchers are critical of the terms *farmers* or *villagers* because, like above, they emphasize mode of production and emphasize a simplistic dichotomous contrast to hunter-gatherers (e.g., forager-farmer differences). Several researchers do not use the term *Bantu* because it is a linguistic term for a particular group of languages and many groups that associate with Congo Basin hunter-gatherers speak non-Bantu languages such as Sudanic or Oubanguian languages.

Using the 80 percent criterion previously described, it seems reasonable to use the terms *farmers, agriculturalists, villagers,* or non-Pygmies to refer to ethnic groups that regularly interact with Congo Basin hunter-gatherers. More than 80 percent practice farming and live in sedentary villages most of the year, and most are taller than their hunter-gatherer neighbors. The term *Bantu* should not be used because (1) it is a linguistic category and several farmer groups do not speak Bantu languages, and (2) it is often contrasted with *Pygmy* (e.g., Pygmy-Bantu social relationships), which gives the impression it is a physical rather than a linguistic category.

Lewis (chapter 8) designed an alternative solution to refer to the different groups in his region by using emic words (i.e., from local people). He uses the forager term *bilo* for farmers or villagers and *Yaka*, a term several forager groups use for themselves. These terms work well with some ethnic groups in ROC and CAR but are not applicable to other areas and groups. For instance, the Baka of Cameroon use the term *kaka* to refer to farming neighbors. *Twa* is also an important emic term that is used by Bantu-speaking farmers in several areas of the Congo Basin, especially in the DRC, to refer to their Pygmy neighbors. Some DRC ethnic groups use this name to refer to themselves, but others do not.

Conventions

The authors of these chapters were encouraged to use the names of specific ethnic groups whenever possible (e.g., Aka, Baka, and Efé), but it was left up to each researcher to decide whether she or he used the term *Pygmy*, *hunter-gatherers*, *forest foragers*, *indigenous*, or *autochtones*. All are considered synonymous unless noted otherwise by an author. The same is true for general terms to refer to ethnic groups associated with the hunter-gatherers; the authors decided whether they preferred to use *non-Pygmies*, *farmers*, *agriculturalists*, or *neighbors*, and all are considered synonymous unless noted otherwise.

Prefixes for Bantu languages and peoples have generally been omitted. For instance, the plural forms of Bantu-speaking forager ethnic groups (Ba)aka, (Ba)mbuti, (Ba)kola, (Ba)bongo are often removed and referred to as Aka, Mbuti, Kola, and Bongo.

Characteristic Features

Anthropologists have tried to identify cultural and biological features that distinguish Congo Basin hunter-gatherers from their neighbors. Bahuchet (chapter 1) and Verdu (chapter 2) review the history of anthropologists' attempts to identify cultural criteria, and Froment (chapter 5) provides a history of how biological anthropologists have characterized Pygmies. Verdu (chapter 2) summarizes studies of cultural criteria and finds that African Pygmy ethnic groups (1) identify themselves as distinct from neighboring groups and the neighboring groups identify the Pygmies as a distinct group; (2) are recognized by neighbors as specialists in forest activities (e.g., hunting, gathering, or fishing) or knowledge (ecological or supernatural); (3) have very mobile settlement patterns; (4) have complex socioeconomic, political, and ritual relationships with specific neighbors; and (5) are often recognized as outstanding musicians, sometimes with specific musical instruments and vocal techniques.

Verdu (chapter 2) and Froment (chapter 5) identify contemporary biological characterizations and indicate that Congo Basin hunter-gatherers (1) are genetically distant from their neighbors, (2) are shorter in stature than their neighbors, and (3) have relatively low faces, wide noses, and thin lips in comparison to their neighbors.

Rupp (chapter 8) is critical of these generalizations, Bahuchet (chapter 1) points out that many Pygmy ethnic groups do not fit the classic cultural criteria (i.e., live in forests or are mobile), Verdu (chapter 2) describes considerable genetic variability, and Froment (chapter 5) emphasizes that morphological features (e.g., height and body shape) are not dichotomous (present or absent physical traits) but clinal in that they are often found to various degrees or gradients across the Congo Basin.

In general, the cultural and biological characteristics described above are especially pronounced among the Efé and Mbuti foragers in the Ituri Forest in the eastern part of DRC; intermediate among Aka and Baka groups in the CAR, ROC, and Cameroon; and least common among the Bongo and Koya of Gabon (i.e., there is an east-to-west gradient of classic features; see figure 1.1 for the locations of these groups). The debate on distinguishing features of Congo Basin hunter-gatherers will continue, but they provide a starting point for discussing the unity and diversity in their biology and cultures.

Unity and Diversity

The subtitle of the book emphasizes diversity of African Pygmy cultures, histories, and biology (genes and morphology) because recent research with more ethnic groups has documented immense intracultural and intercultural diversity. Chapters 1, 10, and 11, in particular, emphasize diversity. Rupp (chapter 10), for instance, suggests Pygmy and non-Pygmy differences have been exaggerated and that the groups share more cultural similarities than differences. Bahuchet (chapter 1) provides an excellent overview of African Pygmy linguistic and cultural diversity. His figures 1.1 and 1.2 identify the locations of over fifteen known ethnic groups and table 1.1 gives the names, approximate population sizes, and linguistic families of the larger or better-documented groups discussed in this book. It is worth pointing out that the population sizes of the ethnic groups are estimates, and we really have a very poor understanding of the total number of forest foragers.

By contrast, chapters 4 and 8 accentuate the commonalities of Congo Basin hunter-gatherer cultures and point out that they are generally, but not always, more egalitarian, share more extensively, value autonomy more highly, and have a more intimate relationship with the forest than do their neighbors. Chapters 2 (genetics) and 5 (morphology) describe immense biological diversity between Pygmy groups and identify several traits that Pygmies share with non-Pygmies, but they also point out that Pygmies' genes and morphology are statistically different from non-Pygmies. As mentioned above, the differences are what geneticists call "clinal" (i.e., exist on a continuum or gradient) rather than dichotomous. Like chapters 2 and 5, chapters 7 (music) and 9 (child development) describe both uniformity and diversity.

Evidence exists for both extensive diversity and some uniformity in African Pygmy cultures, languages, and biology. Congo Basin hunter-gatherers have been around for 70,000 years; currently occupy an area of about 3.7 million km^2, larger than France and India combined; and live in assorted and sundry natural and social environments with different histories of culture contact, so it is not surprising to find profound cultural and biological diversity. Genetic data suggest limited intermarriage with non-Pygmies until relatively recent times, which helps to explain the shared biology, while the shared features of culture that are often identified (chapters 4 and 8) are common to most mobile hunter-gatherer groups in all parts of the world (i.e., egalitarianism, extensive sharing, and high value placed on autonomy). The common cultural characteristics of forest foragers appear to be particularly resilient and adaptive features to a hunter-gatherer way of life in many parts of the world.

Fürniss (chapter 7) uses the metaphor of a "family portrait" to illustrate the links between diversity and commonalities. Everyone in the family is somewhat different from each other genetically and behaviorally, but everyone also shares some resemblances and others recognize them as a family. Profound intercultural and intracultural diversity exists within and between Congo Basin hunter-gatherer ethnic groups, but individuals and ethnic groups also share some cultural and biological features.

Organization of the Book

Serge Bahuchet is one of the most respected and prolific (several books and over fifty book chapters and journal articles) contemporary researchers on African Pygmies and uses chapter 1 to introduce the reader to the pronounced cultural and linguistic diversity between ethnic groups. He provides overviews of extensive variability in hunting and gathering techniques, music, and relations with farmers and offers explanations for the diversity.

Chapters 2–4 use a variety of techniques and approaches to examine the history of Congo Basin hunter-gatherers. Chapter 2, by Paul Verdu, utilizes recent methodological advances in genetics to summarize the deep as well as recent history of Pygmies. Luca Cavalli-Sforza initiated the classic genetic marker studies of African Pygmies in the 1960s that eventually led to his mapping the genes and deep histories of human populations around the world. Verdu summarizes the new developments in genetic sequencing and statistical modeling and describes how they have been applied to understand the origin and distribution of African Pygmy ethnic groups.

Chapter 3, by Karen Lupo, Jean-Paul Ndanga, and Chris Kiahtipes, takes an archaeological and multidisciplinary approach to understanding when, how, and why Congo Basin forager-farmer relationships emerged. They hypothesize that metallurgy played a key role in the emergence of forager-farmer inequalities.

Robert Moïse's chapter 4 utilizes oral and written historical records from the Congo Basin to demonstrate that Pygmies have historical traditions (e.g., commitment to autonomy and entrepreneurialism) in the precolonial period that were substantially different from those among their farming neighbors, described as "equatorial traditions" by historian Vansina (1990).

Chapter 5, by Alain Froment, is longer than other chapters because it covers both the human biology and health of Congo Basin hunter-gatherers. He outlines historical studies of stature and summarizes research on Pygmy morphology (head shape and body proportions) and studies of nutrition and infectious and parasitic diseases. Finally, he reviews and critiques an array of hypotheses used to explain why Pygmies are short-statured.

Chapter 6, by Hiroaki Sato, is a shorter chapter and differs from other chapters in that it reviews one major research question in Congo Basin hunter-gatherer studies: can foragers live in the rainforest without trade relations with farmers who provide agricultural products? He turns to his recent field data and reviews other studies to answer the question. He concludes that forest resources in southeastern Cameroon are able to support a foraging life independent of agricultural products. Chapters 3 and 12 also briefly address the wild yam question.

Chapter 7 shifts attention to more cultural topics with Susanne Fürniss's systematic comparison of the music traditions of the major Pygmy ethnic groups. The chapter clearly illustrates the diversity in the performance, instruments, vocal techniques, and musical features between groups, but at the same time identifies a "family resemblance" in the "family portrait" of Congo Basin hunter-gatherer music.

Chapter 8, by Jerome Lewis, focuses on "remarkable similarities" in egalitarian social organization across Congo Basin hunter-gatherer groups and provides rich ethnographic details of the dynamic processes of demand sharing, play, and other mechanisms that promote egalitarianism and extensive sharing among the Mbendjele Pygmies of the ROC. Lewis concludes that egalitarianism is a core feature of Pygmy life that cuts across political, moral, economic, and other domains of daily life.

Chapter 9 provides an overview of studies conducted with forager infants, children, and adolescents. Features of child development common to several groups include multiple childcare, allomaternal nursing, relatively high involvement by fathers, frequent cosleeping throughout life, and high physical and emotional proximity to parents and others. Intercultural and intracultural diversity exists in these patterns, and explanations are provided to explain both the commonalities and diversity of observed patterns.

Chapter 10, by Stephanie Rupp, critiques the term used to describe peoples of the Congo Basin turns to her own ethnographic experiences and data from Cameroon to demonstrate that existing terms, such as *hunter-gatherers* and *Pygmies*, are misleading because forest people utilize a vast array of identities that cut across subsistence and physical features. She argues that many types of identities exist among forest peoples and communities and that they are always changing in response to modifications in spatial and social circumstances (i.e., they are "multiangular").

Chapter 11, by Kiyoshi Takeuchi, reviews all the research on the nature of forager-farmer relationships in the Congo Basin. He examines ideological (i.e., how they view each other); economic; and social, kinship, and ritual features of these complex relationships. He describes extensive diversity in these relationships and develops a creative system to classify forager-farmer relationships into four general types based on the relative levels of subordinate or autonomous structures in the relationships.

The final chapter, by Mitsuo Ichikawa, addresses a major development question facing forest foragers: how can local indigenous forest foragers maintain access to forest resources they depend on to survive when global conservation groups (e.g., World Wildlife Fund and Wildlife Conservation Society) want to protect tropical forest floral and faunal diversity? Conservation groups have established several parks and reserves in central Africa to protect forest faunal and floral diversity, but this has often led to restricting forager access to traditional lands and essential resources. Ichikawa reviews the political and economic dimensions of the issues, including the bushmeat trade and logging, and recommends that greater emphasis should be placed on community forest utilization of non timber forest products (NTFPs).

Overall, the volume aims to provide overviews of Congo Basin hunter-gatherer research conducted by researchers from all parts of the world in the last forty years. The book also strives to cover a diverse range of topics and is open to alternative theoretical and methodological approaches—cultural and biological methods and data as well as ecological, evolutionary, postmodern, and cultural theoretical orientations. Several of the researchers in this volume disagree on particular issues, but all are supportive of integrating the diverse data and approaches into this volume.

Limitations and Biases

Limitations and biases exist in this book and in Congo Basin hunter-gatherer research. First, most of the research since the late 1980s has been conducted with western Congo Basin groups. Bahuchet (chapter 1) quantitatively demonstrates in figure 1.3 that western Pygmies have received the most academic attention. While research from eastern and southern hunter-gatherers is reviewed in several chapters, little research has been conducted with ethnic groups in these areas since the late 1980s because of the extensive political instability in the DRC. Future research is urgently needed in these areas.

Second, as reviewed elsewhere (Hewlett and Fancher, in press), ecologically oriented scholars dominate research with Congo Basin foragers. The ecological and evolutionary bias has a long history in the anthropology of Congo Basin hunter-gatherers as exemplified in the German *Kulturkreise* (culture circles) school where Pygmies had a special evolutionary position as *Naturvolk* (people in close relationships with nature) (Schmidt 1939). Congo Basin foragers usually (see Bahuchet, chapter 1, for exceptions) identify with the rainforest milieu, are fundamentally shaped by it, and often express a strong preference for forest life (Hewlett 1996), but research in the region has focused on the economic domains of forest life while neglecting other dimensions such as marriage and the family, social and emotional relations, and religion.

Ecological approaches have made significant contributions to our understanding of human-nature relations, but few studies exist that provide us with insights into how Congo Basin foragers think and feel about their lives. We know how many calories of meat they eat each day, how much time they spend hunting, and how much time they spend with infants, but we know little about how forest foragers think and feel about what is important to them—the forest, family relationships, religion, and so on.

Finally, relatively few African anthropologists have conducted research with forest foragers. Jean-Félix Loung (1967) was one of the first African anthropologists to conduct research with forest foragers, and Godefroy Ngima Mawoung (2006) is one of the more recent to publish. African anthropologists' research with forest foragers often takes place in the context of development (e.g., establishing parks or building an oil pipeline), so their work shows up in reports rather than academic publications. Scholars from Congo Basin countries are also usually trained in European universities, and their advisors often impact their research interests and perspectives (i.e., they are often ecologically oriented). But even with these constraints, we need more African scholar research with forest foragers. In the near future it is also possible that forest foragers will complete courses in higher education and begin to contribute to the research literature. Their intimate lived experiences in the Congo Basin will provide alternative and useful views on Congo Basin forest peoples.

References

ACHPR and IWGIA. 2006. *Indigenous Peoples in Africa: The Forgotten Peoples?* African Commission on Human Rights and Peoples' Rights (ACHPR) and International Work Group for Indigenous Affairs (IWGIA). Copenhagen.

Bahuchet, S. 1985. *Les Pygmées Aka et la Foret Centrafricaine*. Paris: SELAF.

Bailey, R. C. 1991. *The Behavioral Ecology of Efe Pygmy Men in the Ituri Forest, Zaire*. University of Michigan Anthropological Papers No. 86. Ann Arbor: University of Michigan.

Barnard, A. 2002. "The Forging Mode of Thought." In *Self- and Other Images of Hunter-gatherers*, edited by H. Stewart, A. Barnard, and K. Omura. Senri Ethnological Studies, No. 60. Osaka: National Museum of Ethnology, 5–24.

Cavalli-Sforza, L. L. 1986. *African Pygmies*. New York: Academic Press.

Hewlett, B. S. 1991. *Intimate Fathers: The Nature and Context of Aka Pygmy Paternal Infant Care*. Ann Arbor: University of Michigan Press.

———. 1996. "Cultural Diversity among African Pygmies." In *Cultural Diversity among Twentieth Century Foragers: An African Perspective*, edited by S. Kent. Cambridge: Cambridge University Press, 215–44.

Hirasawa, A. 2005. "Infant Care among the Sedentarized Baka Hunter-gatherers in Southeastern Cameroon." In *Hunter-gatherer Childhoods*, edited by B. Hewlett and M. Lamb. New Brunswick: Aldine Transaction.

Loung, J-F. 1967. "Le Nom Authentique du Group Pygmée de la Région Côtière Camerounaise." *Revue de Géographie du Cameroun* 7:81–94.

Mawoung Ngima, G. 2006. "Perception of hunting, gathering and fishing techniques of the Bakola of the coastal region, southern Cameroon." *African Study Monographs* Supplement 33, 49–69.

Nobutaka, K. 2013. "Hunting and Gathering Culture and School Education: Comparative Studies on Baka Children in 1990s and 2010s." Paper presented at the International Conference on Hunting and Gathering Societies, Liverpool, UK.

Oishi, T. "Cash Crop Cultivation and Interethnic Relations of the Baka Hunter-gatherers in Southeastern Cameroon." In *African Study Monographs*, supplement 43:115–36.

Robillard M. and S. Bahuchet. 2012. "Les Pygées et les Autres: Terminologie Categorization et Politique." *Journal des Africanistes* 82:167–92.

Schmidt, W. 1939. *The Culture Historical Method of Ethnology*. Translated by S. A. Sieber. New York: Fortuny's.

Soengas, B. 2012. "Des Pygmées Cultivateurs, les Bakoya: Changements Techniques et Sociaux Dans la Foret Gabonaise." *Journal des Africanistes* 82:167–92.

Turnbull, C. M. 1965. *Wayward Servants: The Two Worlds of the African Pygmies*. New York: The Natural History Press.

Yasouka, H. 2012. "Fledgling Agriculturalists? Rethinking the Adoption of Cultivation by the Baka Hunter-gatherers." In *African Study Monographs*, supplement 43:85–114.

1

Cultural Diversity of African Pygmies

Serge Bahuchet

A Brief History

In the second half of the nineteenth century, European and American explorers encountered various groups all over Central Africa that appeared physically different from other peoples with whom they shared the same rainforest. Whenever such a group was discovered, they were commonly recorded under the homogenous category "Pygmy," after the mythical population of short stature described in Greek antiquity. Other, similarly indistinct names were given depending on the nationality of the explorer: Dwarfs, *Zwergen*, *Nains*, or *Négrilles* (little Negroes), to name a few. It became a great endeavor of travelers in the region to find and record "new" Pygmy groups living in the yet hidden corners of the Congo Basin (Bahuchet 1993a). Some sort of competition developed to discover those with the shortest stature and living the most "primitive" lifestyle, that is, nomadic, hunting and gathering, and living in beehive-shaped huts constructed of leaves and branches. The observers of this time were unconcerned with questions of cultural diversity; on the contrary, they were interested in a unity among all Pygmies that would serve as proof of the persistence of some prehistoric way of life, pure and monotheist. Many of the confusions of this early period persist even today. Professional scientific observations of specific groups began quite late. Among the first were those of Paul Schebesta, who in the 1930s lived and worked with the BaMbuti people of the Ituri rainforest, located in the remote eastern part of the Congo Basin. Schebesta wrote a series of immense tomes (1938, 1941, 1948, 1952) detailing the techniques, societies, languages, and vocabulary of the BaMbuti people. He was the first scholar to recognize that within these groups there were various subgroups called by different names—Asua, Efe, BaSua, and Kango—and that discrete languages were spoken by each.

In the 1950s, Schebesta was followed by Colin Turnbull (1965a, 1965b). Critical of Schebesta's methodology, Turnbull developed a summary ethnography of the BaMbuti, emphasizing a strong technological difference dividing two groups: those who practiced net-hunting—those of the subgroup BaMbuti, or BaSua—and those who practiced bow-hunting—the Efe (Turnbull 1965a). He found that beyond the linguistic diversity recognized by Schebesta among these African rainforest populations, there was in fact profound cultural diversity, and he was the first to plead for a careful distinction to be made between them: "The term 'Pygmy' leads far too easily to meaningless generalization" (Turnbull 1965a).

Turnbull did not go far enough. He still considered the people of the Ituri essentially homogenous, without questioning the actual relations between groups that did not speak the same languages (Turnbull 1983). As a matter of fact, after many years of often fine anthropological studies in various regions of the Ituri, we still lack ethnographic descriptions of the social exchanges between, for instance, the Efe and the Mbuti.

Despite the initial advances made by Turnbull and Schebesta in recognizing the tremendous degree of diversity among "Pygmies," their work also contributed strongly to the lasting stereotype of the "true Pygmy" that has permeated much of the literature since. Contrary to the stereotype, there are in fact extensive variations and permutations in virtually every facet of Pygmy life and culture. Rather than in the deep rainforest, some groups have been found to live on the savannahs, some in the forest-savannah ecotone, and some in the mountains or swamps. Others, though short-statured, are not hunter-gatherers; while for some the reverse is true. An excellent illustration of this point comes from the lengthy and important fieldwork done with the BaTwa in Rwanda by Peter Schumacher (1949–1950), a contemporary of Schebesta. This group diverged greatly from the stereotype; they were living in the mountains and were not hunter-gatherers but potters, servants, and musicians in a complex caste society. The books of Schumacher, however, are never used in any discussion about the Pygmy populations.

As time progressed, researchers gradually woke to the fact that the peoples and cultures of the Congo Basin were indeed far from uniform: they populated diverse environments from the rainforest to the savannah and displayed a broad spectrum of physical, linguistic, and cultural traits. To date, the African rainforest foragers collectively known as *Pygmies* have been the subjects of thousands of publications, though they are of varying quality and accuracy (Plisnier-Ladame 1970; Hewlett and Fancher 2010). Although some of these publications have contributed concretely to furthering the anthropological discourse on central African rainforest peoples, the majority of the documentation consists of weakly supported, generalized statements. Altogether, despite the apparent abundance of literature on these foragers, great gaps remain in our knowledge.

Thus, after almost a century of anthropological research in central Africa, we are still confronted with numerous difficulties. As a consequence of the uneven quality and quantity of description and information about the various groups, many important questions remain unanswered. One of the foremost of these regards the validity of the term *Pygmies* to describe African rainforest foragers: Do the various populations to whom it refers bear any meaningful commonalities? Is it perhaps more relevant to compare the differences and similarities among the Pygmies themselves than to compare them as a single group against other African foragers or against all other central African populations?

Names and Locations of Groups

In spite of the immense physical, cultural, and linguistic diversity among Pygmy peoples, an imbalance in the literature has given rise to the popular stereotype based on "forest-oriented" groups. These most famous of Pygmies are typified by their short stature, seminomadism, hunting and gathering, use of temporary camps with domed huts, and regular exchange with neighboring farmers (Hewlett 1996). The forest-oriented groups, in geographic order from the Atlantic coast to the eastern Congo Basin, are as follows:

- Baka (some of them known as Bangombe) (Joiris 1992, 1996; Leclerc 2001, 2012; Sato 1998; Tsuru 1998; Vallois and Marquer 1976)
- BaAka[1] (also Bayaka, also known as Babinga or Bambenga, some of them—the western groups—known as BaMbenzele) (Bahuchet 1972, 1985; Demesse 1978, 1980; Hewlett 1991; Kitanishi 1995; Lewis 2002)
- BaMbuti of Ituri forest, northeastern Democratic Republic of the Congo (DRC, formerly Zaire). This last group is actually divided into at least three ethnic groups: Efe, Asua, and BaSua (or BaMbuti, strictly speaking[2]) (Bailey 1991; Demolin 1993; Harako 1976, 1981; Ichikawa 1978, 1981, 1983; Peacock 1984, 1985; Schebesta 1941; Tanno 1976, 1981; Terashima 1983, 1985, 1998; Turnbull 1965a, 1965b, 1983).

Other groups live in a markedly different socioeconomic situation; having settled in villages, they practice a more or less efficient agriculture in addition to foraging while maintaining relations with neighboring farmers. Although they appear less distinct from farmers than the preceding groups on cultural and often physical bases, they are always distinguished and categorized by farmers as different from themselves. They are, from west to east,

- BaKola (also known as BaGyeli, southwestern Cameroon) (Biesbrouck 1999; Joiris 1992, 1994; Koppert et al. 1997; Ngima Mawoung 1996, 2001; Seiwert 1926)
- BaBongo (also known as Akoa, central Gabon) (Andersson 1983; Annaud and Leclerc 2002; Knight 2003; Le Bomin and Mbot 2012a; Leroy 1928 [1897]; Matsuura 2006, 2011; Mayer 1987)
- BaKoya (sometimes called BaKola, at the border of Gabon-Congo) (Soengas 2009, 2010, 2012; Tilquin 1997)
- BaTwa (in the Equatorial region in western DRC) (Elshout 1963; Pagezy 1975, 1976, 1986, 1988; Schultz 1986; Sulzman 1986)

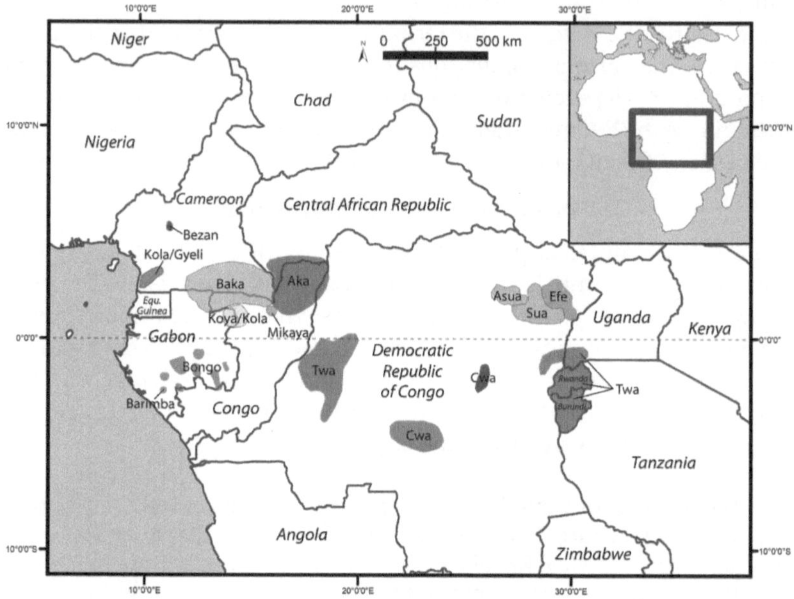

Figure 1.1. Group names and locations.

Other, less well-known Pygmy groups that live on the forest periphery have been documented in only a few papers or reports. These groups are generally less mobile and live in settled villages, sustain a low population, and practice a more or less efficient agriculture. Meanwhile, like seminomadic groups, they maintain relations with neighboring farmers. The groups called Kola, BaBongo, BaKoya, and BaTwa live in the rainforest, while the Bedzan, various groups named Cwa (read "Tshwa") and a scattered group known under the name of BaTwa live on the savannahs at the periphery of the forests in Rwanda and Burundi. These groups are (again, from west to east)

- BaKola, BaBongo, BaKoya, and BaTwa, who live in the rainforest.
- Bedzan[3] living with the Tikar in central Cameroon (Leclerc 1995, 1999; Mebenga Tamba 1998).
- BaCwa (pronounced "Batshwa"), several groups under this name live in the southeastern savannah of Congo Democratic Republic (Kazadi 1981; Vansina 1954).
- BaTwa, the name of scattered groups in Rwanda and Burundi. They are not foragers, but farmers and handworkers (potters), constituting a caste in highly hierarchized societies of Rwanda (Lewis 2000; Lewis and Knight 1996; Schumacher 1949–1950).

I must also mention that there are still other groups in southern central Africa (see figure 1.2), but their documentation is very meager and their status

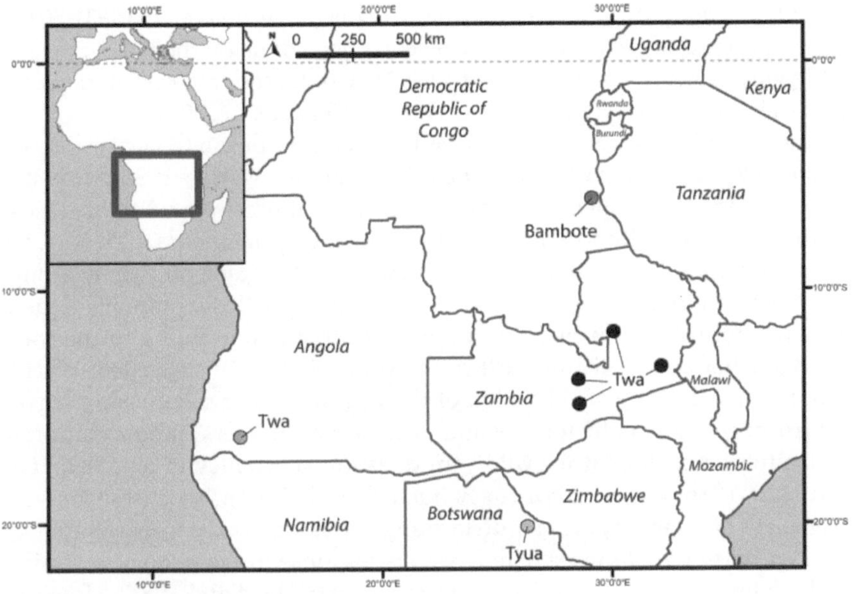

Figure 1.2. Groups with ambiguous status.

remains ambiguous as a result. They are not called "Pygmies" by any of the authors, but their names look similar to those of Pygmy groups in the Congo Basin. These are the Vatwa in southwestern Angola (Estermann 1962, 1976), the different Batwa swamp fishermen in Zambia (Rosen 1925; Macrae 1934; Barham 2006; Haller and Merten 2008), as well as the Bambote of the shores of Lake Tanganyika (Terashima 1980).

Orientation of the Literature

The forest-oriented canonic stereotype of "true Pygmy" originated during the first phase of "classical ethnography," conducted on the BaMbuti of the Ituri Forest in the east of the DRC (Schebesta 1938, 1941, 1948, 1952; Turnbull 1965a, 1965b, 1983). Japanese scholars followed in the 1970s and 1980s, couching their numerous studies in the area within the framework of cultural ecology (Harako 1976, 1981; Ichikawa 1978, 1981; Tanno 1976, 1981; Terashima 1983, 1985; Terashima et al. 1988). American anthropologists then focused on one of the BaMbuti subgroups, the Efe, for a long-term biocultural study (Bailey 1991; Bailey and DeVore 1989; Bailey and Peacock 1988; Ivey 2000; Peacock 1984, 1985).

The research expanded in the late 1970s to include the BaAka (known previously as Babinga, or Bayaka), whose range was in the western part of the Congo Basin. Monographs on this new group, also forest-oriented, by Demesse (1978, 1980) and Bahuchet (1972, 1985) were accompanied by ethnobiological (Bahuchet 1985; Motte 1980) and musicological (Arom 1987) research, and a multidisciplinary study focused on their language (Thomas et al. 1981–2013). Other anthropological studies have recently been done on the Baka, the largest Pygmy group in Cameroon (Sato 1992), with further material and insight offered by the dissertations of Joiris (1998) and Leclerc (2001).

The historical aspects of hunter-gatherer culture began to garner interest in the 1980s (Leacock and Lee 1982; Rottland and Vossen 1986). This was an important step, especially for the study of rainforest populations, whose cultures are deeply intertwined with their sedentary neighbors. Indeed, since Turnbull's book *Wayward Servants* (1965b), the discourse on Pygmy groups has been conditioned by questions concerning this close, mutually dependent relationship, even in publications not directly pertaining to the topic (cf. Bahuchet and Guillaume 1982; Terashima 1998; Vansina 1986; Waehle 1986). In particular, consideration of the Pygmy/farmer relationship stirred controversy over the historicity and necessity of foragers' subordination to agriculturalist groups (the "wild yam question," cf. Bailey et al. 1989; Hart and Hart 1986; and *a contrario* Bahuchet et al. 1991). Because economic exchanges of foraging groups with farming populations exist in all rainforests, it was proposed that the exchanges were a response to the strong constraints of the rainforest ecosystem (cf. Bailey et al. 1989; Headland 1987). However, Pygmy/farmer transactions are not exclusively economic, and some recent papers have analyzed the social ties between the two types of communities (Joiris 2003; Robillard 2012; Rupp 2003; Sawada 1998; Terashima 1987). Three dissertations have examined the relationship from the villagers' perspective: Delobeau (1989) discusses the relationship between the Monzombo and the BaAka in the Central African Republic of the Congo (CAR); Grinker (1994) describes the Lese and Efe in the DRC; and Rupp (2001, 2011) writes about the Bangando and Baka of Cameroon. Combining anthropological inquiries and analysis of oral history archives, three important studies have placed the relationships between Pygmies and farmers into a strong historical framework over the last two centuries (Delobeau 1989; Guillaume 2001; Hardin 1999), while an ethnolinguistic synthesis (Bahuchet 1992, 1993b) attempts to trace the relationship back much further. Klieman (1997, 2003) approached the issue similarly, with glottochronological and linguistic research. In recent years some collective publications, some of which issued from symposia, have offered regional and comparative approaches in gathering chapters of various thematic approaches from research in different regions (Biesbrouck et al. 1999; Kent 1996; Ichikawa 1998; Ichikawa and Kimura 2003).

A rough survey of the serious ethnographical literature (345 titles, papers and books, excluding general papers), some of which is discussed above, reveals a huge disparity in documentation (figure 1.3). A mere five groups (BaMbuti, Efe, BaAka, BaMbenzele, and Baka) provide the topics for 86 percent of all publications on Congo Basin peoples. While studies in several fields have

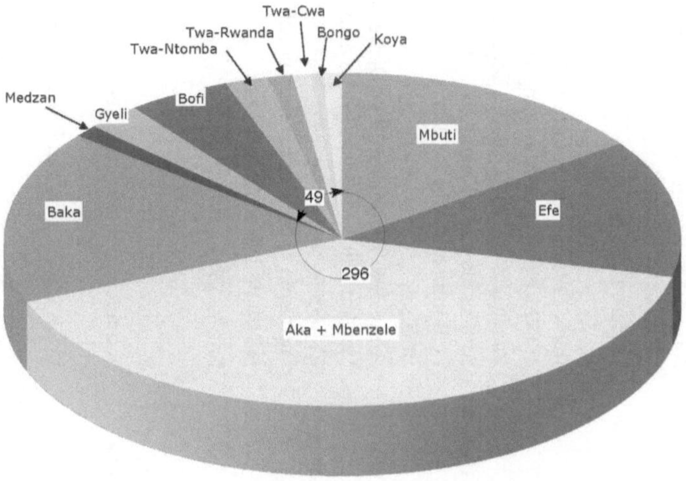

Figure 1.3. Ethnographic publications about Pygmy groups (n = 345).

concentrated on these "classical" forest-oriented groups, the eight other main Pygmy groups—smaller demographically and less mobile—are the subject of comparatively little research, just 14 percent of the literature. In light of this imbalance, we are now at pains to develop an accurate physical, cultural, and linguistic cartography of the Congo Basin rainforests.

Present State of Knowledge

It is very difficult in a short text to give anything approaching an adequate picture of the cultural details of all the groups. Here, I will briefly outline some of what we now know about the range of variability among Pygmy groups, concentrating on some technical examples.

Hunting Techniques

While classically considered to be more or less pure "hunter-gatherers," some Pygmy groups in fact depend no more on hunting, in terms of subsistence, than other forest agriculturalist populations. Those that rely primarily on hunting employ a variety of techniques and weapons according to region. Collective hunting is the notably prevalent technique (Bahuchet 1987; Harako 1976; Tanno 1976; Terashima 1983). Net-hunting is the far more common, with two exceptions: Efe of eastern DRC instead practice collective bow-hunting, and Baka of Cameroon and Gabon practice collective spear hunting without nets. All the groups exhibit much variation in strategy, from the methods used in flushing out game to the particular placement of nets in the undergrowth. In many cases, women are equal partners with men in these collective hunts. This of course has strong social implications.

Table 1.1. An inventory of Pygmy groups (from west to east).

Name	Country	Other Names	Population Size and Environment	Main References
Baka	Cameroon, Congo, Gabon	Bangombe, Bibayak, Babinga	± 30,000–40,000, rainforest	Joiris 1998; Leclerc 2001, 2012: Sato 1992; Tsuru 1998; Vallois and Marquer 1976
Bedzan	Cameroon	Tikar Pygmies	± 400, rainforest, savannah margins	Leclerc 1999; Mebenga Tamba 1998
(Ba)Kola	Cameroon	(Ba)Gyeli	± 4,000, rainforest	Joiris 1994; Koppert et al. 1997; Loung 1959, 1987, 1996; Ngima Mawoung 1996
(Ba)Koya	Gabon, Congo	(Ba)Kola	± 2,600, rainforest	Soengas 2009, 2010, 2012; Tilquin 1997
(Ba)Bongo	Gabon	Akoa, Barimba	± 3,000, rainforest, savannah margins	Andersson 1983; Knight 2003; Matsuura 2006; Le Bomin and Mbot 2012a
Mikaya	Congo	Bambenga	No data, rainforest	Boyi s.d.
(Ba)Aka	CAR, Congo	Bayaka, Biaka, Babinga, Bambenga, BaMbenzele	± 30,000–50,000, rainforest	Arom 1987; Bahuchet 1985; Demesse 1980; Hewlett 1991; Kitanishi 1995; Lewis 2002
(Ba)Twa	DRC	Konda Twa	No data, rainforest, swamps	Elshout 1963; Pagezy 1986, 1988; Schultz 1986; Sulzmann 1986; Van Everbroeck 1974
(Ba)Cwa	DRC	Bushong Twa, Kuba Cwa	± 6,000, savannah	Hiernaux 1966; Vansina 1954
(Ba)Cwa/(Ba) Tembo	DRC	Luba Cwa, Batwa, Bambote	No data, savannah	Kazadi 1981
(Ba)Twa	Uganda	Batwa	± 2000?, mountain forest	Kenrick 2000, UNEP 2005
(Ba)Twa-(Ba) Rhwa	DRC	Batwa, Kivu Twa, western Twa, Barhwa	± 6.000, mountain forest	Schumacher 1949, 1950; Seitz 1993
(Ba)Twa	Rwanda, Burundi	Batwa, eastern Twa	± 70,000?, savannah, mountains	Lewis 2000; Lewis and Knight 1996
(Ba)Sua	DRC	(Ba)Mbuti, Kango	± 26,000, rainforest	Harako 1981; Hart and Hart 1986; Ichikawa 1978; Tanno 1976; Turnbull 1965a
Asua	DRC	(Ba)Mbuti, Akka, Tikki-tikki	± 10,000, rainforest	Demolin 1992; Schebesta 1952
Efe	DRC	(Ba)Mbuti	± 10,000, rainforest	Bailey 1991; Bailey and Peacock 1988; Demolin 1993; Terashima 1983, 1985

Note: CAR = Central African Republic, DRC = Democratic Republic of Congo

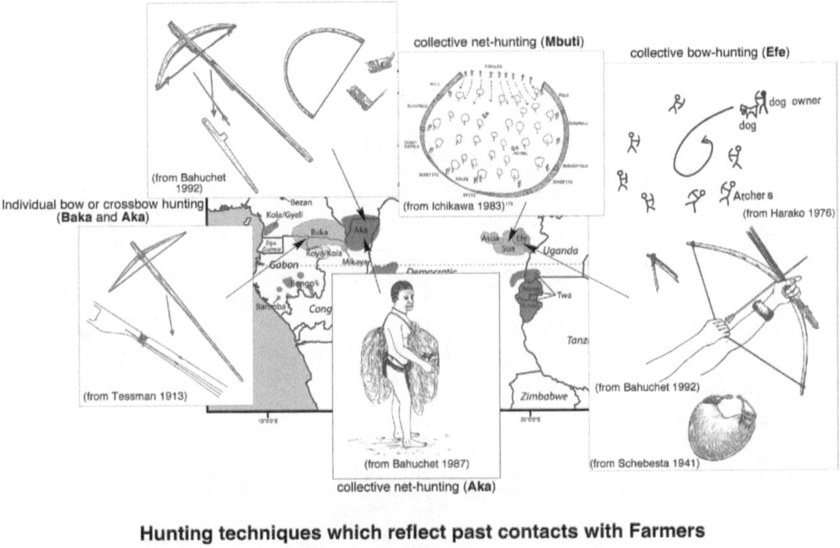

Figure 1.4. Hunting techniques.

Pygmy groups also enjoy remarkable flexibility in what they hunt and how and use different combinations of techniques and weapons according to the game. Duikers, the most commonly hunted animal, are pursued in collective hunts. Arboreal animals, such as monkeys and large birds, are shot with bows and arrows or crossbows. Various types of crossbows have been recorded among groups in the western Congo Basin, and extensive variations in bows are seen in other populations (figure 1.4). Spears affixed with large blades are used in collective hunts to bring down elephants, antelopes, apes, and wild hogs. Various individual techniques—often involving dogs—are also utilized for trapping and catching porcupines or giant rats in their nests.

Considering that all Congo Basin forest populations' food supplies actively depend to one degree or another on hunting, it is difficult to determine any characteristics that are specifically "Pygmy." If some Pygmy specificity should be drawn, however, it may be identified in two aspects: (1) the relatively greater amount of time Pygmies devote to pursuing game, and (2) their designation by other groups as specialists in elephant hunting.

None of the types of weapons are unique to any one particular Pygmy group, and all are in some way similar to the ones used by farmers. Thus, the history of hunting technology in the region appears to be long and involves all the populations of the Congo Basin; hence, it is also a testimony to the historical depth of contact between groups. For instance, bow-and-arrow hunting, which once appeared unique to the Efe Pygmies of the Ituri, is now known to

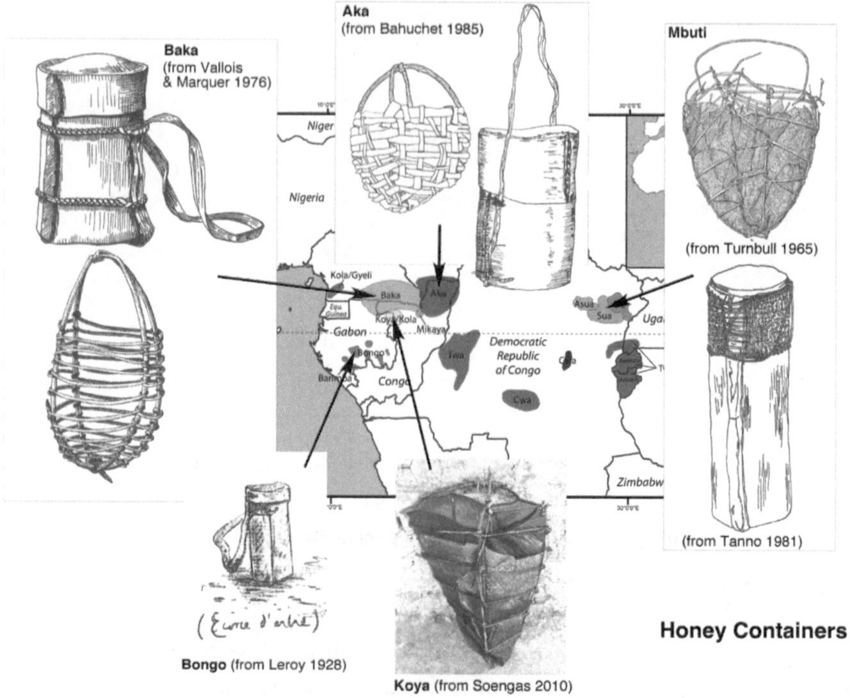

Figure 1.5. Honey containers.

be a technique shared by all the Sudanic language–speaking populations in the region, including the Efe.

Gathering

Common to all Pygmy groups is the practice of gathering yams, nuts, mushrooms, leaves, and insects such as termites and caterpillars. In addition, there is perhaps nothing in the forest more sought after than honey, and Pygmies are universally considered by their neighbors to be experts in its collection. Very few details are available with concern to their techniques, but the process frequently involves climbing what are often extremely tall trees to access the hive's position in the canopy. Remarkably, several groups use very similar tools: a singular axe with an elbow-shaped handle; a temporary collection basket made on the spot with leaves and vines, which is discarded after collection; and a bark container for transporting the honey back to camp (figure 1.5). The three tools have been described among the BaMbuti (Turnbull 1965a; Schebesta 1941; Tanno 1981), the BaAka (Demesse 1980; Bahuchet 1985), and the Baka (Vallois and Marquer 1976; Bahuchet 1992). The use of some, but not all, of these tools has been recorded among certain Pygmy groups in Gabon (BaBongo, drawings in Leroy 1928; BaKoya, in Soengas 2010).

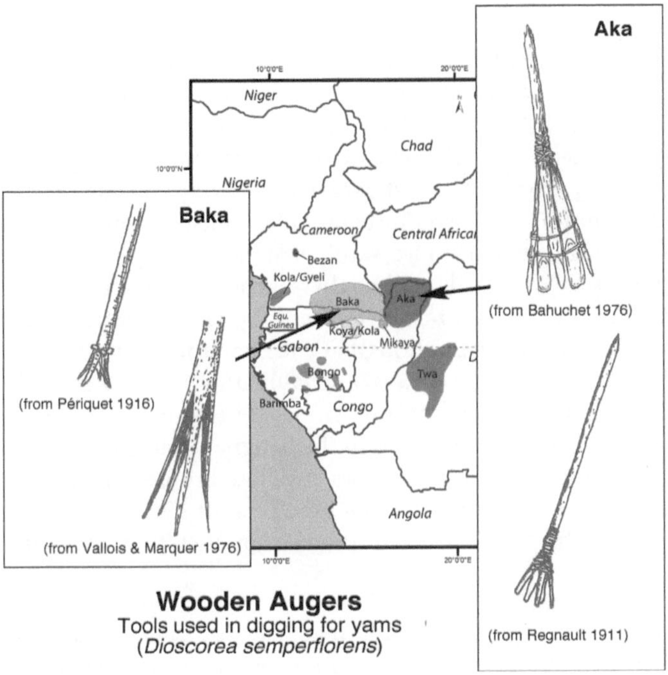

Figure 1.6. Wooden augers.

For the task of procuring wild yam tubers, Pygmies typically use various forms of digging sticks (Dounias 2001a). However, a peculiar drill with a cone at the end of a very long handle (figure 1.6) has been discovered among BaAka of CAR and Baka of Cameroon and is used specially for uprooting *Dioscorea semperflorens* (Bahuchet 2000; Dounias 2001b). To my knowledge, this tool has not been observed in any other population, either Pygmy or farmer.

Other Subsistence Activities

Though fishing is not of prime importance for the majority of the groups, the BaTwa provide an exception. Living in the swamps surrounding Lake Tumba in central Congo, they derive a substantial percentage of their diet from the water, together with Ekonda and Ntomba farmers (Pagezy 1975). In other regions, fishing is practiced seasonally and is a valuable complement to more prevalent modes of subsistence. Many methods have been developed. The BaAka, for example, release ichthyotoxic plant poisons into small rivers. During the dry season, Baka women employ the dam-and-bail method, in which they trap and catch fish by damming a portion of a river and then bailing water from upstream over the barrier. With the depletion of forest game, fishing may begin to assume a more prominent role in subsistence strategies as Pygmy groups seek to maintain their protein intake.

Habitat

"Classical" Pygmies are well-known for their beehive-shaped huts, constructed from plied stems woven together and covered with Marantaceae leaves. As in the case of the BaAka, Baka, and BaMbuti, individual family huts are typically arranged in a circle to form temporary, mobile camps. On a superficial level, there is a strong contrast between the living arrangements of these groups and the more sturdily built rectangular houses of many others, which are often organized linearly into semipermanent villages. However, closer examination of early records and of these groups' seasonal mobility quickly alters this impression. For instance, the BaBongo of central Gabon presently live in sedentary hamlets, though at the time they were first contacted by Paul Du Chaillu in June 1865 (Du Chaillu 1867), they were living in beehive huts in camps. A similar account from 1911 (Regnault) provides a precise description of BaAka forest camps in the Sangha. Semispherical huts situated side by side with rectangular shelters populated those camps. Thus, preferences in terms of housing type can be seen to shift over time and are more a reflection of the historical moment than a sign of cultural diversity.

Music

A pronounced talent among Pygmies for musical performance is widely recognized by their neighbors. Here again, some degree of contrast appears between Pygmies of the canonic stereotype and the other Pygmy groups. The BaMbuti, BaAka, and Baka perform styles of music highly distinct from those of surrounding populations. Huge polyphonic choirs, involving many singers, supported by rhythm instruments and a type of yodeling characterize these styles (Arom 1987; Arom and Fürniss 1992). Some yodeling does exist in the music of other Pygmy populations—and this may signal some hidden similarities—but on the whole their music has more in common with the surrounding farming populations (e.g., less polyphonic and more strophic). The music particular to each of the "classical" groups may, to casual foreign listeners, be practically indistinguishable; but there are, in fact, fundamental differences in their construction (see chapter 7 for detailed discussion).

Melodic instruments, although they play a secondary role in the Pygmy music, are nevertheless present. Examples of such instruments among the BaAka are musical bows, harps, and flutes, which are used in much of their intimate repertoire. In terms of shape, these instruments are not entirely unique to Pygmy groups, but they generally differ from the instruments of their direct neighbors. For instance, the BaAka and Baka share the use of a rare two-stringed musical bow (Guillaume 1982) played only by women. None of their neighbors know this instrument. Strikingly enough, similar instruments were described very far away, at the border of Uganda and the DRC, among one of the Pygmy groups (Efe, Demolin 1990) and among some farmers (Bahuchet 1992). Once again, musical instruments appear like

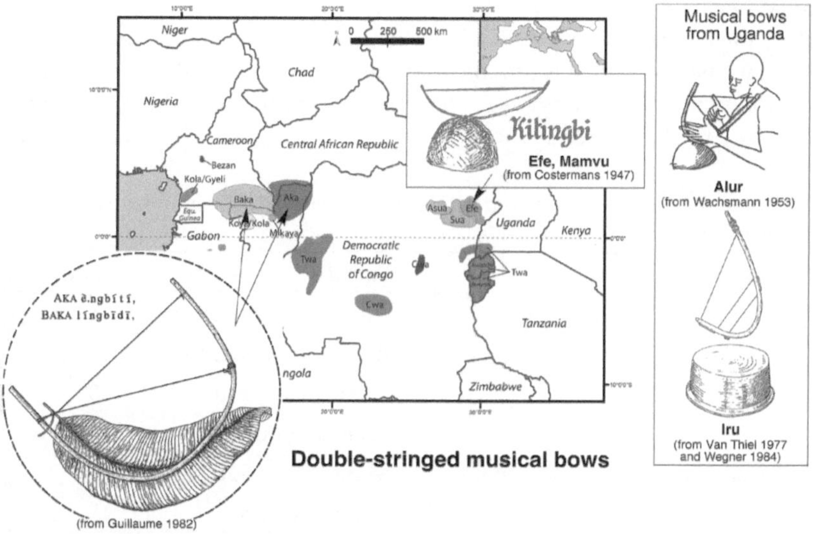

Figure 1.7. Double-stringed musical instruments.

artifacts of past relations difficult to reconstruct (Fürniss and Bahuchet 1995) (figure 1.7).

Finally, we must take care to emphasize the role that music and dance play in the relationships between Pygmy and non-Pygmy communities (cf. Le Bomin and Mbot 2012b). While organizing important festivals such as funerals and initiations, the farmers very frequently ask for the participation of Pygmy musicians, who play the music of the farmers during such occasions, not their own. In past Rwandan society, the BaTwa, in addition to other duties, served as the official musicians of the king (Gansemans 1993; Seitz 1993).

Linguistic Diversity

The linguistic situation among Pygmy populations is definitely diverse, showing strong heterogeneity. Unfortunately, linguistic studies of Pygmy languages are much less available than anthropological ones. The majority of the languages are not documented or described, the exceptions being Efe (Vorbichler 1965; Vorbichler and Brandl 1979), Gyeli (Renaud 1976), Aka (Cloarec-Heiss and Thomas 1978; Thomas 1991), and Baka (Brisson 1984, 2010; Brisson and Boursier 1979; Kilian-Hatz 1989, 1995, 2008; Paulin 2010).

All Pygmy tongues are clearly related to other African languages, belonging to the two *phyla* of central Africa (tables 1.2 and 1.3): Niger-Kordofanian (Niger-Congo stock: Bantu and Adamawa-Ubangian families) and Nilo-Saharan (central Sudanic stock). There is no unique "Pygmy linguistic family" such as the one composed of the "click" languages of southern African "Bushmen" (Khoisan languages, Güldemann and Vossen 2000). The relatively indistinct nature of Pygmy languages shows very clearly that the history of all Pygmy groups is very strongly linked to that of farmers (i.e., non-Pygmy populations). Indeed, it is notable that the socioeconomic—and in many ways phenotypic—distinctions between Pygmies and their neighbors are not paralleled by a linguistic divergence.

Table 1.2. Language groups that include the main Pygmy languages in central Africa.

Phylum	Stock	Family	Number
Niger-Kordofanian	Niger-Congo	Adamawa-Ubangian	1
		Bantoid non-Bantu	1
		Northwest Bantu	8
		Central Bantu	5
Nilo-Saharan	Central Sudanic	Mangbetu-Asua	1
		Mangbutu-Efe	1

Nevertheless, these languages are related in varying ways to farmer (non-Pygmy) languages. Bearing in mind the difference between a dialect and a language—dialects are forms of speech that are mutually intelligible; languages are not (in the absence of learning)—we find two categories: groups that speak a tongue similar to some farmers (the majority) and groups whose tongues are different from farmers. Of the groups in the first category, some (e.g., the BaKola, BaBongo, BaTwa, and BaMbuti/BaSua) are from the Bantu family. The Bedzan speak a dialect of Tikar, a Bantoid language. The Efe speak a tongue that appears to be a dialect very close to the language of Lese farmers (central Sudanic family). The second category consists of three groups that speak languages that are related to, but mutually unintelligible with, farmer languages. These include representatives of different language families. The Baka speak a distinct Ubangian language, the BaAka a distinct Bantu language, and the Asua a distinct central Sudanic language.

Here we must insist upon the obvious: languages that are related to each other share a common language ancestor. In other words, the ancestors of present-day Pygmies acquired their language while living with the ancestors of present-day farmers whose language is in the same family; therefore, they have a (long) common history. With this confirmation, the question is raised of *how* Pygmy communities acquired their languages. As we begin to answer, we can, in light of the linguistic studies, assert what Van Bulck (1952) sensed: there are a few Pygmy languages, but these languages do not form a particular family. Herein lies the "Pygmy paradox": everywhere, Pygmy populations went through a language shift in such a way that it is impossible to find any surviving "Pygmy language." From a linguistic point of view, Pygmies merged into

Table 1.3. Genealogical classification of Pygmy languages.

Pygmy Group	Language Family	Sub-branch	Closest Farmer Language	Status of the Pygmy Language
Niger-Kordofanian, Niger-Congo				
Baka	Ubangi	Gbanzili-Sere	Ngbaka	Language
Bedzan	Bantoid	Non-Bantu	Tikar	Dialect
Kola	Bantoid	Northwest Bantu A80	Mvumbo	Dialect
Koya	Bantoid	Northwest Bantu B20	Ngom	Dialect
Bongo	Bantoid	Northwest Bantu B30	Tsogho	Dialect
Bongo	Bantoid	Northwest Bantu B60	Kaningi	Dialect
Bongo	Bantoid	Northwest Bantu B70	Tege	Dialect
Mikaya	Bantoid	Northwest Bantu C10	Ngundi?	Language
Aka	Bantoid	Northwest Bantu C10	Ngando	Language
Twa	Bantoid	Northwest Bantu C60	Konda	Dialect
Cwa	Bantoid	Central Bantu L30	Luba	Dialect?
Cwa	Bantoid	Central Bantu L30	Hemba	Dialect
Twa	Bantoid	Central Bantu JE10	Kiga	Dialect
Twa	Bantoid	Central Bantu D50	Shi?	Dialect
Twa	Bantoid	Central Bantu JD60	Rundi	Dialect
Sua	Bantoid	Central Bantu D30	Bila	Dialect
Nilo-Saharan, Central Sudanic				
Asua	Moru-Mangbetu	Mangbetu-Asua	Mangbetu	Language
Efe	Moru-Mangbetu	Mangbutu-Efe	Lese	Dialect

Note: Here I follow the classification of Greenberg (1966) for African languages as revised by Ruhlen (1991) and that of Guthrie (1967–71) revised by Maho (2003, 2009) for Bantu languages.

the large families of African languages. In contrast, their ways of life, habitats, technologies, and cultures have usually remained remarkably distinct. For the Pygmies in central Africa, language shift was not linked to a cultural shift (Bahuchet 1993c, 2012).

Relations with Farmers

I must stress that all Pygmy populations, whatever their economic style, live in association with neighboring farmers, but that the nature and degree of the association differs considerably. The long histories of these relations are revealed by linguistic affinities as well as the obvious complementarity of many of their techniques and activities. In some sense, if "Pygmies" do not exist, neither do "farmers." Of course, the term *farmers* is somewhat imprecise; neighboring populations are in many instances hunters or fishermen. Some 155 different ethnic and linguistic communities presently live in the Congo Basin (Joiris and Bahuchet 1994). Thus, *farmers* appears to be a category altogether too general to be very useful, and I agree with Rupp (2003), who criticizes the "homogenization of a great variety of neighbors into one conceptual package of 'villagers' or 'farmers,'" as reinforcing a binary opposition between two categories of people. Central Africa is a cultural and linguistic medley of dozens of populations, each with different languages, cultural settings, and histories, many of them, but not all, having some kind of relationship with Pygmy communities. However, throughout the Congo Basin, the ethnic groups that live in the forest but do not currently interact with Pygmy groups outnumber the groups associated with Pygmies. In the regions where farmers and Pygmies do share the same territories, not all the families in a single village have the option of associating with Pygmies.

Having acknowledged this heterogeneity, it is still possible to recognize certain themes that remain somewhat consistent throughout the relations Pygmies maintain with their neighbors. The latter tend to maintain a strong distinction between themselves and the Pygmy groups, in whom they recognize particularities, both positive and negative. For instance, foragers are often considered great musicians and dancers and are invited to perform during the festivals of the farmers. They are revered as healers, able to cure any kind of mental or physical disease. Patients frequently travel considerable distances for consultation. The respect farmers have for Pygmies has marked limitations, however. For example, there are rarely any occasions when rules prohibiting intermarriage between the foragers and farmers are violated. This results from a deeply entrenched hierarchical social relationship in which the farmer commonly considers the Pygmy to be his property. It is worth reiterating that, while the relationship described above may be typical, or even prevalent, it is in no way universal. The following sections further explore Pygmy-farmer dynamics in their mythological, economic, and social aspects.

Mythology

Perceptions of Pygmies are of a somewhat mixed and ambiguous nature; they are both despised and admired, or even feared—at times simultaneously (Bahuchet and Guillaume 1982; Kazadi 1981). Their significant presence in the mythology of farming peoples throughout the region is amply demonstrated. Several documents attest to the symbolic importance of the Pygmies in some farmers' societies, and the oral traditions of many societies in central Africa grant them a quasi-supernatural symbolic status at the border of the human world (Abega 1997; Arom and Thomas 1974; Bahuchet 1993c; Bahuchet and Guillaume 1982; Delobeau 1989; Grinker 1994; Guillaume 2001; Joiris 2003; Kazadi 1981; Sulzmann 1986; Vorbichler and Brandl 1979; Waehle 1986).

Certain oral traditions tell of encounters with Pygmies during an initial migration. They were guides and initiated the farmers into the forest world, teaching them rites and specific techniques—including some not considered to be typical of hunter-gatherers, such as the forging of iron (for the Ngbaka, Arom and Thomas 1974; for the Beti, Laburthe-Tolra 1981).

In the groups whose sociopolitical system is a monarchy, such as the Tikar (Cameroon) or the Konda (CDR), the Pygmies are considered to be in the same revered category as the sacred king (Bahuchet 1993c; Sulzmann 1986). They are the only persons allowed to touch him during his birth, enthronement, and the period of mourning following his death.

Economy

At the economic level, both societies rely upon the services (food or work) supplied by the other: Pygmies offer forest products (meat, honey, etc.) as well as medicinal treatments. They also assist in slash-and-burn agriculture. In return, farmers procure iron tools, pots, and crops from their fields. This association allows the two partners to better exploit their juxtaposed ecosystems, rainforest on the one hand, fields and secondary forest on the other. The two groups are complementary to one another; they divide their efforts and thereby work less to gain qualitatively more. This is the situation for the groups with a nomadic lifestyle, namely, BaMbuti, Efe, BaAka, and Baka. The nature of exchanges between other villagers and more settled groups (e.g., BaKola, BaBongo) is similar. Other Pygmy groups that rely less on hunting and gathering activities tend to offer more services than products, which, as in the case of the BaTwa of Rwanda, may include musical performance or craftsmanship.

Social Custom, Interaction, and Organization

Like virtually every aspect of their culture, social organization among the peoples of the Congo Basin is highly variable. Patrilineal kinship systems are the most common, though many groups in Gabon are matrilineal. Most groups are organized according to clan structures, with little hierarchy between clans. Conversely, groups such as the Fang are organized in a hierarchical pyramid of lineages and clans. Other groups in the same region, such as the Tikar in central Cameroon, Ntomba in central DRC, and the Kuba in southern DRC, are centralized chiefdoms and kingdoms.

The social dimensions of the interactions between Pygmies and the other populations are important and complex (Bahuchet and Guillaume 1982; Delobeau 1989; Grinker 1994; Joiris 2003; Ngima 2001; Turnbull 1965b; Waehle 1986). They involve, at the same time, internal independence, external dependence, and interactions that are regulated by strong rules.

Both communities are independent as far as social functions are concerned; kinship, social organization, and religion are different between them. Nevertheless, both societies are constructed in such a way that the products of the interrelationship enter into the heart of their networks. To a certain extent, it is possible to consider Pygmies as an integral part of the social system of the farmers: the social system of each needs the other for its reproduction. For instance, among BaAka and Baka groups, iron tools are essential for the bride price. Marriage is the basis of the socioeconomic organization of these groups, and alliance between families leads to strong cooperation between bands. For the villagers, the large amount of meat provided by the Pygmy hunters forms the stock necessary for the ceremonial meetings of the largest communities. These meetings correspond to the end of mourning periods among most of the farmers associated with the BaAka in Central African Republic. In DRC, among the farmers associated with the Efe and the BaSua, large festivals mark the end of the boys' initiation period. This association is sealed by sacred links: members of one group participate in important ceremonies of the other group, thus creating an indestructible fraternity. For instance, when some farmers are initiated into the men's brotherhood of the *Jengi* among the Baka (eastern Cameroon) or when young Efe and BaSua boys are circumcised together with villagers (DRC), the coinitiates become blood brothers—a bond dissolved only by death. This cojoining is also evident in the presence of BaAka people with their most sacred raphia mask of *Ezengi* during the mourning of their villager patrons.

A social hierarchy in which the farmers place themselves above the Pygmies marks external dependency. Though, again, the degree and tightness of this dependency varies, from a very loose exchange relationship to integration into a caste society; for instance, the Ntomba of central DRC is formed with two castes, the BaOto and the BaTwa, and it is the same for the Konda in the same region (Sulzmann 1986; Pagezy 1986). Association can be fixed by various mechanisms of "brotherhood" or "fictive kinship" systems (cf. Joiris 2003; Robillard 2012 for examples among the Baka).

Even if Pygmies and non-Pygmies live together and have many exchanges, avoidance practices are frequent. For instance, in some regions it is impossible for a Pygmy to sit close to a farmer. Of greater salience is the difficulty and rarity of intermarriage. Generally, marriage is permitted between farming males and Pygmy females, but the reverse is almost never true, with a few exceptions in some regions of Gabon (BaBongo).

Polyethnic Systems

As crucial as the Pygmy-farmer relationship is, it is necessary to move beyond this apparent dichotomy to fully appreciate the dynamics that shape Pygmy society and culture. They engage in a complex set of relationships that commonly involves multiple farming groups, as well as many other Pygmy groups, living in the same region. Of great assistance in the analysis of such intricate social webs is the concept of a regional "polyethnic system" (Robillard 2010). Within this framework, the relationship between Pygmies and farmers is only one element among many others, a situation typified by the BaAka, whose range covers much of southern CAR and northern Congo and who are in contact with as many as twenty different ethnic groups. Other examples are found in Cameroon with the Baka, in Ituri with the BaMbuti, and also in central Gabon with the BaBongo.

Conclusion: Are Cultural Differences due to the Relations with the Farmers?

Some twenty distinct ethnic groups are united under the umbrella term *Pygmies*. Though, perhaps the single most unifying factor among them is their very difference, a difference that is manifested in practically innumerable variations but that has somehow marked Pygmies in a somewhat intangible sense as a contiguous group, distinct from neighboring communities. Despite this distinction, and in many ways precisely because of it, Pygmy and farmer groups have also developed and maintained mutually constitutive relationships, each iteration characterized by varying degrees of complexity and intimacy. At present, the issue is to determine whether there are more differences than commonalities among and between Pygmy groups, or whether the reverse is true.

Which aspects, if any, do the people we dubiously refer to as "Pygmies" have universally in common? There have been two long-standing hypotheses: either these groups have nothing in common and are to be considered separately or they have a common origin and their dispersion is to be explained. In the latter case, the location of present groups can be considered as the remnant of an ancient and immense range covering the whole Congo Basin, the population of which was dispersed because of the "invasion" of non-Pygmy groups into the rainforest. Alternatively, there may have been one or several dispersions of the present groups' ancestors from a single, unknown area of origin. Again, this dispersion could have been the result of encounters with non-Pygmy populations.

It is presently impossible to give arguments in favor of either of these hypotheses. Internal comparisons of Pygmy cultural settings, as well as comparative linguistic studies of the Pygmy languages, offer only limited results because neither approach reaches very far into the past. However, two important elements link the history of the Pygmies closely with that of the other populations.

First, there is the specificity of Pygmy linguistics: languages spoken by Pygmy groups are all related to the languages spoken by non-Pygmy groups with whom they are closely associated. This type of binary linguistic relationship exists in all three of the major African linguistic families.

Second, recent genetic studies have yielded important results, confirming both the common origin of the Pygmy populations as well as their genetic divergence. These studies have also demonstrated the various contributions of non-Pygmy populations to the Pygmy genetic pool during episodes of hybridization. Some of the genetic differences between the Pygmy groups reflect their contact with diverse populations throughout history (see Verdu in this volume; Patin et al. 2009; Verdu et al. 2009).

In this context, it is necessary to consider the influence of dispersion, migration and encounter events between populations through time from various demographic, ethnographic, socioeconomic, and ecologic perspectives in order to understand how the cultural diversity observed nowadays among Pygmy populations came to be. Reconstructing these histories will require an immense amount of information of a different nature still lacking at this point in our research and increasingly difficult to gather as the contexts in which Pygmy populations live continue to change and evolve.

Notes

1. African linguistic conventions are to spell the names of languages in a neutral tense, without any mark of plural. In a Bantu name, Ba- is the prefix for the plural form: MoAka (singular) and BaAka (plural). However, to fit with the names more commonly known in anthropological literature, in the following text I will mark prefixes and roots with capital letters: BaAka instead of Aka, BaKola instead of Kola, etc.
2. "Mbuti" is the term used by the Bira people to name the Bantu-speaking Pygmies associated to them.
3. singular: Medzan; plural: Bedzan

References

Abega, S. C. 1997. "Princes et Chimpanzés. Le Pygmée Bedzang Dans les Représentations Mentales des Tikar de Nditam (Cameroun). *Anthropos* 92(4/6):523–34.

Andersson, E. 1983. *Les Bongo-Rimba*. Occasional Papers IX. Uppsala: Uppsala University, 128.

Annaud, M., and C. Leclerc. 2002. *Etude d'impact du PSFE Sur les Populations Rurales et Pygmées and Proposition de Mesures-Mécanismes (Gabon).* Rapport Pour le GEF/World Bank-Gabon. Paris: GEPFE, 90.

Arom, S. 1987 (1978). *Anthologie de la Musique Des Pygmée Aka (Centrafrique).* CD, Paris: OCORA.

Arom, S., and S. Fürniss. 1992. "The Pentatonic System of the Aka Pygmies of the Central African Republic. In *European Studies in Ethnomusicology*, edited by M. P. Baumann. Berlin: International Institute for Traditional Music/Noetzel, 159–73.

Arom, S., and J. M. C. Thomas. 1974. *Les Mimbo, Génies du Piégeage et le Monde Surnaturel des Ngbaka-Ma'bo (R.C.a.).* Paris: SELAF, 153.

Bahuchet, S. 1972. Étude Ecologique d'un Campement de Pygmées Babinga (République Centrafricaine). *Journal d'Agriculture Tropicale et de Botanique Appliquée* 19(12):509–59.

———. 1985. *Les Pygmées Aka et la Forêt Centrafricaine, Ethnologie Ecologique.* Paris: SELAF, 638.

———. 1987. "Le Filet de Chasse des Pygmées Aka (R.C.A.). In *De la Voûte Céleste au Terroir, du Jardin au Foyer: Mosaïque Sociographique. Textes Offerts a Lucien Bernot,* edited by Koechlin, Sigaut, Thomas, Toffin. Paris: Ed. de l'ehess, 209–26.

———. 1992. *Dans la Forêt d'Afrique Centrale; Les Pygmées Aka et Baka.* Paris: Peeters-Selaf, 426.

———. 1993a. "L'invention des Pygmées." *Cahiers d'Etudes Africaines* 33(1):153–81.

———. 1993b. *La Rencontre Des Agriculteurs. Les Pygmées Parmi Les Peuples d'Afrique Centrale.* Paris: Peeters-Selaf. 174 P.

———. 1993c. "History of the Inhabitants of the Central African Rain Forest: Perspectives from Comparative Linguistics." In *Tropical Forests, People and Food,* edited by C. M. Hladik, A. Hladik, O. F. Linares, H. Pagezy, A. Semple, and M. Hadley, "Man and Biophere Series, Vol. 13." Paris/Lancs: Unesco/ Parthenon, 37–54.

———. 2000. "La Tarière a Ignames des Pygmées de L'ouest du Bassin Congolais." In *Aratoires en Afrique, Innovations, Normes et Traces,* edited by C. Seignobos, Y. Marzouk, and F. Sigaut. *Outils* Paris: Karthala-IRD, 237–45.

———. 2012. "Changing Language, Remaining Pygmy." *Human Biology* 84(1): 11–43.

Bahuchet, S., and H. Guillaume. 1982. "Aka-farmer Relations in the Northwest Congo Basin. In *Politics and History in Band Societies,* edited by E. Leacock and R. B. Lee. Cambridge/Paris: Cambridge University Press /M.S.H., 189–211.

Bahuchet, S., D. Mckey, and I. de Garine. 1991. "Wild Yams Revisited: Is Independence from Agriculture Possible for Rain Forest Hunter-gatherers?" *Human Ecology* 19(2):213–43.

Bailey, R. C. 1991. *The Behavioral Ecology of Efe Pygmy Men in the Ituri Forest, Zaire.* Ann Arbor: Museum of Anthropology, University of Michigan, 144.

Bailey, R. C., and I. De Vore. 1989. "Research on the Efe and Lese Populations of the Ituri Forest, Zaïre." *American Journal of Physical Anthropology* 78:459–71.

Bailey, R. C., G. Head, M. Tenike, B. Owen, R. Rechtman, and E. Zechenter. 1989. "Hunting and Gathering in Tropical Rainforest: Is It Possible?" *American Anthropologist* 91:59–82.

Bailey, R. C., and N. R. Peacock. 1988. "Efe Pygmies of Northeast Zaïre: Subsistence Strategies in the Ituri Forest." In *Coping with Uncertainty in Food Supply,* edited by I. de Garine and G. A. Harrison. Oxford: Oxford University Press, 88–117.

Barham, L. 2006. "Batwa in the Mist." *Before Farming* 4(9):1–4.

Biesbrouck, K. 1999. "Agriculture among Equatorial African Hunter-gatherers and the Process of Sedentarization: The Case of the Bagyeli in Cameroon." In

Challenging Elusiveness: Central African Hunter-gatherers in a Multidisciplinary Perspective, edited K. Biesbrouck, S. Elders, and G. Rossel. Leiden: CNWS, 189–206.
Biesbrouck, K., S. Elders, G. Rossel, eds. 1999. *Central African Hunter-Gatherers in a Multidisciplinary Perspective: Challenging Elusiveness*. Leiden: Research School for Asian, African and Amerindian Studies (CNWS) Universiteit Leiden, 332.
Boyi, J. *Les Mikaya*. Mémoire de DEA, Unpublished, EHESS, Paris.
Brisson, R. 1984. *Lexique Français-Baka*. Douala: BP 5351, 396.
———. 2010a. *Lexique Français-Baka, Sud-Cameroun*. Paris: L'Harmattan, 274.
———. 2010b. *Petit Dictionnaire Baka-Français, Sud-Cameroun*. Paris: L'Harmattan, 640.
Brisson, R, and D. Boursier. 1979. *Petit Dictionnaire Baka-Français*. Douala: B.P. 1855, 506.
Cloarec-Heiss, F, and J. M. C. Thomas. 1978. *L'aka, Langue Bantoue des Pygmées de Mongoumba (Centrafrique): Introduction a L'étude Linguistique, Phonologie*. Paris: SELAF, 204.
Costermans, B. 1947. "Muziëk-Instrumenten Van Watsa-Gombari en Omstreken." *Zaïre, Revue Congolaise* 6, Vol. 1, 629–63.
Delobeau, J. M. 1989. *Yandenga et Yamonzombo. Les Relations Entre les Villages Monzombo et Les Campements Pygmées Aka Dans la Sous-Préfecture de Mongoumba (Centrafrique)*. Paris: Peeters-SELAF, 268.
Demesse, L. 1978. *Changements Techno-Economiques et Sociaux Chez les Pygmées Babinga (Nord-Congo et Sud-Centrafrique)*. Paris: SELAF, 262.
———. 1980. *Techniques et Economie des Pygmées Babinga*. Paris: Institut d'Ethnologie, 302.
Demolin, D. 1990. *Chants de l'orée de la Forêt. Polyphonies des Pygmées Efe*. CD; Fonti Musicali Fmd 185.
———. 1991–1992. *Le Mangbetu. Etude Phonétique et Phonologique* [Phd]. Bruxelles: Université Libre De Bruxelles, 331+ Annexes.
———. 1993. "Les Rêveurs de la Forêt. Polyphonies des Pygmées Efe de l'Ituri (Zaïre)." *Cahiers de Musiques Traditionnelles* 6:139–51.
Dounias, E. 2001a. "The Management of Wild Yam Tubers by the Baka Pygmies in Southern Cameroon. *African Study Monographs*, supplement 26:135–56.
———. 2001b. "Les Tarières à Ignames Sauvages des Pygmées Aka et Baka d'Afrique Centrale. *Techniques et Culture* 37:127–54.
Du Chaillu, P. B. 1867. *A Journey to Ashango-Land and Further Penetration into Equatorial Africa*. London, Murray, 501.
Elshout, P. 1963. *Les Batwa des Ekonda*. Tervuren: Musée Royal De l'Afrique Centrale, 65.
Estermann, C. 1962. "Les Twa du Sud-Ouest de l'Angola." *Anthropos* 57:465–74.
———. 1976. "The Southern Twa." In *The Ethnography of Southwestern Angola*, edited by G. D. Gibson. Vol. 1, *The Non-Bantu Peoples and the Ambo Ethnic Group*. New York: Africana Publishing Co., 31–5.
Fürniss, S., and S. Bahuchet. 1995. "Existe-t-il des Instruments de Musique Pygmées?" In *Ndroje Balendro, Musiques, Terrains et Disciplines (Textes Offerts à Simha Arom*, edited by V. Dehoux, S. Fürniss, S. Le Bomin, E. H. R. Olivier, and F. Voisin. Paris: Peters-Selaf, 87–109.
Gansemans, J. 1993. *Polyphonies des Twa du Rwanda*. Bruxelles: Fonti Musicali, CD FMD196.

Greenberg, J. H. 1966. *The Languages of Africa.* Bloomington/The Hague: Indiana University/Mouton & Co., 180.
Grinker, R. R. 1994. *Houses in the Rainforest: Ethnicity and Inequality among Farmers and Foragers in Central Africa.* Berkeley: University of California Press, 226.
Guillaume, H. 1982. *Chasseurs Pygmées: Pygmées Aka d'Afrique Centrale.* Paris: ORSTOM/SELAF, Disque 33t./30 cm., CETO795.
———. 2001. *Du Miel au Café, de l'ivoire à l'acajou. La Colonisation de l'interfluve Sangha-Oubangui et l'évolution des Rapports Entre Chasseurs-Collecteurs Pygmées Aka et Agriculteurs (Centrafrique, Congo) 1880–1980.* Leiden: Peeters-Selaf, 786.
Güldemann, T., and R. Vossen. 2000. Khoisan. In *African Languages,* edited by B. Heine and D. Nurse. Cambridge: Cambridge University Press, 99–122.
Guthrie, M. 1967–1971. *Comparative Bantu.* Hants: Gregg.
Haller, T, and S. Merten. 2008. "'We Are Zambian—Don't Tell Us How to Fish!' Institutional Change, Power Relations and Conflicts in the Kafue Flats Fisheries in Zambia." *Human Ecology* 36:699–715.
Harako, R. 1976. "The Mbuti as Hunters: A Study of Ecological Anthropology of the Mbuti Pygmies (I)." *Kyoto University African Studies* 10:37–99.
———. 1981. "The Cultural Ecology of Hunting Behavior among the Mbuti Pygmies in the Ituri Forest, Zaire." In *Omnivorous Primates,* edited by R. Harding and G. Teleki.. New York: Columbia University Press, 499–555.
Hardin, R. 1999. *Translating the Forest: Tourism, Trophy Hunting and the Transformation of Forest Use in Central African Republic (CAR).* [Phd]: Yale University, 676.
Hart, T. B., and J. A. Hart. 1986. "The Ecological Basis of Hunter-gatherer Subsistence in African Rain Forests: The Mbuti of Eastern Zaire." *Human Ecology* 14(1):29–55.
Headland, T. 1987. "The Wild Yam Question: How Well Could Independent Hunter-gathers Live in a Tropical Rainforest Ecosystem?" *Human Ecology* 15:463–91.
Hewlett, B. S. 1996. "Cultural Diversity among African Pygmies." In *Cultural Diversity among Twentieth-century Foragers: An African Perspective,* edited by S. Kent. Cambridge: Cambridge University Press, 215–44.
———. 1991. *Intimate Fathers: The Nature and Context of Aka Pygmy Paternal Infant Care.* Ann Arbor: University of Michigan Press, 202.
Hewlett, B. S., and J. M. Fancher. 2010. *Bibliography of Congo Basin Hunter-gatherer Research.* International Conference on Congo Basin Hunter-gatherers, Montpellier, France, Sept. 2010, 89.
Hiernaux, J. 1966. "Les Bushong et les Cwa du Royaume Kuba: Pygmées, Pygmoïdes et Pygméisation; Anthropologie, Linguistique et Expansion Bantoue." *Bulletins et Mémoires de la Société d'Anthropologie de Paris* 9:299–336.
Ichikawa, M. 1978. "The Residential Groups of the Mbuti Pygmies." *Senri Ethnological Studies 1 (Africa 1)*:131–88.
———. 1981. "Ecological and Sociological Importance of Honey to the Mbuti Net Hunters, Eastern Zaire." *African Study Monographs* 1:55–68.
———. 1983. "An Examination of the Hunting-dependant Life of the Mbuti Pygmies, Eastern Zaïre. *African Studies Monographs* 4:55–76.
———, ed. 1998. "Man and Nature in Central African Forests. Kyoto." *African Study Monographs.* supplement N 25.

Ichikawa, M., and D. Kimura, eds. 2003. "Recent Advances in Central African Hunter-gatherer Research. Kyoto" *African Study Monographs,* supplement N 28:158.

Ivey, P. K. 2000. "Cooperative Reproduction in Ituri Forest Hunter-gatherers: Who Cares for Efe Infants?" *Current Anthropology* 41(5):856–66.

Joiris, D. V. 1992. "Entre le Village et la Forêt: Place des Femmes Bakola et Baka dans des Sociétés en Voie de Sédentarisation." In *Relations de Genre et Développement, Femmes et Sociétés,* edited by F. Pinton and F. Lecarme. Paris: ORSTOM, 125–48.

———. 1994. "Elements of Techno-economic Changes among the Sedentarized Bagyeli Pygmies (South-west Cameroon)." *African Study Monographs* 15(2):83–95.

———. 1996. "Ce Que 'Bien Manger' Veut Dire chez les Pygmées Kola (Gyeli) et Baka du Sud-Cameroun." In *Bien Manger et Bien Vivre. Anthropologie Alimentaire et Développement en Afrique Intertropicale,* edited by A. Froment, I. de Garine, C. Binam Bikoi, and J. F. Loung. Paris: ORSTOM-L'Harmattan, 365–70.

———. 1998. *La Chasse, la Chance, le Chant: Aspect du Système Rituel des Baka du Cameroun* [Thèse de Doctorat]: Université Libre de Bruxelles, 448.

———. 2003. "The Framework of Central African Hunter-gatherers and Neighbouring Societies." *African Study Monographs,* supplement 28:57–79.

Joiris, D. V., and S. Bahuchet. 1994. "Afrique Équatoriale." In *Situation des Populations Indigènes des Forêts Denses et Humides,* edited by S. Bahuchet, 387–448. Bruxelles: Commission Européenne.

Kazadi, M. 1981. "Méprisés et Admirés: L'ambivalence des Relations entre les Bacwa (Pygmées) et les Bahemba (Bantu)." *Africa* 51(4):837–47.

Kent, S., ed. 1996. *Cultural Diversity among Twentieth-century Foragers: An African Perspective.* Cambridge: Cambridge University Press, 344.

Kilian-Hatz, C. 1989. *Contes et Proverbes des Pygmées Baka.* Avec Une Introduction au Peuple et a la Langue Baka par R. Brisson et D. Boursier. Paris: ACCT, 262.

———. 1995. *Das Baka: Grundzüge Einer Grammatik aus der Grammatikalisierungs—Perspektive.* Afrikanistische Monographien (AMO) 6. Köln: Universität Zu Köln.

———. 2008. *Contes des Pygmées Baka du Cameroun.* Schriften Zur Afrikanistik–Research in African Studies 14. Frankfurt: Peter Lang.

Kitanishi, K. 1995. "Seasonal Changes in the Subsistence Activities and Food Intake of the Hunter-gatherers in Northeastern Congo." *African Study Monographs* 16(2):73–118.

———. 1997. *Hunters and Farmers of the Equatorial Rainforest: Economy and Society, C. 3000 BC to 800 AD* [Phd]. Los Angeles: University of California.

———. 2003. *"The Pygmies Were Our Compass": Bantu and Batwa in the History of West Central Africa, Early Times to C. 1900 C.E.* Portsmouth: Heinemann, 254.

Knight, J. 2003. "Relocated to the Roadside: Preliminary Observations on the Forest Peoples of Gabon. *African Studies Monographs,* supplement 28:81–121.

Koppert, G., A. Froment, S. Bahuchet, and G. Ngima Mawoung. 1997. *Survey of Pygmy Populations: Lolodorf to Kribi Area, Republic of Cameroon.* Paris: GEPFE/LACITO, 44.

Laburthe-Tolra, P. 1981. *Les Seigneurs de la Forêt. Essai sur le Passé Historique, L'organisation Sociale et les Normes Ethiques des Anciens Beti du Cameroun.* Paris: Publications de la Sorbonne, 490.

Leacock, E, R. B. Lee, eds. 1982. *Politics and History in Band Societies.* Paris/ Cambridge: CUP/MSH, 500.
Le Bomin, S., and J. E. Mbot. 2012a. "Sur les traces de l'histoire des Pygmées du Gabon: résultats de cinq ans de prospections." *Journal des Africanistes* 82(1–2):277–318.
———. 2012b. "The Musical Heritage of the Gabon Bongo: Heritage under Influence." *Before Farming* [Online Version] 2012/1 Article 3.
Leclerc, C. 1995. *Le Rapport à la Nature Comme Rapport Social. Les Pygmées Bedzan: Entre la Forêt, la Savane et les Tikar (Cameroun)* [DEA]. Nanterre: Paris X, 197.
———. 1999. "Des Pygmées Entre la Forêt et la Savane: Etude Comparative de Deux Campements Medzan (Vallée du Mbam, Cameroun)." In *Challenging Elusiveness: Central African Hunter-gatherers in a Multidisciplinary Perspective,* edited by K. Biesbrouck, S. Elders, G. Rossel.. Leiden: CNWS, 169–86.
———. 2001. *En Bordure de Route: Espace Social, Dynamisme et Relation a l'environnement Chez les Pygmées Baka du Sud-Est Cameroun* [Thèse Nouveau Régime]. Nanterre: Paris X, 372.
———. 2012. *L'adoption de l'agriculture Ches les Pygmées Baka du Cameroun. Dynamique Sociale et Continuité Structurale.* Paris, Versailles: Editions de la MSH/Editions Quae, 244.
Leroy, A. 1928 (1897). *Les Pygmées: Négrilles d'Afrique et Négritos d'Asie.* Paris: Procure Générale des Pères du St-Esprit, 367.
Lewis, J. 2000. *The Batwa Pygmies of the Great Lakes Region.* London: Minority Rights Group International, 32.
———. 2002. *Forest Hunter-gatherers and Their World: A Study of the Mbendjele Yaka Pygmies of Congo-Brazzaville and Their Secular and Religious Activities and Representations* [Phd]. London: London School of Economics and Political Science, 312.
Lewis, J., and J. Knight. 1996. *Les Twa du Rwanda. Rapport d'évaluation de la Situation des Twa et Pour la Promotion des Droits des Twa Dans le Rwanda d'après-Guerre*: World Rainforest Movement, IWGIA, and Survival International, 118.
Loung, J. F. 1959. "Les Pygmées de la Forêt de Mill. Un Groupe de Pygmées Camerounais en Voie de Sédentarisation." *Les Cahiers d'Outre-Mer* 48:362–79.
———. 1987. "Le Nom Authentique du Groupe Pygmée de la Région Côtière Camerounaise". *Revue de Géographie du Cameroun* 7(2):81–94.
———. 1996. "L'insuffisance des Féculents Sauvages Comestibles et Ses Conséquences Chez les Pygmées Bakola du Cameroun." In *Bien Manger et Bien Vivre. Anthropologie Alimentaire et Développement en Afrique Intertropicale,* edited by A. Froment, I. de Garine, C. Binam Bikoi, J. F. Loung. Paris: ORSTOM-L'Harmattan, 173–94.
Macrae, F. B. 1934. "The Lukanga Swamps." *The Geographical Journal* 83(3):223–7.
Maho, J. F. 2003. "A Classification of the Bantu Languages: An Update of Guthrie's Referential System." In *The Bantu Languages,* edited by Derek Nurse and Gérard Philippson. Routledge Language Family Series 4. London: Routledge, 639–51.
———. 2009, *NUGL Online.* The Online Version of the New Updated Guthrie List, a Referential Classification of the Bantu Languages. goto.glocalnet.net/ mahopapers/nuglonline.pdf.
Matsuura, N. 2006. "Sedentary Lifestyle and Social Relationship among Babongo in Southern Gabon." *African Studies Monographs,* supplement 33:71–93.

———. 2011. "Historical Changes in Land Use and Interethnic Relations of the Babongo in Southern Gabon." *African Studies Monographs* 32(4):157–76.

Mayer, R. 1987. "Langues des Groupes Pygmées au Gabon: Un Etat des Lieux." *Pholia* 2:111–24.

Mebenga Tamba, L. 1998. "Les Campements Pygmées-Bedzan du Pays Tikar." In *Paléo-Anthropologie en Afrique Centrale: Un Bilan de l'Archéologie au Cameroun*, edited by M. Delneuf, J. M. Essomba, A. Froment. Paris: L'Harmattan, 323–30.

Motte, E. 1980. *Les Plantes Chez les Pygmées Aka et les Mozombo de la Lobaye (Centrafrique). Contribution a Une Ethnobotanique Comparative Chez des Chasseurs-Cueilleurs et des Pêcheurs-Agriculteurs Vivant Dans Un Même Milieu Végétal.* Paris: SELAF, Bibl. 80–82, 546.

Ngima Mawoung, G. 1996. "Economie-Consommation des Produits-Artisanat-Budgets Familiaux Chez les Pygmées Bakola de la Région Côtière du Cameroun." In *Bien Manger et Bien Vivre. Anthropologie Alimentaire et Développement en Afrique Intertropicale: Du Biologique au Social*, edited by A. Froment, I. de Garine, C. Binam Bikoi, and J. F. Loung. Paris: L'Harmattan-ORSTOM, 195–200.

———. 2001. "The Relationship between the Bakola and the Bantu Peoples of the Coastal Regions of Cameroon and Their Perception of Commercial Forest Exploitation." *African Studies Monographs*, supplement 26:209–35.

Pagezy, H. 1975. "Les Interrelations Homme-Faune de la Forêt du Zaïre." In *L'homme et l'animal. 1er Colloque d'ethnozoologie.* Paris: Institut International d'Ethnosciences, 63–88.

———. 1976. "Quelques Aspects du Travail Quotidien des Femmes Oto et Twa Vivant en Milieu Forestier Equatorial (Lac Tumba, Zaïre). *L'Anthropologie* 80(3):465–90.

———. 1986. "The Food System of the Ntomba of Lake Tumba, Zaïre." In *Food Systems in Central and Southern Africa*, edited by J. Pottier.. London: SOAS, 61–79.

———. 1988. *Contraintes Nutritionnelles en Milieu Forestier Equatorial Liées a La Saisonnalité et la Reproduction: Réponses Biologiques et Stratégies de Subsistance Chez les Ba-Oto et les Ba-Twa (Lac Tumba, Zaire)* [Thèse de Doctorat d'État]. Aix-Marseille: Université d'Aix-Marseille, Faculté des Sciences, 500.

Patin, E. et al. 2009. "Inferring the Demographic History of African Farmers and Pygmy Hunter-gatherers Using a Multilocus Resequencing Data Set." *Plos-Genetics* 5(4):2–13.

Paulin, P. 2010. *Les Baka du Gabon Dans Une Dynamique de Transformations Culturelles Perspectives Linguistiques et Anthropologiques.* Thèse, Université Lumière Lyon 2, 626.

Peacock, N. R. 1984. "The Mbuti of Northeast Zaïre: Women and Subsistence Exchange." *Cultural Survival Quarterly* 8(2):15–17.

———. 1985. *Time Allocation, Work and Fertility among Afe Pygmy Women of Northeastern Zaire* [PhD]. Cambridge: Harvard University, 157+39.

Périquet, 1915–1916. *Rapport Général Sur la Mission de Délimitation A.E.F.-Cameroun* (1912–1914), III. Flore, Faune, Cultures, Animaux Domestiques, 3 Vol.

Plisnier-Ladame, F. 1970. *Les Pygmées, Enquête Bibliographique.* Bruxelles: CEDESA, XXIX + 258.

Regnault, D. R. 1911. "Les Babenga (Négrilles de la Sanga)." *L'Anthropologie* XXII–3:261–88.

Renaud, P. 1976. *Description Phonologique et Eléments de Morphologie Nominale d'une Langue de Pygmées du Sud-Cameroun: Les Bajèlè (Bipindi)* [Thèse 3e Cycle]: Paris III-Sorbonne Nouvelle, 375.

Robillard, M. 2010. *Pygmées Baka et Voisins Dans la Tourmente des Politiques Environnementales en Afrique Centrale*. Thèse, Paris: Muséum National D'histoire Naturelle, 490.

———. 2012. "De la Nécessité d'étudier les Relations Interethniques Pour Appréhender la Dynamique du Changement: Le Cas des Baka et des Fang-Mvè de Minvoul (Gabon)." *Journal des Africanistes* 82(1–2):137–165.

Rosen, E. 1925. "The Swedish Rhodesia-Congo Expédition, 1911–192." *Geografiska Annaler* 7:9–17.

Rottland, F., and R. Vossen, eds. 1986. "Afrikanische Wildbeuter. African Hunter-gatherers. Chasseurs et Cueilleurs en Afrique." Internationales Symposion. Hamburg: *Sprache Und Geschichte in Afrika*. Vol. 7–1/2:458+465.

Ruhlen, M. 1991. *A Guide to the World's Languages*. Vol. 1, *Classification*. Stanford: Stanford University Press, 464.

Rupp, S. K. 2001. *I, You, We, They: Forests of Identity in Southeastern Cameroon* [PhD]. New Haven: Yale University, 573.

———. 2003. "Interethnic Relations in Southeastern Cameroon: Challenging the 'Hunter-gatherer'— 'Farmer' Dichotomy." *African Studies Monographs*, supplement 28:37–56.

———. 2011. *Forests of Belonging: Identities, Ethnicities, and Stereotypes in the Congo River Basin*. Seattle: University of Washington Press, 306.

Sato, H. 1992. "Notes on the Distribution and Settlement Pattern of Hunter-gatherers in Northwestern Congo." *African Study Monographs* 13(4):203–16.

———. 1993. "Food Restriction on Elephants among the Baka Pygmies in the Northwestern Congo." *Reports of Liberal Arts*. Hamamatsu University School of Medicine 7:19–30.

Sawada, M. 1998. "Encounters with the Dead among the Efe and the Balese in the Ituri Forest: Mores and Ethnic Identity Shown by the Dead." *African Studies Monographs*, supplement 25:85–104.

Schebesta, P. 1938. *Die Bambuti-Pygmäen Von Ituri*, Bd I. *Geschichte, Geographie, Umwelt, Demographie Und Anthropologie Der Ituri-Bambuti*. Bruxelles, 438.

———. 1941. *Die Bambuti-Pygmäen Von Ituri*, Bd II. *Ethnographie Der Ituri-Bambuti, 1 Teil : Die Wirtschaft Der Ituri-Bambuti (Belgisch Kongo)*. Bruxelles, 284.

———. 1948. *Die Bambuti-Pygmäen Von Ituri*, Bd II– *Ethnographie Der Ituri-Bambuti, Teil II: Das Soziale Leben*. Bruxelles, 551.

———. 1952. *Les Pygmées du Congo Belge*. Bruxelles, 432.

Schultz, M. 1986. "Economic Relations Between the Batua and Baoto of Bibelo Village, Bikoro Zone, Zaïre: A Preliminary Report on New Fieldwork." *Sprache Und Geschichte in Afrika* 7(2):339–59.

Schumacher, P. 1949–1950. *Expedition Zu Den Zentralafrikanischen Kivu Pygmäen (Twiden)*. I. *Die Physische Und Sociale Umwelt Der Kivu-Pygmäen*. II. *Die Kivu-Pygmäen*. Brussels: Mémoires de l'Institut Royal Colonial Belge, 2 Vol.

Seitz, S. 1993. *Pygmées d'Afrique Centrale*. Translated by L. Bouquiaux and G. Lex. Paris: Peeters/SELAF, 358.

Seiwert, J. 1926. "Die Bagielli, Ein Pygmäenstamm des Kamerun Urwaldes." *Anthropos* 21(1–2):127–47.

Soengas, B. 2009. "Preliminary Ethnographic Research on the Bakoya in Gabon." *African Studies Monographs* 30(4):187–208.

———. 2010. *La Subsistance des Pygmées Bakoya à L'épreuve de l'agriculture: Dynamique des Savoirs Ethnobotaniques et des Pratiques. (Département de la Zadié, Ogooué-Ivindo, Gabon).* Thèse, Paris: Muséum National d'histoire Naturelle.

———. 2012. "Des Pygmées cultivateurs, les Bakoya: changements techniques et sociaux dans la foret gabonaise." *Journal des Africanistes* 82(1–2):167–192.

Sulzmann, E. 1986. "Batwa Und Baoto—Die Symbiose Von Wildbeutern Und Pflanzern Bei Den Ekonda Und Bolia (Zaïre)." *Sprache Und Geschichte in Afrika* 7(1):369–89.

Tanno, T. 1976. "The Mbuti Net-Hunters in the Ituri Forest, Eastern Zaire: Their Hunting Activities and Band Composition." *Kyoto University African Studies* 10:101–35.

———. 1981. "Plant Utilization of the Mbuti Pygmies." *African Studies Monographs* 1:1–53.

Terashima, H. 1980. "Hunting Life of the Bambote: An Anthropological Study of Hunter-gatherers in a Wooded Savanna." *Senri Ethnological Studies* 6:223–68.

———. 1983. "Mota and Other Hunting Activities of the Mbuti Archers." *African Study Monographs* 3:71–85.

———. 1985. "Variation and Composition Principles of the Residence Group (Band) of the Mbuti Pygmies." *African Study Monographs*, supplement 4:103–20.

———. 1987. "Why Efe Girls Marry Farmers?: Socio-ecological Backgrounds of Inter-ethnic Marriage in the Ituri Forest of Central Africa." *African Studies Monographs*, supplement 6:65–83.

———. 1998. "Honey and Holidays: The Interactions Mediated by Honey between Efe Hunter-gatherers and Lese Farmers in the Ituri Forest." *African Studies Monographs*, supplement 25:123–34.

Terashima, H., M. Ichikawa, and M. Sawada. 1988. "Wild Plant Utilization of the Balese and the Efe of the Ituri Forest (Zaïre)." *African Study Monographs*, supplement 8:1–78.

Tessman, G. 1913. *Die Pangwe*. Ernst Wasmuth A.-G., Berlin.

Thomas, J. M. C. 1991. "La Langue des Aka." In *Encyclopédie des Pygmées Aka*, Livre I, Fascicule 4, "Les Pygmées Aka," edited by J. M. C. Thomas and S. Bahuchet. Paris: Selaf-Peeters, 1–182.

Thomas, J. M. C., and S. Bahuchet, eds. 1981. *Encyclopédie des Pygmées Aka—Techniques, Langage et Société des Chasseurs-Cueilleurs de la Forêt Centrafricaine*. Paris: SELAF. 18 Vol.

Tilquin, O. 1997. *Esquisse Ethnoécologique et Sociologique d'une Population de Pygmées Sédentarisés. (Les Bakola de l'axe Mbomo-Mbandza, Nord Congo).* Rapport, Bruxelles: APFT/ULB, 1–97.

Tsuru, D. 1998. "Diversity of Ritual Spirit Performances among the Baka Pygmies in Southeastern Cameroon." *African Study Monographs*, supplement 25:47–84.

Turnbull, C. M. 1965a. *The Mbuti Pygmies: An Ethnographic Survey*. New York: American Museum of Natural History, 140–282.

———. 1965b. *Wayward Servants: The Two Worlds of the African Pygmies*. New York: The Natural History Press, 392.

——. 1983. *The Mbuti Pygmies: Change and Adaptation*. New York: Holt, Rinehart, and Winston, 162.
UNEP. 2005. *Bwindi Impenetrable National Park, Uganda*. United Nations Environment Programme—World Conservation Monitoring Centre, 7.
Vallois, H. V., and P. Marquer. 1976. *Les Pygmées Baka du Cameroun: Anthropologie et Ethnographie*. Paris: Muséum National d'Histoire Naturelle, 196.
Van Bulck, G. V. 1952. "Existe-t-il Une Langue des Pygmées en Afrique Centrale?" *Kultur Und Sprache* 9:365–96.
Van Everbroeck, N. 1974. *Ekond'e Mputela. Histoire, Croyances, Organisation Clanique, Politique, Sociale et Familiale des Ekonda et de Leurs Batoa*. Tervuren: Musée Royal de l'Afrique Centrale, Archives d'anthropologie N 21, 306.
Vansina, J. 1954. "Note Sur les Twa du Territoire de Mweka (Kasaï)." *Zaïre* 7:729–32
——. 1986. "Do Pygmies Have a History?" *Sprache Und Geschichte in Afrika* 7(1):431–45.
Van Thiel, P. 1977. *Multi-Tribal Music of Ankole*, Tervuren, Annales Sciences Hum. M.R.A.C., 234.
Verdu, P., F. Austerlitz, A. Estoup, R. Vitalis, M. Georges, S. Théry, A. Froment, S. Le Bomin, A. Gessain, J-M Hombert et al. 2009. "Origins and Genetic Diversity of Pygmy Hunter-gatherers from Western Central Africa." *Current Biology* 19:312–8.
Vorbichler, A. 1965. *Die Phonologie Und Morphologie des Balese (Ituri-Urwald, Zaïre)*. Glückstadt: Afrikanistische Forschungen N 2.
Vorbichler, A., and R. M. Brandl. 1979. *Die Oralliteratur Der Balese-Efe Im Ituri-Wald (Nordost Zaïre)*. St. Augustin: Anthropos Institut, 349.
Wachsmann, K. P. 1953. "The Sound Instruments." In *Tribal Crafts of Uganda*, edited by Trowell and Wachsmann. London, O.U.P., 311–417.
Waehle, E. 1986. "Efe (Mbuti Pygmy) Relations to Lese-Dese Villagers in the Ituri Forest, Zaïre: Historical Changes during the Last 150 Years." *Sprache Und Geschichte in Afrika* 7(2):375–411.
Wegner, U. 1984. *Afrikanische Saiten-Instrumente*. Berlin: Museum Für Völkunde, 306.

2

Population Genetics of Central African Pygmies and Non-Pygmies[1]

Paul Verdu

> *In reality, Pygmies do not exist. The humans that do exist are called Baka, BaBongo, BaKola, BaAka, BaSua, Efe, Asua, BaTwa . . .*
> —Serge Bahuchet, "L'invention des Pygmees" (1993)

Anthropologists and ethnographers have described extensively the cultural diversity of central African Pygmy societies over the past seventy years. However, little is known about the origins of Pygmies and neighboring non-Pygmies, mainly because of a lack of paleoarchaeological, historical, and demographic data concerning the Congo Basin. In this chapter, we present how human population geneticists since the 1960s have shed new light on the previously unknown peopling history of central Africa. In particular, we will show that interdisciplinary research between biologists and cultural anthropologists, rooted in joint fieldwork, is a powerful tool to understand human biological diversity.

First, we introduce the main paradigms underlying population genetics studies of biological diversity and briefly illustrate several findings concerning human origins and migrations. Second, we highlight the lack of knowledge about the peopling history of central Africa and describe the complex cultural criteria underlying the "Pygmy and non-Pygmy" population categorization in a multiethnic context. Third, we show how population genetics and ethnographic definitions of *a population* can be reconciled through interdisciplinary approaches and joint fieldwork. Thus we establish the framework for studying central African genetic diversity and use population genetics methods to reconstruct the peopling history of Pygmies and non-Pygmies. In particular, we show how population geneticists have investigated questions regarding (1) the common or independent origin of central African Pygmy and non-Pygmy populations, (2) migration patterns among central African populations, and (3) the evolution of differential adult height between Pygmies and non-Pygmies.

Population Genetics Approaches of Human Origins and Past Migrations

Where do I come from? Why do we all look so similar and yet so different? Most people may find these questions vertiginous, and trying to address them is the fundamental challenge for human biologists. Since the 1950s and the discovery of the chemical structure of DNA, one of the major molecular bases of biologic inheritance, genetics provided rapidly growing sets of tools to reconstruct the history of mankind, especially when archaeological data are lacking. Any individual's DNA was inherited from half the DNA of each one of his or her parents, who themselves inherited their DNA from their respective parents, and so on at each generation. New mutations, that is, new variants at a given location in the genome often called "genetic markers," arise randomly in the parental DNA and are transmitted to the offspring who will in turn transmit them to their descendants.

An individual's DNA is thus a complex mosaic of DNA chunks inherited with mutations through the whole parental genealogy. This is why the DNA sequence of every existing individual is different[2] and why certain mutations are found in biological families and populations while absent in others. In a sense, the DNA sequence of an individual carries the coded history of his or her whole biological parenthood since the emergence of *Homo sapiens*. Therefore, by studying the distribution of given mutations across individuals, families, populations, and groups of populations, population geneticists try to reconstruct the history of human lineages that gave birth to existing individuals.

Finally, a small molecule of DNA is present in the mitochondria, a cell compartment responsible, among other functions, for producing the energy necessary to cell metabolism. This mitochondrial DNA is transmitted almost exclusively by mothers across generations. Therefore, by studying the mitochondrial DNA mutations, population geneticists focus specifically on the history of maternal lineages. In addition to the twenty-two pairs of DNA molecules (chromosomes) present in each cell nucleus of the organism, individuals carry a pair of sex chromosomes. Males carry an X and a Y chromosome, the Y chromosome being transmitted exclusively from fathers to sons. Therefore, as opposed to mitochondrial DNA, mutations on the Y chromosome carry the genealogical history of the whole paternal lineages.

The distribution of genetic variation across individuals is not only determined by how and when new mutations emerge or by the way parents transmit half of their DNA to their offspring. It is also influenced by natural selection forces. For an instance of positive natural selection, imagine a new mutation increasing the reproductive success of his or her carrier. This individual will have more children than individuals who do not carry this mutation and his or her children will in turn transmit this mutation to their numerous offspring. Generation after generation, the frequency of this mutation will increase in the whole population. Conversely, for negative natural selection, a mutation decreasing the reproductive success of individuals carrying them will decrease in population frequency over time. Therefore, population geneticists can study the distribution of genic mutations among existing individuals to identify potential signatures of past natural selection forces that have influenced the evolution and adaptation of our species.

Demographic forces also strongly influence the distribution of mutations within and among populations. For instance, imagine that an individual dies accidentally before reproducing. If the population is of small census size, the unique mutations he or she carried may thus disappear from the whole population, whether these mutations provided a better reproductive success to the carrier or not. Conversely, in a large population, several individuals are likely carrying some mutations inherited from their common ancestors. Therefore, the same "accidental" loss of individuals and mutations will affect the population frequency of mutations more slowly than in a small census population. This stochastic process of loss or fixation of mutations in a population, called *genetic drift*, thus strongly influences the evolution of population mutation frequencies over time in interaction with population size and demographic expansions or contractions. Finally, migration and admixture between previously isolated populations are other major demographic forces influencing observed patterns of genetic diversity. For instance, genetic variants specific to a given population can appear in another population from where they were previously absent through the migration and mating of individuals. Therefore, by studying how neutral[3] genetic variation is distributed across individuals among families and populations, population geneticists can infer the history of divergences, isolations, migrations, and demographic expansions or contractions that have likely given birth to existing human populations. Finally, by dating when mutations appear in individuals and genealogies, it is possible to date the occurrence of these various demographic events.

In this context, population genetics studies have vastly furthered our understanding of human origins and biological evolution. Among the major findings of the past decades, it was shown that the worldwide genetic diversity patterns were consistent with an African origin of all existing human populations. The most up to date model suggests that the current peopling of the world results from serial migration "Out of Africa" over the past 100,000 years. In this model, mankind expanded throughout the world via a series of migrations increasingly distant from the African source, each migration followed by founding events of new populations (Cavalli-Sforza and Feldman 2003; DeGiorgio et al. 2009). Interestingly, population geneticists found that an eastward migration initiated roughly 60,000 years ago via a coastal route out of the Horn of Africa, in addition to an earlier migration route out of northeast Africa, could jointly explain the genetic patterns found in populations living nowadays in southwest, south-central, and southeast Asia (Macaulay et al. 2005; Quintana-Murci et al. 1999).

Importantly, population geneticists demonstrated that patterns of worldwide human genetic diversity strongly contradicted the existence of population-based biological races among humans. Indeed, the rapid colonization of the world from a common ancestral African population and numerous subsequent historical migrations and admixture between populations locally have shaped the genetic diversity of mankind: the majority of human genetic variation (more than 95 percent on average across the genome) is found between individuals within populations rather than between individuals across populations (Cavalli-Sforza and Feldman 2003; Rosenberg et al. 2002). This shows that most of the phenotypic diversity observed between groups of populations, such as skin color, in fact grounds its genetic diversity basis within a very small proportion of the total genetic diversity found in mankind. In this context, the reader should constantly bear in mind that when population geneticists claim to have found "genetically differentiated populations," this differentiation between individuals from different populations only stems from a very small proportion of the total genetic variation found in the whole sample.

The Multiethnic Context of Central Africa: Pygmy and Non-Pygmy Populations

African hunter-gatherer Pygmies are often seen as living examples of a lifestyle that dominated over 135,000 years of our evolution since the emergence of *Homo sapiens* roughly 150,000 years ago. Furthermore, Pygmies are currently neighboring populations with a widely different subsistence, agriculture, acquired much more recently in our evolutionary history, about 5,000 years ago in central Africa. In this context, biological anthropologists have studied central African populations over the past fifty years. In particular, they tried to evaluate the influence of lifestyles and ecologies (e.g., parasitic or nutritional environments) on the biological and genetic diversity of human populations (Campbell and Tishkoff 2010; Froment 1993; Perry and Dominy 2009).

Archaeological records demonstrated that the Congo Basin has been densely and continuously inhabited by human populations for at least 45,000 years (Cornelissen 2002; Mercader 2003; Phillipson 2005). Nevertheless, how these past populations relate to the existing Pygmy and non-Pygmy populations remains largely unknown. This is essentially because of the lack of ancient human remains in this region because the acidic soil of the equatorial forest rapidly degrades bones. Therefore, biological anthropologists are challenged by a tremendous lack of historical and demographic knowledge, even recent, in central Africa. Indeed, trying to understand how environmental pressures influenced the biological diversity observed across populations requires the prior knowledge of whether the studied populations share a common or independent origin and when the population divergence events occurred. For instance, knowing whether Pygmy populations have a common or independent origin is essential to understanding whether biological features specific to several Pygmy populations evolved recently and separately in each population or whether these features were inherited from a common ancient ancestral population.

The binary categorization of central African populations into Pygmies and non-Pygmies is deeply rooted in the Western culture. The ancient Greek poet Homer first introduced the word "Pygmy" to refer to an anthropomorphic mythical population in the epic narrative *The Iliad* (Song 3, v. 1–6). He briefly describes the Pygmies as a population of short-statured individuals (the word *Pygmy* refers to a length or volume measurement, such as a cubit or a handful), living somewhere overseas and at war with migrating cranes. Western explorers traveling across central Africa during the second half of the nineteenth century encountered numerous short-statured populations and designated them as Pygmies in a scholarly reference to Homer (Bahuchet 1993; Du Chaillu and Owen 1867; Schweinfurth 1873).

During the second half of the twentieth century, anthropologists described *a posteriori* the numerous and complex cultural criteria underlying the historical Pygmy or non-Pygmy categorization of central African populations (see other chapters in this volume and Bahuchet 1993; Bahuchet 2012; Froment 1993; Hewlett 1996; Hewlett et al. 1982; Joiris 2003). They showed that populations could be distinctly designated as Pygmies from neighboring non-Pygmies if (1) they designate themselves with a different ethno-name than neighboring ethnic groups; (2) they are designated as a different ethnic group by neighbors; (3) they are recognized by outsiders as specialists of forest activities, such as hunting, gathering, and fishing and possess an extensive knowledge of forest environments (both ecological and esoterical); (4) they are very mobile in their environment; (5) they maintain complex socioeconomic relationships with specific neighbors, including the trading of goods, exchanges of workforce, and sometimes participation in social and ritual events; and (6) they are recognized as outstanding musicians, sometimes with specific musical instruments and vocal techniques. Additionally, physical anthropologists demonstrated that populations categorized as Pygmies based on these cultural criteria were among the shortest worldwide, despite a tremendous variability in average stature across Pygmy populations (Froment 1993; Hiernaux 1974).

Most importantly, ethnographers and anthropologists extensively demonstrated that cultural features were highly variable across Pygmy populations.

Therefore, the Pygmy or non-Pygmy categorization is highly reductive of an often much more complex reality. For instance, the various groups historically designated as Pygmies do not share a common language today. In fact, they all speak different languages including languages from different linguistic families (Bahuchet 2012). Furthermore, Pygmies do not have a common myth of origin and rarely know of other populations also called Pygmies by outsiders. The wide cultural and morphological diversity across Pygmy populations *a priori* challenges the common view that all Pygmy populations throughout central Africa share a common heritage or origin, a view *de facto* assumed by the widespread use of the umbrella term *Pygmy*.

In summary, the peopling history of central Africa and the origins of Pygmy and non-Pygmy populations remain widely unknown. This is mainly because of the lack of historical and archaeological data as well as the complex cultural features underlying the Pygmy and non-Pygmy categories in the multiethnic context of central Africa. In this context, population genetics approaches provide a promising set of methods and tools to try to indirectly reconstruct the past origins of existing central African Pygmy and non-Pygmy populations.

Reconciling the Definitions of a Population between Research Fields

We have previously shown that population genetics methods can be used to infer the past demographic history of populations by studying patterns of genetic diversity within and among existing populations. Similarly, as in cultural anthropology disciplines, all the observations and conclusions drawn by geneticists about the diversity and origins of human groups rely on how a population is defined. Population geneticists, because they are studying the history of the genes carried by existing individuals (their genealogies, literally), intuitively define a population as a group of individuals more likely to mate with each other than to mate with other groups, albeit not necessarily strictly.

The multiethnic context of central Africa exemplifies how the complex and dynamic cultural-based definition of ethnic groups and populations can be at odds with this genetic-based definition. As previously described, numerous culturally heterogeneous and geographically isolated populations are gathered under the term *Pygmy*. Merging *a priori* all DNA samples obtained from various Pygmy populations in a single population labeled "Pygmies" can thus be strongly misleading for population genetics studies because the individuals merged may belong to distinct, mutually exclusive, reproductive units or genetics "populations." To apply population genetics methods in central Africa it is therefore necessary to precisely assess how the DNA samples collected from individuals relate to (1) existing populations as defined *a priori* in population genetics approaches and (2) the complex Pygmy or non-Pygmy cultural categorization established by anthropologists and ethnographers.

This can be achieved by interdisciplinary fieldwork between ethnographers and population geneticists and jointly sampling DNA and ethnographic data such as ethnicity, family structures, marriage customs and mobility of spouses, lifestyle, and so on. Familial anthropology data as well as knowledge of marriage customs and the mobility of spouses allow geneticists to identify *a priori* groups of sampled individuals more likely to mate with each other than they are to mate with other communities. Furthermore, detailed ethnographic data on cultural features such as endogenous and exogenous ethnic representations, lifestyles, or socioeconomic interactions between neighboring communities further allow population geneticists to accurately categorize sampled individuals and populations in Pygmies or non-Pygmies based on ethnographic criteria. Such an interdisciplinary approach was first developed to study central African Pygmies and non-Pygmies by Hewlett and Cavalli-Sforza, starting in the 1960s to 1970s. These authors established a powerful framework to study central African genetic diversity. It further allowed studying the complex relationships between a variety of cultural and biological phenomenon such as how the widespread cultural barriers against marriages between Pygmies and non-Pygmies translate (or not) into genetic isolation between neighboring communities.

First, we present what population genetics approaches rooted in interdisciplinary fieldwork revealed about the origins and biological diversity of central African Pygmy and non-Pygmy populations. Despite ethnographers and anthropologists repeated efforts, the overly simplistic myth of the short-statured nomadic forest hunter-gatherer Pygmies isolated from the rest of the world still overwhelms the imagination of Westerners.[4] We then present how population geneticists faced these myths to study patterns of genetic admixture among central African populations and the differential biological evolution and adaptation of Pygmies and non-Pygmies through the example of height.

The Genetic Diversity and Origins of Pygmies

Starting in the late 1960s, Cavalli-Sforza and his collaborators were the first population geneticists to systematically study the genetic diversity of several central African Pygmy populations. Using a reduced number of genetic markers underlying the various blood groups present in western (CAR and Cameroon) and eastern (Western Democratic Republic of Congo) Pygmy groups, these authors first found that Pygmy populations were highly differentiated from other central African non-Pygmy populations, although Pygmies indisputably exhibited a genetic pattern of African origin. Second, these authors identified major genetic differences between the various eastern and western Pygmy populations as well as among western Pygmy groups (Cavalli-Sforza et al. 1969; Santachiara-Benerecetti et al. 1980). Nevertheless, all these groups shared specific blood types and other protein-related genetic patterns that led Cavalli-Sforza and his colleagues to conclude in favor of a likely ancient common origin of all Pygmy populations, followed by a more recent split between eastern and western Pygmy populations. In this model, the differential influence of genetic drift after the east-west split and recent admixture between Pygmies and neighboring non-Pygmies explained the genetic differentiation found among Pygmies today (Cavalli-Sforza 1986; Cavalli-Sforza et al. 1977; Cavalli-Sforza et al. 1969; Daiger and Cavalli-Sforza 1977; Santachiara-Benerecetti et al. 1980).

Cavalli-Sforza and his colleagues could not formally test these hypotheses or obtain precise estimate for the divergence times and admixture proportions because of the limited number of genetic markers and the lack of statistical and methodological tools available at the time. Nevertheless, these studies posed the initial questions for all further population genetics studies of central Africa: when did Pygmy populations diverge from non-Pygmy populations? When did eastern and western Pygmy populations split? When and how did western Pygmy populations start their genetic differentiation? Are Pygmy populations genetically isolated from neighboring non-Pygmies?

The Aka (CAR) and Mbuti (Eastern DRC) Pygmy DNA samples collected by Hewlett and Cavalli-Sforza in the 1980s during interdisciplinary fieldwork were later included in wider worldwide genetic databases.[5] They were studied alongside other African and non-African populations in the emerging systematic investigation of human worldwide genetic diversity (Cann et al. 2002; Cavalli-Sforza 2005). Despite the growing number of studies showing that these two Pygmy populations exhibited largely different genetic patterns, a very limited number of studies tried to tackle the question of the origins of Pygmies until the twenty-first century, perhaps because of the belief that this question had already been definitively solved by the extensive work of Cavalli-Sforza and his colleagues.

Giovani Destro-Bisol and his colleagues were among the first to rediscover the issue of the origins of Pygmies with a series of articles published between 2000 and 2004 and tried to precisely characterize the genetic diversity among the available population samples (Destro-Bisol et al. 2000; Destro-Bisol et al. 2004a; Destro-Bisol et al. 2004b). They found that the three Pygmy populations under study (the Aka and Mbuti samples previously gathered by Cavalli-Sforza and Hewlett and a third sample of Aka Mbenzele Pygmies from southwestern CAR) had widely differentiated mitochondrial DNA, but that they shared a number of markers indicating an ancient common maternal origin (Destro-Bisol et al. 2004a). Assuming a common origin of eastern and western Pygmy populations, the authors simulated genetic data and found that the ancestral maternal lineage split roughly 70,000 years ago into two lineages respectively giving birth to Pygmy and non-Pygmy populations. They further estimated that eastern and western Pygmy groups of populations diverged from the ancestral Pygmy population between 3,000 and 18,000 years ago. However, these studies did not fundamentally question the initial hypothesis of a common origin of Pygmies. In fact, the lack of both population samples and genome wide genetic markers made it difficult to test this hypothesis.

Genetic Differentiation among Central African Pygmy Populations

To overcome these limitations, numerous interdisciplinary studies, following Hewlett and Cavalli-Sforza's initial approach, were mainly conducted by French and American researchers. Between 2002 and 2009, researchers sampled DNA from more than twenty populations in western and eastern central Africa (Batini et al. 2011; Patin et al. 2009; Tishkoff et al. 2009; Verdu et al. 2009, see figure 2.1 and table 2.1). These populations were categorized as Pygmies and non-Pygmies based on the numerous cultural criteria set by ethnographers and anthropologists.

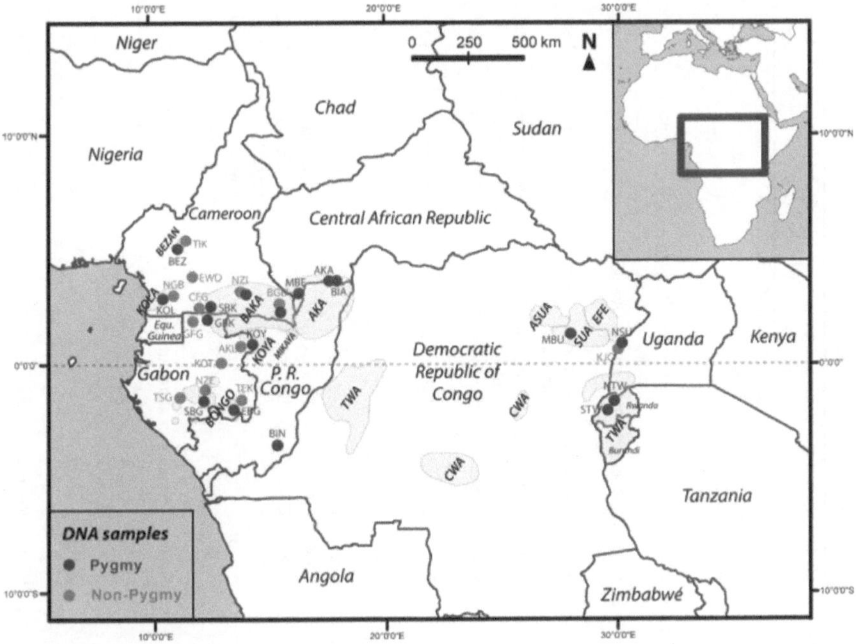

Figure 2.1. Location of Pygmy and non-Pygmy population samples in recent population genetics studies focusing on the peopling history of central Africa. *Note:* See table 2.1 for population codes and publication references. Pygmy and non-Pygmy categories were established using numerous cultural criteria following anthropology and ethnology studies (see Introduction). Peopling areas on the map were inferred from ethnographic fieldwork and presented in Bahuchet 2012.

Table 2.1 Pygmy and non-Pygmy population samples in some recent population genetics studies.

	Population Name	Code	Country	Linguistic Family	Sampling	Autosomal Data References	mtDNA references	Sample Sizes Used for This Study
Pygmies	Bezan	BEZ	Cameroon	Bantoïd Non-Bantu, Tikar	A. Froment, P. Verdu	Verdu et al. 2009; Tishkoff et al. 2009	Quintana-Murci et al. 2008	29 (Verdu et al. 2009)
	Kola	KOL	Cameroon	Bantu, A80	A. Froment	Verdu et al. 2009; Patin et al. 2009; Tishkoff et al. 2009	Quintana-Murci et al. 2008	31 (Verdu et al. 2009)
	Central Baka	CBK	Cameroon	Adamawa-Ubangian, Ngbaka	A. Froment	Verdu et al. 2009; Patin et al. 2009; Tishkoff et al. 2009	Quintana-Murci et al. 2008	29 (Verdu et al. 2009)
	Eastern Baka	EBK	Cameroon	Adamawa-Ubangian, Ngbaka	A. Froment	Verdu et al. 2009	*nd*	29 (Verdu et al. 2009)
	Southern Baka	SBK	Cameroon	Adamawa-Ubangian, Ngbaka	A. Froment	Verdu et al. 2009	Quintana-Murci et al. 2008	29 (Verdu et al. 2009)
	Gabonese Baka	GBK	Gabon	Adamawa-Ubangian, Ngbaka	J-M Hombert, L. Van Der Veen	Verdu et al. 2009; Patin et al. 2009	Quintana-Murci et al. 2008	30 (Verdu et al. 2009)
	Koya	KOY	Gabon	Bantu, B20	S. Le Bomin	Verdu et al. 2009	Quintana-Murci et al. 2008	29 (Verdu et al. 2009)
	Eastern Bongo	EBG	Gabon	Bantu, B70	P. Verdu	Verdu et al. 2009	Quintana-Murci et al. 2008	30 (Verdu et al. 2009)
	Southern Bongo	SBG	Gabon	Bantu, B30–B50	P. Verdu	Verdu et al. 2009	*nd*	30 (Verdu et al. 2009)
	Babinga	BIN	PRC	*nd*	*nd*	*nd*	Batini et al. 2007; 2011	*nd*
	Aka	AKA	CAR	Bantu, C10	B. Hewlett	*This study*	*nd*	27 (*This study*)
	Biaka	BIA	CAR	Bantu, C10	L.L. Cavalli-Sforza, B. Hewlett	Rosenberg et al. 2002; Patin et al. 2009; Tishkoff et al. 2009	Destro-Bisol et al. 2004; Quintana-Murci et al. 2008; Batini et al. 2011	36 (Rosenberg et al. 2002)
	Mbenzele Aka	MBE	CAR	Bantu, C10?	*nd*	Destro-Bisol et al. 2004	Destro-Bisol et al. 2004; Batini et al. 2007; 2011	*nd*
	Nsua	NSU	Uganda	Sudanic?	P. Verdu	*This study*	*nd*	18 (*This study*)
	Mbuti	MBU	DRC	Sudanic, Mangbutu-Efe/Bantu?	L.L. Cavalli-Sforza, B. Hewlett	Rosenberg et al. 2002; Patin et al. 2009; Tishkoff et al. 2009	Destro-Bisol et al. 2004; Quintana-Murci et al. 2008; Batini et al. 2011	15 (Rosenberg et al. 2002)
	Northern Twa	NTW	Rwanda	*nd*	*nd*	Patin et al. 2009	*nd*	*nd*
	Southern Twa	STW	Rwanda	*nd*	*nd*	Patin et al. 2009	*nd*	*nd*

	Tikar	TIK	Cameroon	Bantoïde Non-Bantu, Tikar	A. Froment	Verdu et al. 2009; Tishkoff et al. 2009	*nd*	30 (Verdu et al. 2009)
	Nzime	NZI	Cameroon	Bantu, A80	A. Froment	Verdu et al. 2009; Tishkoff et al. 2009	*nd*	31 (Verdu et al. 2009)
	Bangando	BGD	Cameroon	Adamawa-Ubangian, Bangandu, Gbaya	A. Froment	Verdu et al. 2009	*nd*	30 (Verdu et al. 2009)
	Ngumba	NGB	Cameroon	Bantu, A80	A. Froment	Verdu et al. 2009; Patin et al. 2009; Tishkoff et al. 2009	Quintana-Murci et al. 2008	30 (Verdu et al. 2009)
	Ewondo	EWD	Cameroon	Bantu, A70	A. Froment	Verdu et al. 2009	Quintana-Murci et al. 2008	24 (Verdu et al. 2009)
	Fang	CFG	Cameroon	Bantu, A70	A. Froment	Verdu et al. 2009; Tishkoff et al. 2009	Quintana-Murci et al. 2008	30 (Verdu et al. 2009)
Non-Pygmies	Fang	GFG	Gabon	Bantu, A70	J-M Hombert, L. Van Der Veen	Verdu et al. 2009	Quintana-Murci et al. 2008	30 (Verdu et al. 2009)
	Akele (Bongomo)	AKL	Gabon	Bantu, B20	S. Le Bomin	Verdu et al. 2009	*nd*	13 (Verdu et al. 2009)
	Teke	TEK	Gabon	Bantu, B70	J-M Hombert, L. Van Der Veen	Verdu et al. 2009	Quintana-Murci et al. 2008	30 (Verdu et al. 2009)
	Nzebi	NZE	Gabon	Bantu, B50	J-M Hombert, L. Van Der Veen	Verdu et al. 2009	Quintana-Murci et al. 2008	30 (Verdu et al. 2009)
	Kota	KOT	Gabon	Bantu, B20	J-M Hombert, L. Van Der Veen	Verdu et al. 2009	Quintana-Murci et al. 2008	30 (Verdu et al. 2009)
	Tsogho	TSG	Gabon	Bantu, B30	J-M Hombert, L. Van Der Veen	Verdu et al. 2009	Quintana-Murci et al. 2008	30 (Verdu et al. 2009)
	Konjo	KJO	Uganda	Bantu, ?	P. Verdu	*This study*	*nd*	28 (*This study*)

Note: Pygmies and non-Pygmies are categorized as such based on numerous cultural criteria following anthropology and ethnology studies (see Introduction). Recent population genetics studies studying autosomal and mitochondrial DNA (mtDNA) genetic data are given separately in two different columns. Linguistic affiliations were determined following the classification in Greenberg 1966 and Guthrie 1967. Where no data was available is indicated by "*nd*." The column "Sample Sizes Used for This Study" indicates the number of family unrelated individuals considered for the original population genetics analyses of genetic differentiation performed in this chapter (see above and figure 2.2). We studied the distribution of genetic diversity within and among central African populations using twenty-six independent neutral autosomal microsatellite markers, that is, twenty-six locations in the genome (not sex-specific) far from one another and far from any known or predicted genes. These genomic regions exhibit known mutations genetically differentiating individuals and potentially populations. Original publications corresponding to the genetic data used here are given between parentheses. The mention "*This study*" indicates that genetic data concerning these populations are reported in this chapter for the first time and were obtained following the protocol and normalization procedures described in Verdu et al. (2009) and Becker et al. (2010). Genotype data and technical aspects of the genotyping and normalization procedures are available upon demand to the author. French, US, Ugandan, CAR, Gabonese, and Cameroonian official ethical and research review boards delivered appropriate research and sampling authorizations for this study. DNA was originally extracted from blood or saliva samples obtained from informed voluntary donors according to standard protocols. Oral, video-recorded, or written informed consents were provided by each donor.

Considering twenty-six central African populations (728 individuals, table 2.1 and figure 2.1), we study here the genetic diversity patterns using twenty-six independent neutral markers genome-wide (see table 2.1 and figure 2.2 legend). We found that certain Pygmy and non-Pygmy pairs of populations exhibit the largest genetic differentiation levels in our data set, even when considering only Cameroonian and Gabonese populations (figure 2.2). Furthermore, most of the genetic variation explained by differences among populations (1.66 percent of the total genetic variation AMOVA F_{ST}, Excoffier et al. 1992) was accounted for by the major differentiation among Pygmy populations (2.25 percent of the total genetic variation when considering only thirteen Pygmy populations). Conversely, non-Pygmy populations exhibited small, often not significant levels of population genetic differentiation (0.46 percent of the total genetic variation when considering only thirteen non-Pygmy populations). Finally, eastern and western groups of Pygmy populations exhibit major genetic differentiation, followed by high levels of pairwise population differentiation respectively among western and eastern Pygmy groups (see figure 2.2 and Patin et al. 2009; Tishkoff et al. 2009; Verdu et al. 2009). These results significantly expanded previous observations by Cavalli-Sforza, Destro-Bisol, and their collaborators and echo the vast cultural diversity found by ethnographers among populations designated as Pygmies. Furthermore, while some Pygmy populations showed large genetic distances with non-Pygmy populations, some others, such as the two Gabonese Bongo populations, were found resembling genetically non-Pygmy neighbors (figure 2.2 and Verdu et al. 2009). These results further jeopardized the traditional view that all Pygmies are biologically homogeneous and widely different from their neighbors.

Common Genetic Origins of Central African Pygmies

In this context, we formally tested for the first time whether genetic patterns observed among numerous Pygmy populations from Cameroon and Gabon were consistent with a common origin of Pygmy populations or with a historical scenario where each existing Pygmy population diverged independently from the lineage that gave birth to non-Pygmy populations (Verdu et al. 2009). To address this question considering highly complex historical and demographic models and numerous population samples, we used new statistical inference methods called Approximate Bayesian Computation (ABC; Beaumont et al. 2002; Tavaré et al. 1997). Briefly, these methods proceed by simulating numerous genetic data sets under a given complex historic and demographic model. We then summarize the genetic information by calculating population genetics statistics for each simulated data set and compare these summary statistics to the same statistics calculated on the real observed data. We infer *a posteriori* the distribution of the model parameters, such as divergence times and population sizes, that were used to produce

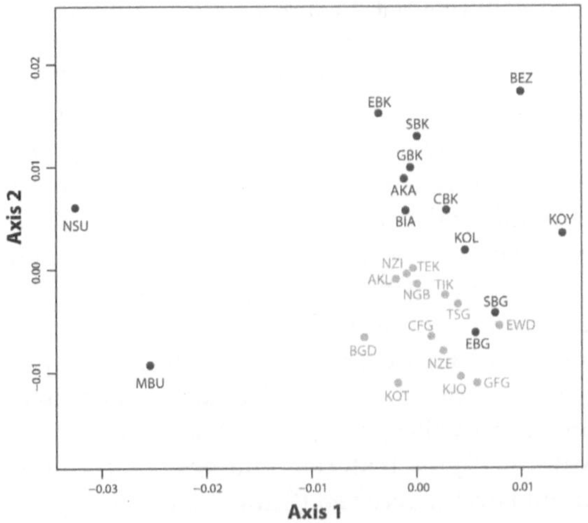

Figure 2.2. Two-dimensional multidimensional scaling representation of central African population pairwise genetic differentiation levels.
Note: Pairwise levels of genetic differentiation between populations were calculated as FST (Weir and Cockerham 1984) using twenty-six independent tetranucleotide microsatellite markers (see table 2.1). Nonsignificant FST values based on 10,000 permutations of individuals among populations were set to 0. Pairwise FST values are represented in two dimensions using a metric multidimensional scaling approach. To evaluate how well the two-dimensional plot represented the pairwise genetic differentiation levels, we calculated the Spearman correlation between Euclidean distances among pairs of populations on the plot and the corresponding pairwise FST levels of genetic differentiation. We obtained a value of Spearman rho equal to 0.84 indicating a very high correlation between pairwise distances on the plot and the corresponding genetic differentiation levels. Pygmy populations are represented in red and non-Pygmies in blue. See table 2.1 for population codes, sample sizes, and publication references.

the simulations (e.g., 0.1 percent) with summary statistics closest to our real data set.

Using ABC methods, we demonstrated for the first time that the genome-wide neutral genetic diversity of numerous Pygmy populations from western central Africa was much more consistent with scenarios of a common Pygmy origin than with scenarios considering an independent origin of each Pygmy population (relative posterior probability of the common origin scenario was 96 percent; Verdu et al. 2009). We estimated in this study that the ancestral Pygmy population diverged from the ancestral non-Pygmy population in a remote past, roughly 50,000–90,000 years ago. Furthermore, we found that the ancestral western central African Pygmy population split much more recently, about 3,000 years ago, to give birth to the various Pygmy populations found today in Cameroon and Gabon. Finally, our genetic data further indicated that the rapid genetic diversification among western Pygmy population during the last 3,000 years could be explained by reproductive isolation and genetic drift in each Pygmy population separately.

Importantly, Patin and colleagues further applied similar ABC methods to formally test whether western and eastern Pygmy groups of populations had a common or an independent origin (Patin et al. 2009). They found a likely common origin between eastern and western Pygmy groups roughly 20,000 years ago, consistently with previous estimates. Furthermore, they estimated that the original divergence between ancestral Pygmy and non-Pygmy populations occurred roughly 60,000 years ago. Finally, using full mitochondrial DNA sequences, Batini and his colleague (2011) further explored the common or independent origin of eastern and western Pygmy maternal lineages and detailed scenarios of their respective past demography. These authors expanded Patin and his colleagues results, finding a likely common origin of eastern and western Pygmy maternal lineages about 27,000 years ago, and an ancient common origin between Pygmies and non-Pygmies about 70,000 years ago (Batini et al. 2011). They also found that Pygmies likely experienced a decrease in maternal effective population sizes since about 4,000 years ago until recently, and that, conversely, non-Pygmies likely experienced a strong increase in effective maternal population size starting shortly after the original split (about 65,000 years ago) until the present day. It is worth mentioning here that all the divergence times reported here have potentially wide credibility intervals and will need to be more accurately estimated in the future using more genetic makers and individual samples.[6] A schematic population genetics-based model for the peopling history of central African Pygmy and non-Pygmy populations is presented in figure 2.3.

It is difficult, if not impossible, for population geneticists to determine the external causes for the divergence and demographic changes experienced by central African ancestral populations, especially the ancestral Pygmy and non-Pygmy original divergence about 70,000 years ago. Nevertheless, a number of archaeological and paleoclimatological data provide interesting sets of hypotheses to build a more comprehensive evolutionary framework for the divergence history in the ancestral Pygmy lineages. Patin, Batini, and their colleagues found that the western-eastern ancestral Pygmy divergence event coincided with the Last Glacial Maximum. This period is characterized by a fragmentation of the forest and the formation of small forest refuges throughout the Congo Basin. Therefore, these authors hypothesized that the reduction of forest-covered areas potentially triggered the isolation of ancestral western and eastern Pygmy populations (Batini et al. 2011; Patin et al. 2009). However, paleoclimatic and paleoecological data are scarce to precisely map the fluctuations of the forest coverage in the Congo Basin over the last 30,000 years, and this hypothesis makes a number of yet unproven assumptions. In particular, it assumes that current eastern and western Pygmy populations settle the same areas as their ancestors and that these areas indeed formed forest refuges more than 20,000

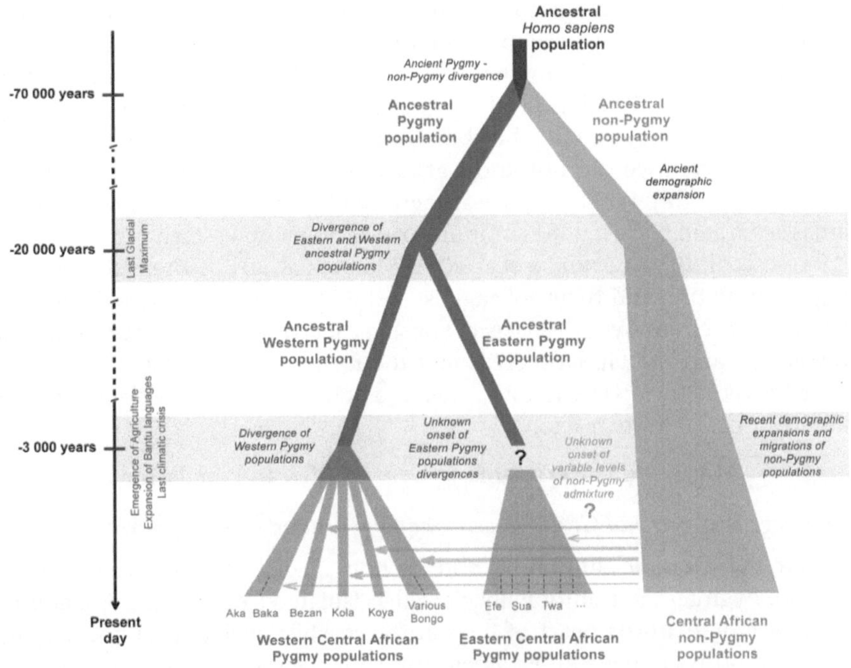

Figure 2.3. Synthetic model for the genetic origins of central Pygmy and non-Pygmy populations.
Note: This historical peopling model of central Africa is based mainly on population genetics studies and Approximate Bayesian Computation approaches to reconstruct the evolutionary history of central African genetic diversity by Verdu et al. (2009), Patin et al. (2009), and Batini et al. (2011).

years ago. It also assumes that Pygmy populations are and have always been confined to forest environments, which remains to be demonstrated.

Concerning the western Congo Basin, archaeological and archaeolinguistic data respectively showed that agriculture and Bantu languages emerged about 4,000–5,000 years ago and rapidly spread in central and southern Africa (Phillipson 2005; Vansina 1995). In addition, population geneticists showed that the diffusion of Bantu languages may have been accompanied by human population migrations, as attested by the demographic expansion of non-Pygmy Bantu-speaking populations and the high levels of non-Pygmy migrations found using genetic data[7] (Berniell-Lee et al. 2009; de Filippo et al. 2011; Montano et al. 2011; Quintana-Murci et al. 2008). Pygmy and non-Pygmy populations currently share complex socioeconomic relationships that have been shown, at least in some Aka and Baka groups, to substantially influence the mobility and potential isolation of Pygmy populations (Bahuchet 1992; Hewlett et al. 1982; MacDonald and Hewlett 1999). Therefore, we proposed that the emergence of agriculture and migrations of Bantu-speaking non-Pygmies may have triggered profound social and demographic changes in the ancestral western Pygmy population. This may in turn have led to the fragmentation and isolation of the various Pygmy populations currently settling CAR, Cameroon, and Gabon over the past 3,000 years (Verdu et al. 2009). However, we did not know at the time that a recent climatic crisis occurring roughly 4,000–5,000 years ago resulted in the contraction of forest areas separated by vast zones of more arid savanna in western central Africa (Hessler et al. 2010; Lebamba et al. 2009; Maley et al. 2012). Therefore, following Patin, Batini, and their colleagues, and with the caution invoked above, we can further our hypothesis by proposing that the agricultural and cultural revolution and the climatic crisis and the formation of forest refuges may have triggered the divergence and isolation of western central African Pygmy populations 3,000 years ago.

Migration Patterns among Central African Populations

Marriage Practices and Gene Flow among Pygmies and Non-Pygmies

In both the scientific community and the non-specialized Western audience, African Pygmies are traditionally viewed as isolated from other populations, living remotely in the heart of the rainforest while agricultural non-Pygmy populations live outside forest areas. This somewhat romantic view has been frequently contradicted by the extensive work of ethnographers and anthropologists (see other chapters in this volume and Bahuchet 1992; Hewlett 1996; Joiris 2003): (1) most Pygmy populations also often practice agriculture and eat agricultural produces on daily bases; (2) non-Pygmy populations practice a slash-and-burn agriculture inside the forest and also often eat forest goods such as game or wild honey; (3) numerous Pygmy groups settle in the vicinity or even inside non-Pygmy villages, if not always permanently, at least for long periods of time; and (4) Pygmies and non-Pygmy neighbors interact socially (including with intermarriages) and economically on daily bases. In addition, the work of Cavalli-Sforza and his collaborators already identified genetic signatures of migrations between Aka Pygmies and non-Pygmies (Cavalli-Sforza 1986), further casting doubt on the representation that Pygmies are isolated from any other central African population.

Since the late 1990s, population geneticists studying sex-specific genetic markers from the male-inherited Y chromosome and the female-inherited mitochondrial DNA further refined sex-specific patterns of migration in sub-Saharan Africa. In particular, Salas and his colleagues found some levels of sharing of mitochondrial DNA diversity between Pygmy and non-Pygmy populations. Population geneticists attributed these patterns to traces of maternal gene flow between Pygmies and non-Pygmies through patrilocal marriage customs,[8] which would have occurred during the migrations of Bantu-speaking populations (Salas et al. 2002; Seielstad et al. 1998). This hypothesis was expanded by later findings from Destro-Bisol, Batini, Quintana-Murci, and their respective colleagues considering numerous Pygmy population samples (Batini et al. 2007; Destro-Bisol et al. 2004a; Quintana-Murci et al. 2008). Furthermore, Hammer and colleagues found that the diversity of the male-inherited Y chromosome, showing small among-population genetic differentiation, was also consistent with extensive male migrations among sub-Saharan populations, a surprising result at odds with patrilocal customs in central Africa (Hammer et al. 2001).

In a founding article, Destro-Bisol and colleagues (2004b) used the interdisciplinary framework of Cavalli-Sforza and Hewlett to study complex interactions between cultural and biological phenomena. They reconciled the apparent discrepancies previously found between male and female signatures of interpopulation migrations by further showing how complex sociocultural behaviors could trigger movements of spouses and admixed offspring. Indeed, ethnologists previously showed that while marriages between a Pygmy male and a non-Pygmy female are often strongly prohibited, marriages between a Pygmy female and a non-Pygmy male can occur more frequently (Bahuchet and Guillaume 1982; Hewlett 1996; Joiris 2003). Nevertheless, social discriminations against the newly wedded Pygmy female that moved to the village of her in-laws often lead to divorce. The Pygmy wife then returns to her community of origin, potentially with her admixed offspring. Importantly, such mobility of Pygmy females and admixed offspring in intermarriages also occurs with the death of the non-Pygmy husband. This culture-driven mobility of spouses and admixed offspring could thus explain male-mediated gene flow from non-Pygmies into Pygmy populations in a context of patrilocality and social discrimination against Pygmies.

We further used our extensive population sample set and genome-wide genetic markers to more precisely estimate admixture levels and assess the potential variability of admixture patterns across numerous Pygmy populations (Verdu et al. 2009). Results showed that all western central African Pygmy populations considered in our study exhibited substantial levels of asymmetric admixture from the non-Pygmy populations, hence expanding previous findings described above. Surprisingly, we found that average levels of admixture were largely variable across Pygmy populations. For instance, Baka and Bezan Pygmies were significantly less admixed than Bongo populations. Using ethnographic data we observed that the heterogeneous levels of admixture found across western Pygmy populations were inversely proportional to the strength of barriers against intermarriages: strong sociocultural prohibition of intermarriages and discriminatory endogenous representations against Pygmies could be the cause of lower genetic admixture rates (Verdu et al. 2009). Considering eastern Pygmy population samples, Patin and his colleagues (2009), followed by Tishkoff and his colleagues (2009), further found that eastern Pygmy populations were also substantially admixed with non-Pygmies. Altogether, these findings showed that a Pygmy population with virtually no trace of admixture with non-Pygmies remains yet to be found, if such a population even exists today.

Dispersal Behavior and Reproductive Isolation between Neighboring Pygmy Populations

We have seen that the various Pygmy populations were genetically differentiated despite sharing a recent common genetic origin in the west and a more ancient one between eastern and western groups of populations. The reproductive isolation is not surprising between geographically distant eastern and western Pygmy populations. In the light of the known exploratory mobility of Pygmy populations, such isolation is more unexpected locally, for instance, between Baka and Koya neighboring Pygmy populations (figure 2.1). Indeed, Pygmies are traditionally described as being extremely mobile in the forest environment (Bahuchet 1993; Hewlett 1996). Ethnographers demonstrated that this mobility was mainly because of social interactions such as search for spouses, social and ritual events among the Pygmy communities, or socioeconomic interactions with non-Pygmy neighbors (Bahuchet 1992; Hewlett et al. 1982; MacDonald and Hewlett 1999). If such extended exploratory mobility translates into extended reproductive mobility, it would be surprising to identify genetic isolation signals between neighboring Pygmy populations.

When trying to assess the extent to which exploratory mobility relates to reproductive mobility or effective dispersal, namely, how far children are born from the birthplace of their parents, detailed demographic data on the birthplaces of biologically related family members are needed but are not readily available in Pygmy populations. The correlation between genetic differentiation levels and geographic distances between birthplaces can indirectly provide accurate estimates of effective dispersal ranges among generations. This was previously confirmed by population genetics studies of New Guinean populations where extensive transgenerational demographic data were available (Long et al. 1987; Rousset 1997; Wood 1987). Applying these population genetics methods to Cameroonian Baka Pygmy populations, we found a significant pattern of genetic isolation by distance corresponding to very short effective dispersal ranges between 12.4 and 63.2 km2 on average (Verdu et al. 2010). In other words, two Baka individuals born far away from each other are likely to be genetically more differentiated than two individuals born close-by. Following Cavalli-Sforza and Hewlett's (1982) assumptions about the dispersal shape among Aka Pygmies, these effective dispersal ranges indicated that, on average, half of the Baka individuals were born at most between 1.7 and 3.9 km away from their parents respective birthplaces (Verdu et al. 2010). Such a limited reproductive dispersal behavior, if found in other Pygmy populations, could very well explain the rapid and extensive levels of genetic differentiation found across neighboring Pygmy populations from western central Africa. In any case, this study showed that detailed quantitative demographic and ethnographic data on the transgenerational mobility behaviors of Pygmy populations would provide key elements to understand how and why geographically neighboring Pygmy populations are currently genetically isolated. Future population genetics work needs to be conducted to assess more precisely when such limited effective dispersal behavior started among the Baka.

The Stature of African Pygmies

The widely different lifestyles and ecology of Pygmies and non-Pygmies could have triggered different biological and morphological adaptation processes between groups (see other chapters in this volume). Understanding the differences in adult height between Pygmies and non-Pygmies best exemplifies this research paradigm (Perry and Dominy 2009).

Pygmies are found in the lower part of the adult average height distribution of African or worldwide populations. Nevertheless, the extensive work of Froment and Hiernaux (Froment 1993; Hiernaux 1974) established that there was no discontinuity in height between central African Pygmies and non-Pygmies, some non-Pygmy populations neighboring Pygmy groups also being of short stature.[9] Therefore, these authors demonstrated that considering an arbitrary threshold of height could be extremely misleading to categorize populations as Pygmies or non-Pygmies in the multiethnic context of central Africa, thus contradicting Cavalli-Sforza's proposal to set the threshold for Pygmy height to 160 cm (1986). Furthermore, these authors found tremendous variance in adult height within and across Pygmy populations, ranging from an average height of 143 cm among Efe men in eastern DRC to an average 161 cm for Twa men from the Lake Tumba in western DRC.

Based on these observations, Hiernaux proposed that the short stature of African Pygmies was because of convergent adaptation processes to ecological, parasitological, and nutritional constraints in the Congo Basin forest. Cavalli-Sforza, finding clues for a common origin of Pygmy populations, conversely proposed that the short stature of Pygmies was inherited from a common ancestral population that had gone through an adaptation to the hot and humid environment of the equatorial forest (Cavalli-Sforza et al. 1969; Merimee et al. 1972). Indeed, thermoregulation is hard in a hot and humid environment, and being short lowers the energy requirements for thermoregulation, thus providing an adaptive advantage over taller individuals. Diamond further proposed that Pygmies' short stature could reflect an adaptation to hunting and gathering in the dense rainforest environment (Diamond 1991). In this hypothesis, being shorter would allow easier mobility in the rainforest and thus lower energy expenditure while looking for food. Furthermore, Bailey proposed that Pygmies' short stature allowed reduced caloric requirements, which could be advantageous in the nutritionally scarce forest environment (Bailey 1991; Bailey et al. 1989). More recently, Migliano and his colleagues

(2007) proposed that, in the context of high mortality in the equatorial forest, it could be advantageous for the survival of the population to start reproduction earlier, which would in turn result in a reduction in stature of the population. However, none of these hypotheses was formally proven to be true, and all of them were challenged by other anthropology and ethnography work. Furthermore, anthropologists are still debating about whether the short stature of African Pygmies is fully due to environmental phenotypic plasticity rather than stemming from a biological adaptation to natural selection pressures in a challenging environment (Bahuchet 1988; Bahuchet et al. 1991; Froment 1993; Becker et al. 2010; Perry and Dominy 2009).

Physiologists attributed the short stature of Pygmies as compared to non-Pygmies to potential differences between populations in the metabolic pathway determining in part cell and tissue growth: the growth hormone receptor (GHR) and insulin growth factor (IGF) pathway (Merimee et al. 1972; Merimee et al. 1987; Rimoin et al. 1969). Recently, Bozzola and his colleagues (2009) found that African Pygmies had reduced expression levels of certain hormones and receptors of the GHR-IGF pathway as compared to non-Pygmies. However, several studies focusing on the genes encoding some of the proteins of this pathway failed to identify genetic mutations significantly differentiating Pygmies from non-Pygmies (Bowcock and Sartorelli 1990).

Most of these studies implicitly or explicitly assumed the common origin of Pygmy populations, and hence a possible common inheritance of the genetic determinant of height from the ancestral Pygmy population. Although we recently proved that genetic data indeed advocated for a common origin of central African Pygmy populations, further issues concerning the evolution of height emerge. In the lack of ancient human remains in central Africa, we do not know if an ancient population of particularly short stature existed in the region or not. Therefore, we do not know whether Pygmies became shorter in the course of evolution or if neighboring non-Pygmies conversely became taller. This major issue has been somewhat overlooked by some biological anthropologists. However, the answer could fundamentally change the way the question of the genetic determination of central African differences in stature and the identification of potential signatures of natural selection influencing this phenotype should be addressed.

In this context, Becker and colleagues (2011) recently looked for indirect evidences of genetic determination of the differences in height between Pygmies and non-Pygmies. Using numerous Pygmy and neighboring non-Pygmy population samples, categorized as such based on cultural criteria disregarding *a priori* height information, these authors found a significant positive correlation between individual's height and levels of genetic admixture among Pygmies and non-Pygmies. A central African individual categorized as Pygmy based on cultural criteria and found to be highly genetically admixed with non-Pygmies was likely to be significantly taller than a less admixed Pygmy individual. This showed for the first time that yet unidentified genetic factors could be involved in determining height differences between Pygmies and non-Pygmies. This study highlighted that the diversity of Pygmy populations should imperatively be considered and genetic substructures, such as admixture patterns, controlled for when trying to identify genetic determinants associated with height among central Africans. Recently, Jarvis and his collaborators, following the path opened by Becker and his collaborators, released a genome-wide analysis using more than a million genetic markers to identify mutations determining height differences among central Africans (Jarvis et al. 2012). These authors identified a number of mutations spread out across the genome showing signatures of differential adaptation between Pygmies and non-Pygmies. In particular, they found that genomic regions coding for proteins involved in immunological functions were enriched in mutations strongly differentiating Pygmies and non-Pygmies and showing signatures of differential natural selection as compared to the rest of the genome. These results were very recently further confirmed by the study of whole genome sequences of fifteen individuals from three Cameroonian Pygmy populations (Lachance et al. 2012). Therefore, Jarvis, Lachance, and their collaborators proposed that the short stature of central African Pygmies may be resulting from an adaptation to challenging parasitological environments. However, they could not identify genetic mutations significantly associated to height differences in Central Africa, mainly because of a lack of statistical power when considering a very limited number of individuals from both groups of populations. These studies nevertheless paved the way for future promising studies trying to solve the long-standing and still unsolved mystery of central African Pygmies' and non-Pygmies' differential stature.

Perspectives and Conclusions

Perspectives on the Origins of Central African Populations

Population genetics approaches unveiled the common origin of numerous culturally, morphologically, and genetically diverse Pygmy populations and further proposed a time line for the various divergence events among central African populations. However, numerous questions remain unsolved. First, because population geneticists focused on the history of genes carried by individuals, the historical peopling model proposed is not geographically rooted. Indeed, we still do not know where the ancestral Pygmy and non-Pygmy populations settled or when Pygmy and non-Pygmy populations first settled in the locations that they currently occupy. Moreover, while numerous samples from various regions of sub-Saharan Africa are currently available, central Africa is still overall poorly represented in population genetics databases, in particular southern Sudan, CAR, northwestern Uganda, Rwanda,

Burundi, and, most importantly, DRC. Second, the detailed divergence history of the various Pygmy populations remains unknown: the population genetics models here presented could not assess, for instance, whether the Cameroonian Baka Pygmies diverged from the common ancestral western Pygmy population before or after the Bedzan. Furthermore, the divergence history of existing eastern Pygmy populations has never been precisely assessed. Finally, the studies described here highlight a fundamental characteristic of the genetics of African Pygmies: the substantial levels of genetic differentiation among Pygmy populations. In this context, it is difficult to predict how additional samples from other Pygmy populations such as the Twa from Lake Tumba in western DRC, the Asua from northeastern DRC, or the Cwa from southern DRC, will modify our current understanding of the peopling of central Africa.

Perspectives on Culture and Gene Interactions among Pygmies and Non-Pygmies

Determining since when Pygmies and non-Pygmy neighbors share complex socioeconomic relationships has been an important debate animating the ethnologist community in the past decades. The main underlying question is whether the social interactions predated the beginning of the Western colonization of central Africa in the seventeenth century or whether the profound social changes triggered by these colonization events subsequently provoked the emergence of new social interactions between Pygmies and non-Pygmies. Although not all sociocultural interactions lead to intermarriages resulting in genetic admixture between existing populations, estimating the time of the admixture events could at least provide a conservative lower bound for the historical onset of social interactions between communities. Finally, the complex social behaviors triggering the mobility of spouses and admixed offspring proposed by population geneticists for explaining observed admixture patterns remain to be precisely quantified in the field. Indeed, patrilocality also implies that children live with the husband's family. If such behavior was strongly prevalent in Pygmy to non-Pygmy marriages and separations, an alternative explanation of the genetic admixture patterns here found should be considered. Therefore, the mobility of admixed Pygmy and non-Pygmy children in intermarriages, as well as their social status and reproductive success in either community respectively, imperatively needs to be further explored.

Concluding Remarks

In the last fifty years, population genetics studies rooted in the interdisciplinary work conducted by Cavalli-Sforza, Hewlett, and their collaborators tremendously increased our understanding of the peopling history and biological evolution of central African Pygmy and non-Pygmy populations. In particular, they highlighted extremely complex patterns of genetic diversity and admixture among Pygmies and non-Pygmies potentially determined by marriage behaviors. This demonstrates the importance of considering the diversity of populations designated under the umbrella term *Pygmy* in order to understand the evolution of central African populations. Following in the footsteps of ethnologists and anthropologists, population geneticists further deconstructed the pregnant myths and widespread misconceptions surrounding central African Pygmy populations. Population geneticists showed that (1) despite sharing a common origin, Pygmy populations are not biologically homogeneous but are substantially genetically differentiated from one another; (2) Pygmy populations are not isolated from other African populations but are all substantially admixed genetically with non-Pygmy neighbors; and (3) despite an extensive exploratory mobility, at least one Pygmy group has very reduced effective dispersal ranges, a demographic behavior that could trigger reproductive isolation among neighboring Pygmy populations. Finally, population geneticists showed that the differential stature between Pygmies and non-Pygmies was at least partly due to multiple and complex genetic factors as opposed to being caused exclusively by environmental factors experienced by Pygmies during growth. Nevertheless, these genetic factors yet remain to be precisely identified.

While the different approaches and definitions of what is "a population" can create tremendous divisions between geneticists, anthropologists, and ethnologists, the example of the complex Pygmy or non-Pygmy categorization of central African populations shows that successful interdisciplinary efforts establishing common grounds for studying the same groups of individuals from largely different points of view can be infinitely fruitful. Indeed, interdisciplinary approaches rooted in joint fieldwork allowed us to not only rediscover and revisit unsolved questions but also to build a historical and evolutionary framework for studying with a novel eye the cultural and biological evolution of central African populations.

Notes

1. Acknowledgments: The author would like to warmly thank Evelyne Heyer and Serge Bahuchet for giving him the opportunity to study the genetic history of central African populations. Original results presented here were funded in part by the French CNRS, ACI Prosodie, by the French ANR program C3A, and the US NIH grant RO1 GM081441. The author would like to warmly thank all voluntary participants from central Africa as well as ethical and research institutions from Gabon, Cameroon, Central African Republic, and Uganda having provided research authorizations. Finally, the author would like to thank Barry Hewlett for his help and numerous comments on this chapter.
2. Except in the case of monozygotic twins, who carry virtually the same DNA sequence.
3. Genetic markers, or mutations, located far away from any known genes or functional genomic regions, a priori assumed to have evolved neutrally: i.e., without being direct targets of natural selection pressures but only influenced by demographic forces.

4. Illustrated for instance by the first Tarzan movie among many others: *Tarzan the Ape Man* (dir. W. S. Van Dyke, 1932). See also chapters 1 and 10 in this book.
5. The Human Genome Diversity Project—Centre d'Etude du Polymorphisme Humain (HGDP-CEPH)—encompasses more than a thousand individual DNA samples from more than fifty populations across Africa, the Middle East, Europe, Asia, Oceania, and the Americas.
6. Detailed discussions on these divergence time estimates using ABC can be found in Verdu et al. (2009), Patin et al. (2009), and Batini et al. (2011).
7. These mechanisms consistently explain the low levels of genetic differentiation found among non-Pygmy Bantu-speaking populations, as such demographic expansions and migrations would homogenize the gene pools among populations.
8. When marriage customs dictate that the bride moves to live with the husband's family after marriage.
9. For example, the Ba.Konjo neighboring Efe Pygmies, at the western Ugandan border with DRC's Nord Kivu, measure less than 160 cm high on average.

References

Bahuchet, S. 1988. "Food Supply Uncertainty among the Aka Pygmies (Lobaye, C.A.R.). In *Coping with Uncertainty in Food Supply*, edited by I. D. Garine and G. A. Harrison. Oxford: Oxford University Press, 118–49.

———. 1992. "Spatial Mobility and Access to the Resources among the African Pygmies.", In *Mobility and Territoriality: Social and Spatial Boundaries among Foragers, Fishers, Pastoralists and Peripatetics*, edited by Casimir and Rao. New York/Oxford: Berg, 205–57.

———. 1993. "L'invention des Pygmées." *Cahiers d'etudes Africaines* 33:153–81.

———. 2012. "Changing Language, Remaining Pygmy." *Hum Biol* 84:11–43.

Bahuchet, S., and H. Guillaume. 1982. *Aka-Farmer Relations in the Northwest Congo Basin.* In *Politics and History in Band Societies*. Cambridge/Paris: Cambridge University Press/M S H.

Bahuchet, S., D. Mckey, and I. De Garine. 1991. "Wild Yams Revisited: Is Independence from Agriculture Possible for Rain Forest Hunter-gatherers?" *Human Ecology* 19:213–43.

Bailey, R. C. 1991. "The Comparative Growth of Efe Pygmies and African Farmers from Birth to Age 5 Years. *Ann Hum Biol* 18:113–20.

Bailey, R. C., G. Head, M. Tenike, B. Owen, R. Rechtman, et al. 1989. "Hunting and Gathering in Tropical Rainforest: Is It Possible?" *American Anthropologist* 91:59–82.

Batini, C., V. Coia, C. Battaggia, J. Rocha, M. M. Pilkington, et al. 2007. "Phylogeography of the Human Mitochondrial L1c Haplogroup: Genetic Signatures of the Prehistory of Central Africa." *Mol Phylogenet Evol* 43:635–44.

Batini, C., J. Lopes, D. M. Behar, F. Calafell, L. B. Jorde, et al. 2011. "Insights into the Demographic History of African Pygmies from Complete Mitochondrial Genomes." *Mol Biol Evol* 28:1099–110.

Beaumont, M. A., W. Zhang, and D. J. Balding. 2002. "Approximate Bayesian Computation in Population Genetics." *Genetics*. 162:2025–35.

Becker, N. S., P. Verdu, A. Froment, S. Le Bomin, H. Pagezy, et al. 2011. "Indirect Evidence for the Genetic Determination of Short Stature in African Pygmies." *Am J Phys Anthropol* 145:390–401.

Becker, N. S. A., P. Verdu, B. Hewlett, and S. Pavard. 2010. "Can Life History Trade-offs Explain the Evolution of Short Stature in Human Pygmies? A Response to Migliano and Colleagues." *Human Biology*, Vol. 82.

Berniell-Lee, G., F. Calafell, E. Bosch, E. Heyer, L. Sica, et al. 2009. "Genetic and Demographic Implications of the Bantu Expansion: Insights from Human Paternal Lineages." *Mol Biol Evol* 26:1581–9.

Bowcock, A., and V. Sartorelli. 1990. "Polymorphism and Mapping of the IGF1 Gene, and Absence of Association with Stature among African Pygmies." *Hum Genet* 85:349–54.

Bozzola, M., P. Travaglino, N. Marziliano, C. Meazza, S. Pagani, et al. 2009. "The Shortness of Pygmies Is Associated with Severe Under-Expression of the Growth Hormone Receptor." *Mol Genet Metab* 98:310–3.

Campbell, M. C., and S. A. Tishkoff. 2010. The Evolution Of Human Genetic And Phenotypic Variation In Africa. *Curr Biol* 20:R166–73.

Cann, H. M., C. De Toma, L. Cazes, M. F. Legrand, V. Morel, et al. 2002. A Human Genome Diversity Cell Line Panel. *Science* 296:261–2.

Cavalli-Sforza, L. L. 1986. *African Pygmies*. Orlando, Fl: Academic Press.

———. 2005. The Human Genome Diversity Project: Past, Present And Future. *Nat Rev Genet* 6:333–40.

Cavalli-Sforza, L. L., S. P. Daiger, and D. P. Rummel. 1977. Detection Of Genetic Variation With Radioactive Ligands. I. Electrophoretic Screening Of Plasma Proteins With A Selected Panel Of Compounds. *Am J Hum Genet* 29:581–92.

Cavalli-Sforza, L. L., and M. W. Feldman. 2003. The Application Of Molecular Genetic Approaches To The Study Of Human Evolution. *Nat Genet.* 33 Suppl:266–75.

Cavalli-Sforza, L. L., and B. Hewlett. 1982. Exploration And Mating Range In African Pygmies. *Ann. Human Genetics* 46:257–270.

Cavalli-Sforza, L. L., L. A. Zonta, F. Nuzzo, L. Bernini, W. W. De Jong, et al. 1969. Studies On African Pygmies. I. A Pilot Investigation Of Babinga Pygmies In The Central African Republic (With An Analysis Of Genetic Distances). *Am J Hum Genet* 21:252–74.

Cornelissen, E. 2002. Human Responses To Changing Environments In Central Africa Between 40,000 And 12,000 B.P. *Journal Of World Prehistory* 16:197–235.

Daiger, S. P., and L. L. Cavalli-Sforza. 1977. Detection Of Genetic Variation With Radioactive Ligands. Ii. Genetic Variants Of Vitamin D-Labeled Group-Specific Component (Gc) Proteins. *Am J Hum Genet* 29:593–604.

De Filippo, C., C. Barbieri, M. Whitten, S. W. Mpoloka, E. D. Gunnarsdottir, et al. 2011. Y-Chromosomal Variation In Sub-Saharan Africa: Insights Into The History Of Niger-Congo Groups. *Mol Biol Evol* 28:1255–69.

Degiorgio, M., M. Jakobsson, and N. A. Rosenberg. 2009. Explaining Worldwide Patterns Of Human Genetic Variation Using A Coalescent-Based Serial Founder Model Of Migration Outward From Africa. *Proc Natl Acad Sci U S A* 106:16057–62.

Destro-Bisol, G., I. Boschi, A. Caglia, S. Tofanelli, V. Pascali, et al. 2000. Microsatellite Variation In Central Africa: An Analysis Of Intrapopulational And Interpopulational Genetic Diversity. *Am J Phys Anthropol* 112:319–37.

Destro-Bisol, G., V. Coia, I. Boschi, F. Verginelli, A. Caglia, et al. 2004a. The Analysis Of Variation Of Mtdna Hypervariable Region 1 Suggests That Eastern And Western Pygmies Diverged Before The Bantu Expansion. *American Naturalist* 163:212–26.

Destro-Bisol, G., F. Donati, V. Coia, I. Boschi, F. Verginelli, et al. 2004b. Variation Of Female And Male Lineages In Sub-Saharan Populations: The Importance Of Sociocultural Factors. *Mol Biol Evol.* 21:1673–82.

Diamond, J. M. 1991. Anthropology. Why Are Pygmies Small? *Nature* 354:111–2.

Du Chaillu, P. B., and R. Owen. 1867. *A Journey To Ashango-Land: And Further Penetration Into Equatorial Africa.* New York: D. Appleton And Company.

Excoffier, L., P. E. Smouse, and J. M. Quattro. 1992. Analysis Of Molecular Variance Inferred From Metric Distances Among Dna Haplotypes: Application To Human Mitochondrial Dna Restriction Data. *Genetics.* 131:479–91.

Froment, A. 1993. Adaptation Biologique Et Variation Dans L'espèce Humaine: Le Cas Des Pygmées D'afrique. *Bulletin Et Mémoires De La Société D'anthropologie De Paris* 5:417–48.

Greenberg, J. H. 1966. *The Languages Of Africa.* Bloomington/The Hague: Indiana University/Mouton & Co.

Guthrie, M. 1967. *Comparative Bantu; An Introduction To The Comparative Linguistics And Prehistory Of The Bantu Languages.* Farnborough: Gregg.

Hammer, M. F., T. M. Karafet, A. J. Redd, H. Jarjanazi, S. Santachiara-Benerecetti, et al. 2001. Hierarchical Patterns Of Global Human Y-Chromosome Diversity. *Mol Biol Evol.* 18:1189–203.

Hessler, I., L. Dupont, R. Bonnefille, H. Behling, C. Gonzalez, et al. 2010. Millennial-Scale Changes In Vegetation Records From Tropical Africa And South America During The Last Glacial. *Quaternary Science Reviews* 29:2882–99.

Hewlett, B. 1996. Cultural Diversity Among African Pygmies, In *Cultural Diversity Among Twentieth-Century Foragers: An African Perspective,* Vol. 1. Edited By S. Kent, Pp. 215–44. Cambridge: Cambridge University Press.

Hewlett, B., J. Van De Koppel, and L. L. Cavalli-Sforza. 1982. Exploration Ranges Of Aka Pygmies Of The Central African Republic. *Man* 17:418–30.

Hiernaux, J. 1974. *The People Of Africa.* London: Weidenfeld And Nicolson.

Jarvis, J. P., L. B. Scheinfeldt, S. Soi, C. Lambert, L. Omberg, et al. 2012. Patterns Of Ancestry, Signatures Of Natural Selection, And Genetic Association With Stature In Western African Pygmies. *Plos Genet* 8:E1002641.

Joiris, D. V. 2003. The Framework Of Central African Hunter-Gatherers And Neighbouring Societies. *African Study Monographs, Suppl.* 28:57–79.

Lachance, J., B. Vernot, C. C. Elbers, B. Ferwerda, A. Froment, et al. 2012. Evolutionary History And Adaptation From High-Coverage Whole-Genome Sequences Of Diverse African Hunter-Gatherers. *Cell* 150:457–69.

Lebamba, J., A. Ngomanda, A. Vincens, D. Jolly, C. Favier, et al. 2009. Central African Biomes And Forest Succession Stages Derived From Modern Pollen Data And Plant Functional Types. *Climate Of The Past* 5:403–29.

Long, J. C., P. E. Smouse, and J. W. Wood. 1987. The Allelic Correlation Structure Of Gainj- And Kalam-Speaking People. Ii. The Genetic Distance Between Population Subdivisions. *Genetics* 117:273–83.

Macaulay, V., C. Hill, A. Achilli, C. Rengo, D. Clarke, et al. 2005. Single, Rapid Coastal Settlement Of Asia Revealed By Analysis Of Complete Mitochondrial Genomes. *Science.* 308:1034–6.

Macdonald, D. H., and B. S. Hewlett. 1999. Reproductive Interests And Forager Mobility. *Current Anthropology* 40:501–23.
Maley, J., P. Giresse, C. Doumenge, and C. Favier. 2012. Comment On "Intensifying Weathering And Land Use In Iron Age Central Africa." *Science* 337:1040; Author Reply 1040.
Mercader, J. Editor. 2003. *Under The Canopy: The Archaeology Of Tropical Rain Forests.* New Brunswick: Rutgers University Press.
Merimee, T. J., D. L. Rimoin, and L. L. Cavalli-Sforza. 1972. Metabolic Studies In The African Pygmy. *J Clin Invest* 51:395–401.
Merimee, T. J., J. Zapf, B. Hewlett, and L. L. Cavalli-Sforza. 1987. Insulin-Like Growth Factors In Pygmies: The Role Of Puberty In Determining Final Stature. *N Engl J Med* 316:906–11.
Migliano, A. B., L. Vinicius, and M. M. Lahr. 2007. Life History Trade-Offs Explain The Evolution Of Human Pygmies. *Proc Natl Acad Sci U S A* 104:20216-9.
Montano, V., G. Ferri, V. Marcari, C. Batini, O. Anyaele, et al. 2011. The Bantu Expansion Revisited: A New Analysis Of Y Chromosome Variation In Central Western Africa. *Mol Ecol* 20:2693–708.
Patin, E., G. Laval, L. B. Barreiro, A. Salas, O. Semino, et al. 2009. Inferring The Demographic History Of African Farmers And Pygmy Hunter-Gatherers Using A Multilocus Resequencing Data Set. *Plos Genet* 5:E1000448.
Perry, G. H., and N. J. Dominy. 2009. Evolution Of The Human Pygmy Phenotype. *Trends Ecol Evol* 24:218–25.
Phillipson, D. W. 2005. *African Archaeology*, 3rd Edition. Cambridge: Cambridge University Press.
Quintana-Murci, L., H. Quach, C. Harmant, F. Luca, B. Massonnet, et al. 2008. Maternal Traces Of Deep Common Ancestry And Asymmetric Gene Flow Between Pygmy Hunter-Gatherers And Bantu-Speaking Farmers. *Proc Natl Acad Sci U S A.* 105:1596–601.
Quintana-Murci, L., O. Semino, H. J. Bandelt, G. Passarino, K. Mcelreavey, et al. 1999. Genetic Evidence Of An Early Exit Of *Homo Sapiens* From Africa Through Eastern Africa. *Nat Genet.* 23:437–41.
Rimoin, D. L., T. J. Merimee, D. Rabinowitz, L. L. Cavalli-Sforza, and V. A. Mckusick. 1969. Peripheral Subresponsiveness To Human Growth Hormone In The African Pygmies. *N Engl J Med* 281:1383–8.
Rosenberg, N. A., J. K. Pritchard, J. L. Weber, H. M. Cann, K. K. Kidd, et al. 2002. Genetic Structure Of Human Populations. *Science.* 298:2381–5.
Rousset, F. 1997. Genetic Differentiation And Estimation Of Gene Flow From F-Statistics Under Isolation By Distance. *Genetics* 145:1219–28.
Salas, A., M. Richards, T. De La Fe, M. V. Lareu, B. Sobrino, et al. 2002. The Making Of The African Mtdna Landscape. *Am J Hum Genet.* 71:1082–111.
Santachiara-Benerecetti, A. S., M. Beretta, M. Negri, G. Ranzani, G. Antonini, et al. 1980. Population Genetics Of Red Cell Enzymes In Pygmies: A Conclusive Account. *Am J Hum Genet* 32:934–54.
Schweinfurth, G. 1873. *Im Herzen Von Afrika.* Leipzig: Brockhaus.
Seielstad, M. T., E. Minch, and L. L. Cavalli-Sforza. 1998. Genetic Evidence For A Higher Female Migration Rate In Humans. *Nat Genet.* 20:278–80.
Tavaré, S., D. J. Balding, R. C. Griffiths, and P. Donnelly. 1997. Inferring Coalescence Times From Dna Sequence Data. *Genetics* 145:505–18.

Tishkoff, S. A., F. A. Reed, F. R. Friedlaender, C. Ehret, A. Ranciaro, et al. 2009. The Genetic Structure And History Of Africans And African Americans. *Science* 324:1035–44.
Vansina, J. 1995. New Linguistic Evidence And 'The Bantu Expansion'. *Journal Of African History* 36:173–95.
Verdu, P., F. Austerlitz, A. Estoup, R. Vitalis, M. Georges, et al. 2009. Origins And Genetic Diversity Of Pygmy Hunter-Gatherers From Western Central Africa. *Curr Biol* 19:312–8.
Verdu, P., R. Leblois, A. Froment, S. Thery, S. Bahuchet, et al. 2010. Limited Dispersal In Mobile Hunter-Gatherer Baka Pygmies. *Biol Lett* 6:858–61.
Weir, B. S., and C. C. Cockerham. 1984. Estimating F-Statistics For The Analysis Of Population-Structure. *Evolution* 38:1358–70.
Wood, J. W. 1987. The Genetic Demography Of The Gainj Of Papua-New-Guinea. 2. Determinants Of Effective Population-Size. *American Naturalist* 129:165–87.

3

On Late Holocene Population Interactions in the Northwestern Congo Basin: When, How, and Why Does the Ethnographic Pattern Begin?

Karen Lupo
Alfred Jean-Paul Ndanga
Christopher Kiahtipes

Introduction

The ethnographically observed interrelationships between forest foragers and Bantu agriculturalists found throughout the Congo Basin were traditionally cast as necessary responses to rainforest resource insufficiency. These interrelationships were viewed as an exclusively Late Holocene phenomenon characterized by a set of homogenous processes such as the migration of Bantu-speaking populations, the appearance of "neolithic" toolkits, the adoption of domesticated foods, and the rise of "iron-age" villages that were causally linked to climatic perturbations and anthropogenic environmental degradation. Although resource insufficiency as a constraining factor on human occupation of the rainforests is no longer widely accepted, questions about the timing, nature, and catalysts for these interrelationships persist. We begin this paper by reviewing evidentiary lines from different intellectual domains that shed light on these questions. These data show that, although some portions of central Africa underwent similar prehistoric processes, heterogeneous and localized events shaped the overall trajectory of development. We argue that irrespective of when forager-Bantu interrelationships emerged, they were responsive to heterogeneous sociocultural and natural events spanning the last few thousand years and that each of these potentially impacted the nature of these interactions in different ways. However, among all of these events, we nominate the advent of iron tools and iron manufacturing as having a lasting pivotal effect on changing the sociopolitical landscape that gave rise to inequalities among these populations. Historically, iron implements were expensive to manufacture and some items had prestige value, but ironworker specialists controlled all aspects of the production, including manufacturing knowledge and access to finished items. The introduction of high-value prestige items set the stage for amplifying sociopolitical asymmetries between forest foragers and agricultural populations across the Congo Basin that continued into the ethnohistoric period. While data supporting this hypothesis are limited, we suggest that the amplification of social inequalities resulting from the introduction of iron can explain anomalies in the archaeological record such as the so-called "Stone to Metal Age." We end this paper by suggesting ways that this hypothesis could be tested in the archaeological record.

Traditional Views of Forest Foragers and Agriculturalists

Ethnographic forest foragers and agriculturalists maintain complex interrelationships that are based, in part, on the economic exchange of village goods (such as manioc) produced by farmers for forest products (notably meat and honey) and labor provided by the foragers. For decades, questions about how these complex interrelationships emerged were narrowly focused on the idea that the economic exchange was a functional response to the insufficient availability of resources in the rainforest critical to human survival, especially starches (Bailey et al. 1989; Bailey and Headland 1991). Existing wild starches (i.e., *Dioscorea* sp.) are often seasonally restricted, spatially dispersed, or required intensive processing that resulted in high handling costs (Bailey et al. 1989). The fact that many contemporary forest foragers relied on domesticated resources obtained via exchange from agricultural populations seemed to support this overarching explanation (Bailey et al. 1989). Some concluded that rainforests could not permanently support full-time foragers before the advent of food production or elaborate processing technologies (Bailey et al. 1989; Bailey and Headland 1991). This idea cast the interrelationship between foragers and farmers as symbiotic and serving a critical nutritional need. It also impacted archaeological reconstructions of the peopling of the rainforest because it meant that human occupation of the rainforest did not extend into remote periods of time (Mercader 2002).

Conventional archaeological reconstructions of the evolution of the interrelationships between foragers and farmers and the prehistoric peopling of the central African rainforest closely tracked the anthropological view and posited a late colonization associated with the spread of domesticated plants and animals. Prehistoric foragers before the advent of domesticated starches only intermittently used peripheral forest-savanna zones. Bantu farming populations with superior technology (e.g., ceramics, ground stone tools, and sometimes metallurgy) and a reliable subsistence base (domesticated plants and animals) spread in response to a dry climatic interval ca. 3,500–2,000 years BP (Kadomura and Kiyonaga 1995). Historic linguistic models identified migration(s) of Proto-Bantu–speaking farming populations from a common homeland in eastern Nigeria and western Cameroun some 3,000–3,500 years BP that seemed to support this scenario (Bouquiaux 1980; Guthrie 1962; Greenberg 1949, 1955). A dry interval associated with the demographic spread changed forest boundaries, fragmented blocks of impenetrable forest and created savanna corridors that facilitated the spread of human populations between 2,800–2,000 years BP (Schwartz et al. 1990). Foragers permanently occupied deep rainforest habitats after the appearance of new technologies and domesticated resources (Phillippson 1985). Slash-and-burn agriculture and the fuel demands of metallurgy, in concert with climate change, underwrote the development of the interdependent forager–farmer relationship found in different portions of the Congo Basin today (Bailey et al. 1989; Oliver 1966; Vansina 1990). These traditional reconstructions often couched prehistoric forager-Bantu interrelationships within a model derived from the classic ethnographic cases of the Mbuti (see Turnbull 1962; Schebesta 1936) and Efe (Bailey, 1991) and de-emphasized regional and temporal variation in forager-farmer interrelationships and lifeways. A return to moister conditions after 2,000 years BP resulted in forest resurgence that isolated populations and cemented the interdependency between forager and farmer.

Despite the widespread acceptance of the view that rainforests lacked sufficient wild starches to support full-time foragers, a number of scholars challenged this basic assumption (e.g., Bahuchet et al. 1991; Bahuchet 1993; Hladik and Dounais 1993; Vansina 1993/1994). Botanical surveys in concert with ethnobotanical studies conducted in forested west central and central Africa identified high densities of edible wild yams (*Dioscorea* spp.) in some areas (Bahuchet et al. 1991; Hladik et al. 1984; Hladik and Dounais 1993; Sato 2001). At the same time, ethnographic studies show that some forest foragers practice a type of paracultivation or incipient management of wild yams by replanting stems after harvesting (see Dounias 2001 and the references therein). Management practices, ritual beliefs, and traditional knowledge of wild yams imply an established longtime depth of exploitation. Experimental studies conducted among forest foragers demonstrate that foraging for wild yams can seasonally yield high returns that rival those derived from other domesticated starches (Caudell 2011; Dounais 2001). Although domesticated foods are clearly important, wild yams comprised a consistent proportion of the diet for many forest foragers and in some locations were seasonally the primary starch source (Kitanishi 1995). For example, Kitanishi (1995) found that the proportion of wild yams consumed by Aka in northeastern Congo varied seasonally but during the late wet season greatly exceeded the amount of domesticated foods exploited. More recently, Yasouka (2006) documented forest-foraging trips (so-called *molongo's*) spanning several months during which Baka foragers in Cameroun subsisted entirely on forest products, including wild yams (also see Sato in this volume).

In addition to these studies, archaeological research conducted in the central African rainforest and in tropical forested regions throughout the world shows occupation by human foragers that predates the emergence of domesticated food sources (Barker et al. 2007; Barton et al. 2011; Mercader 2002; Mercader et al. 2003; Sémah and Sémah 2012). Cumulatively, these archaeological and ethnographic studies weaken the underlying assumption of a nutritional basis for these relationships and reopen lingering questions about the timing, nature, and catalysts that gave rise to these interrelationships.

Forager-Bantu Population Interactions in the Congo Basin: Timing, Nature, and Catalysts

Although the insufficiency of starches is no longer viewed as a primary catalyst for the emergence of the forager-farmer dyad, the key questions about when, how, and why remain. Here we summarize and review current evidentiary lines from different domains that shed light on these questions.

When Did Forager-Bantu Interrelations Begin?

The question of when forager-Bantu dyads emerged rests, in part, on establishing the timing and trajectory of human occupation in the central African forest. Limited archaeological evidence suggests that hominins occupied portions of the central African rainforest beginning in the mid to late Pleistocene (Gottilogue 2000; Mercader 2002). Early Stone Age (ESA) sites are extremely rare, not well preserved, and reveal very little about how these early populations lived. Archaeological evidence from the Middle Stone Age (MSA) represented by the Lupemban industrial complex, dates to approximately 300,000 years BP (Barham 2001; Taylor 2011) and is often viewed as reflecting the earliest evidence of sustained hominin occupation with a specialized technological adaptation to the rainforest (Barham 2002; Taylor 2011). However, Cornelissen (2002) views the Lupemban as a more generalized adaptation. Although Lupemban tools persist until 25,000 years ago, the gradual appearance of Tshitolian microlithic assemblages in central Africa at roughly 45,000 years ago marks the beginning of the Late Stone Age (LSA), which persists until around 10,000 years BP (Cornelissen 2002).

LSA microliths appear to reflect a flexible technological adaptation allowing individuals to access a variety of habitats within increasingly variable Pleistocene climatic regimes (Cornelissen 2003). However, quartzite microlithic technologies persist until at least 3,000 years BP and perhaps later in west and central Africa (Barham and Mitchell 2008) and until after

AD 1200 in some areas such as the Ituri (Mercader and Brooks 2001). Thus the terminus of the LSA is loosely defined by the beginning of the "Stone to Metal Age. The latter is a largely transitional stage between the LSA and the Early Iron Age, which is characterized by the use of ceramics and well-defined ground stone tools and temporally overlaps with the Early Iron Age (de Maret 1994/1995; Lavachery 2001). There is limited evidence of a cultural continuity between the MSA, LSA, and Middle to Late Holocene assemblages before the widespread use of iron (Taylor 2012), but some long and nearly continuous sequences have been reported. For example, Mercader et al. (2000a) report a continuous sequence of C14 dates from different archaeological sites in the Ituri Forest spanning from 18,800 C14 BP through the Late Holocene (also see Lavachery 2001).

It is not clear how or if modern forest-forager populations are related to these early forest dwellers (but see Hammer et al. 2011). Time of most recent common ancestor (TMRCA) estimates from biomolecular studies of contemporary peoples suggest that forest foragers shared a common ancestor with Bantu populations before 71,000 to 90,000 years ago (Batini et al. 2007, 2011, 2012; Patin et al. 2009; Verdu et al. 2009). These populations diverged some 30,000–70,000 years ago, and the daughter populations possibly became isolated from one another by adapting to separate ecological habitats (e.g., rainforest versus grasslands) or perhaps via sociocultural barriers (but see Destro-Bisol et al. 2004). Biomolecular and historical linguistic analyses identify a subsequent further separation of proto-foragers into two populations occupying the eastern and western parts of the Congo Basin (Bahuchet 1992; Batini et al. 2011; Destro-Bisol et al. 2004a). Bahuchet's (1993) historical linguistic analyses of contemporary Aka and Baka revealed a split between eastern and western groups of forest foragers dating some 40,000 years ago. TMRCA estimates of forest foragers place these events at approximately 10,000–15,000, 21,900 to 27,000 years ago (Batini et al., 2011; Destro-Bisol et al. 2004a). Although Batini et al. (2011) found no evidence for subsequent maternal lineage exchange between these populations, analyses of Y chromosomes and genome-wide variation suggest sustained direct or indirect male-to-male interaction between eastern and western forest foragers. Thus, a backdrop of nearly continuous rainforest occupation over several millennia by groups of mobile forest foragers who, in some areas, had contact with outlying forager populations characterizes the rainforest before the arrival of Bantu populations.

There is ample evidence for a change in the material record that corresponds with biomolecular and linguistic evidence for the migration of Bantu-speaking peoples dating within the last 2,000–5,000 years (Berniell-Lee et al. 2012; Eggert 2002; Holden 2002). Current opinion holds that the heartland of Bantu-speaking populations was the Grassfields in western Cameroon (see Lavachary 2001), and most scholars view these migrations as protracted events characterized by waves of immigrations rather than a single demographic event (Berniell-Lee et al. 2012; Mercader et al. 2000b).

Mercader et al. (2000b) propose prolonged interaction between forest foragers and Bantu farmers spanning almost 1,000 years based on ceramic characteristics from archaeological sites in the Ituri. Oslisly (1999) describes a continuous sequence of LSA, Neolithic (i.e., ground stone and ceramics), and Iron Age occupations spanning from 9,000–1,000 years BP in the Ogooué Valley of Gabon. The appearance of "le stade néolithique" and "l'âge du fer ancien" (Oslisly 1999) at 3,500 and 2,500 years BP, respectively, is described as "*une arrivée en plus grand nombre de nouveaux groups néolithiques*" (1999), which implies that these new arrivals completely displaced LSA hunter-gatherers. The disparate records of technological and demographic change in the Ituri Forest and the Ogooué Valley underscores the need to approach the emergence of the forager-farmer dyad as a localized phenomenon.

Because direct evidence of interaction between the Bantu immigrants and indigenous populations is quite rare in the archaeological record, some of the strongest evidence currently available for early interactions among these populations comes from biomolecular studies that identify the demographic and genetic consequences stemming from the expansion of Bantu populations. Differences between food-producers and hunter-gatherers in the ratios of fast- and slow-evolving polymorphisms of MtDNA and Y chromosomes lead Destro-Bisol et al. (2004) to suggest that expanding Bantu populations inadvertently pushed forest foragers into less desirable habitats, reducing gene flow and ultimately reducing effective population size. Similarly, biomolecular studies conducted by Verdu et al. (2009) found evidence of diversification in western forest-forager groups at 2,625 and 2,900 years BP, which they attribute to genetic drift resulting from forager population contraction and isolation in response to the spread of Bantu farmers. Several studies reveal important purported evidence of the emergence of the well-known ethnographic marriage/mating pattern whereby forager women marry farmer men (i.e., hypergyny), usually as second wives (Destro-Bisol et al. 2004; Quintana-Murci et al. 2008). Conversely, and with few exceptions (see Verdu et al. 2009), forager males do not have access to farmer women as brides/mates (discussed further below). Recently, Verdu et al. (2009) identify heterogeneous levels of admixture from foragers to Bantu populations among western foraging groups that they argue emerged within the last 2,000–5,000 years (also see Bruges Armas et al. 2003). Quintana-Murci et al. (2008) interpret the substantial overlap in ancestral L1c clades in forager and farmer mtDNA as evidence of an early population split sometime before 70,000 years BP. The sharing of derived L1c haplotypes that appeared as early as 40,000 years ago and as recent as 3,000 years ago makes a strong case for recurring gene flow from forest-forager populations to proto-Bantu populations (Quintana-Murci et al. 2008). These results imply a much earlier and sustained link among these populations that significantly predates the archaeological and linguistic evidence for the Bantu expansion. Because the evidence suggests comparatively less maternal gene flow from foragers to Bantu populations in the last few thousand years, it implies a different mating pattern existed in the remote past in comparison to the ethnographic record.

Although biomolecular data provides intriguing evidence for the timing and nature of interactions between ancestral foragers and Bantu populations, linkages of these events to the archaeological material record are problematic because of the large error ranges associated with estimates of TMRCA and haplogroup coalescence times (see table 3.1; Henn et al. 2008; Ho and Larson 2006). For instance, while several analysts have noted that there is temporal overlap between the biomolecular dates for the forager-Bantu population split and the Lupemban industrial complex (e.g., Batini et al. 2011), the appearance of the latter actually predates the emergence of anatomically modern humans and the putative split by approximately 150,000 years. Others note the general correspondence between evidence for asymmetrical introgression in ancestral Bantu populations and foragers and the onset on the Late Stone Age (Quintana-Murci et al. 2008) or the "Neolithic" (Verdu et al. 2009). However, finger-matching between demographic events encoded in the genomes of modern populations and the somewhat arbitrary boundaries of the LSA or the Neolithic (see Lane 2005) can only result in very generalized correlations. Linkages between biomolecular data and archaeological phenomenon, especially where it concerns the emergence of the forager-farmer dyad, will be difficult to make without more detailed archaeological chronologies and improvements in estimates of TMRCA and coalescence times.

Nature of Ethnographic and Prehistoric Forager-Bantu Dyads

Traditionally, the classic ethnographic cases of the Mbuti and Efe were thought to epitomize contemporary and prehistoric forager-Bantu interrelationships. But within the last few decades, ethnographic, historical linguistic, biomolecular, and archaeological analyses provide revealing evidence suggesting a wide range of variation in the nature of forest forager lifeways, their interrelationships with settled agriculturalists and articulation with larger spheres of influence (Barham and Mithell 2008; Lupo 2011; Kleiman 1999; and Verdu in this volume).

Ethnographically, forager-Bantu dyads have important ritual, social, political, and economic dimensions (Hewlett 1990, 1991). Despite these close and complex relationships, social distances and inequalities between populations are maintained by differences in dress, material culture, access to goods and education, residential segregation, marriage patterns, and overall societal beliefs (e.g., Joiris 2008; Schmitt and Lupo 2008). There is, however, a wide range of variability in how these dimensions are manifested (see Verdu in this volume). Even though forager-Bantu interrelationships may be widespread within a specific population, Hewlett (1991) found that some Aka foragers did

Table 3.1. Molecular analyses on contemporary central African populations and TMRCA estimates.

Topic/ Population	Analysis/ Material	Dates	Source
Population divergence Bantu/ forager	Multilocus	53,995 (20,625–121,475, 95% CI) 89,675 (23,025–123,275, 95% CI)	Verdu et al. (2009)
	Multilocus	56,000 (25,800–130,500, 95% CI)	Patin et al. (2009)
	MtDNA	57,100 +/-7,900	Quintana-Murci et al. (2008)
Bantu/ western forager	MtDNA (L1c)		
	L1c1a (forager)	29,250 +/-12,700	Batini et al. (2007)
	L1c4 (forager)	36,700+/-15,600	
	L1c5 (forager)	36,600+/-14,450	
	L1c1b (Bantu)	58,800+/-12,650	
	L1c1c (Bantu)	46,350 +/-10,900	
	L1c2 (Bantu)	55,350 +/-13,600	
Western/ eastern forest foragers	Y-chromosome (B2b2 & B2b4b/B2b3)	10,695 (3,534–17,143, 95% CI) 10,478 (6882–16523, 95% CI) 14,322 (9300–22,909, 95% CI) 15,221 (10,013–23,932, 95% CI)	Batini et al. (2011)
	MtDNA	27,000 (10,000–57,000)	Batini et al. (2010)
	Multilocus	21,900 (14,200–66,300, 95% CI)	Patin et al. (2009)
Western foragers	Multilocus	2,625 (725–35,275, 95% CI) 2,900(850–30,050,95%CI)	Verdu et al. (2009)

not maintain these relationships (see also Lupo and Schmitt 2004, 2005). This could be attributable to contemporary circumstances such as the advent of wage labor and decline of traditional societal structure, but it remains unknown whether a similar range of variation characterized past forager-Bantu dyads.

Kleiman (1999) draws on historical linguistic analysis and historical evidence to demonstrate that forest foragers in parts of central Gabon (Ebongo-Tsogho) and southern Republic of Congo (Ibongo-Iyaa) participated in wider economic trade networks acting as autonomous specialists and procurers of forest goods (see also Bahuchet 1979, 1993). Albeit very limited, historical sources from the seventeenth century (Kleiman 1999:99) and mid- to late nineteenth century (de Quatrefages 1895) add further support to this view.

These sources portray forest foragers as primarily big game hunters adept at procuring elephants (*Loxodonta africana*). Battel in AD 1625 (de Quartrefages 1895) and Dapper in AD 1686 (Bahuchet and Guillaume1979) describe them as elephant hunters who transported and exchanged ivory with middlemen on the coast of Congo. De Quatrefages (1895; Masui 1897) cites historical sources that depict forest foragers as autonomous, occupying independent villages in remote regions of the forest and having chiefs. According to Schweinfurth (as cited in de Quatrefages 1895), forest foragers served as a warrior group for King Munza. Although the historical evidence is anecdotal, it suggests different kinds of forager lifeways and interactions than those recapitulated in the ethnographic record.

Biomolecular analyses of maternal profiles provide evidence for important differences in the demographic histories of forager groups who occupy the eastern and western portions of the Congo Basin (Quintana-Murci et al. 2008). For instance, these analyses demonstrate that eastern groups of foragers have maternal profiles that are more similar to eastern African nonforager populations, while those in the west are more allied with western agriculturalists (Patin et al. 2009). The degree of introgression between foragers and Bantu populations also significantly differed between eastern and western forest foragers. Patin et al. (2009) found that gene flow from western forest foragers to Bantu populations was three to seven times higher than it was for eastern forest foragers and their neighboring agriculturalists (also see Verdu et al. [2009]). These differences may reflect differences in the intensity of social inequalities as manifested by hypergyny. There are also differences among forest-forager populations in past effective population sizes. Patin et al. (2009) found that the effective population size for Bantu speakers increased before 65–30,000 years ago, while eastern forager foragers underwent a bottleneck some 20,000–4,000 years ago with an estimated 90–95 percent decrease in population. Western pygmies also underwent a reduction in population size 4,000–650 years ago, with an estimated 80 percent reduction population followed by a recovery period (also see Batini et al. 2010; Verdu et al. 2009). Differences in the timing of forager population reductions could reflect asynchronous responses to spatial and reproductive constraints imposed by immigrant and growing neighboring populations (*sensu* Verdu et al. 2009) and other anthropogenic factors such as habitat loss and destruction. But other kinds of factors can also cause bottlenecks such as a reduction in foraging efficiency, disease, and competitive resource or sociopolitical exclusion.

Recent archaeological studies show that the appearance of material remains, such as ceramics and a shift to villages, often thought to index the Bantu expansion, have independent histories and do not appear to be concordant traits or a uniform response to population migration and the adoption of food production (Casey 2005; de Maret 2005; Stahl 1993). To complicate matters further, the appearance of certain artifacts, such as ceramics, were traditionally viewed as evidence for the spread of farming, even in the absence of direct and independent indicators of the use of domesticated foods. For example, the spread of ceramics along the west coast, in west-central and into central Africa some 2,500–2,000 years ago is often taken as evidence of slowly expanding Bantu farmers (Diamond and Bellwood 2003; Holden 2002). But as many now point out the assumption that specific artifacts equate to farming cannot be supported without corresponding early direct evidence for food production in the archaeological record (Eggert 1993, 2002). Linguistic evidence indicates a common origin for certain domesticated foods (Vansina 1994/1995; Rossel 1999), but actual physical archaeological evidence for the early use of domesticated foods is slim. For example, domesticated yams are often assumed to have been among some of the earliest domesticates because their wild counterparts thrive in forests and ethnographic populations relied heavily on yams in western and west-central Africa (e.g., Daniell and Latham 1848; Mignouna and Dansi 2003). But actual evidence for the use of domesticated yams in prehistoric contexts can only be inferred by the presence of certain artifacts. Limited archaeological evidence indicates the use of goats and plantains (*Musa* sp.) in southern Cameroon between 2,600–2,200 BP (Mbida et al. 2000, 2001), but this evidence may require additional assessment (see Neumann and Hilderbrand 2009). More recently, Eggert et al. (2006) report the recovery of pearl millet seeds (*Pennisetum glaucum*) from Bwambé Hill and Abang Minko'o dating to approximately 2,200 BP and Bambara groundnut seeds (*Vigna subterranean*) from Akonétye dating to 1,700 BP in southern Cameroon. These remains are particularly significant because these plant types are not usually found in forested habitats and are adapted to arid savanna biomes. Dramatic increases in the pollen of the oil palm (*Elaeis guineenis* Jacq.) in Late Holocene pollen records has been argued to be the result of widespread arboriculture or paracultivation of indigenous oleaginous oil palm trees beginning around 5,000 and intensifying by 3,000 to 2,500 years ago (Sowunmi 1999). Although this hypothesis has been challenged (Maley and Chepstow-Lusty 2001), charcoal identification from sites in the middle Ogooué Valley, Gabon, show clear and enduring preferences for specific kinds of trees and intensive habitat management post-1,400 BP (Oslisly 1999). Thus, it is possible that arboriculture characterized some areas and populations during particular time frames, but not others. Arboriculture has distinct and different cultivation requirements than many other kinds of domesticated crops (especially root crops) (Lantis 2000) and does not necessarily equate to sedentism or the dependency on domesticated foods; there are ethnographic cases of hunter-gatherers practicing incipient arboriculture in other parts of the world (Hill and Baird 2003). Though scant, the direct evidence suggests a variety of plants and cultivation techniques were utilized, and some of these (e.g., pearl millet) may have been acquired via trade or cultivated on forest and savanna fringe zones (Kahlheber et al. 2009). But even if direct evidence of food production is found associated with ceramics or artifacts in certain areas, it does not necessarily reflect the widespread dispersal and use of domesticated foods.

Despite new and emerging research, the exact nature of the lifeways of early foragers and Bantu populations remains unclear. What is clear is that many of the assumptions researchers originally held about how these populations lived, and by extension interacted, are not supported by independent evidence. Researchers, such as Kleiman (1999), have pointed to the fallacy of retrodicting ethnographic patterns of forager lifeways into the past, but a similar problem concerns the assumption that all early Bantu populations were farmers or had identical lifeways. Clearly, it is possible that some groups of early Bantu speakers engaged in a variety of economic pursuits (i.e., pure farming, mixed farming and foraging, or foraging with no farming). It is also clear that innovations (such as ceramics) and beliefs move independently of people via exchange and transmission (Eggert 2005). If a high degree of variation characterizes the interrelationships of foragers and Bantu famers in the ethnographic record, then it is possible that a similar degree of variation also characterizes these relationships in the past.

Homogenous "Push Factors"

The final question concerns the circumstances that purportedly catalyzed the development of these relationships. These circumstances usually include climate-driven population migration and anthropogenic environmental impacts resulting from the cumulative effects of the onset of slash-and-burn agriculture and ironworking. Probably the most common "push factor" cited by researchers is the onset of an arid interval during the Late Holocene that purportedly launched the spread of the Bantu-speaking peoples (Ngomanda et al. 2009). Proxy evidence including fossil pollen, diatoms, and geochemical elements from west and west-central Africa identify a widespread, prolonged, and in some cases abrupt arid phase spanning from approximately 4,500 to 2,000 years BP (figure 3.1). This climatic interval fragmented forest boundaries, regressed lakes and wetlands, and is marked in some places by changes in hydraulic budgets, an increase in grass species, or increased dust deposition (Delègue et al. 2001; De Menocal et al. 2000; Elenga et al. 2004; Maley and Brenac 1998; Ngomanda et al. 2005; Nguetsop et al. 2004; Reynaud-Farrera et al. 1996; Vincens et al. 1999). In some areas, forest fragmentation and regressions during the Late Holocene were as severe as those associated with the climatic interval at the end of the Late Pleistocene (Elenga et al. 1996). This arid interval is often viewed as part of a continent-wide pattern following the African Humid Period (10,000 to 6,000 years ago) and manifested by spatial variation in precipitation patterns. Some link this interval to the southward retreat of the Intertropical Convergence Zone at approximately 4,500–4,100 years ago that kicked off hyper-aridity and cooling in the Sahara and opened

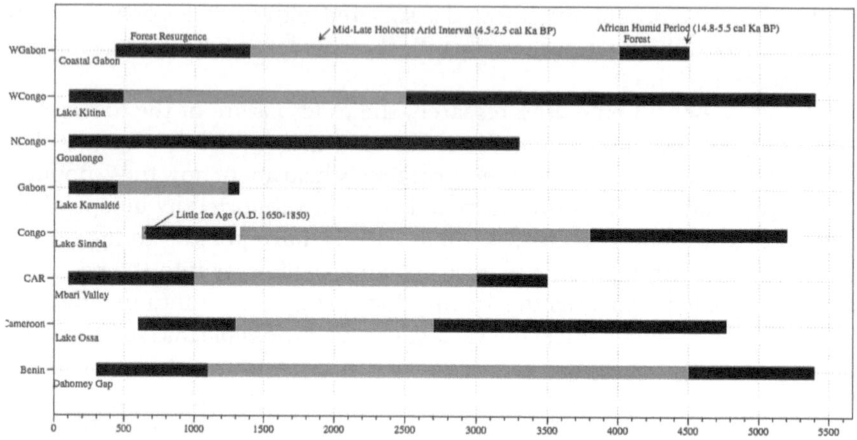

Figure 3.1. Simplified paleovegetation sequence spanning the last 5,000 years in west and west-central Africa.
Sources: Brncic et al. (2007); Elenga et al. (2004); Maley and Brenac (1998); Ngeutsop et al. (2004); Reynaud-Farrera et al. (1996); and Runge (2002).

forested areas in west Africa (Bonnefille and Chalié 2000; Salzmann and Hoelzmann 2005).

But localized factors such as differences in precipitation regimes, forest sensitivity, vegetation communities, topographic features, edaphic environment, and historical-environmental circumstances influence the local manifestations of large-scale climate events (Elanga et al. 1996; Hassan 2002; Vincens et al. 1999). As recently demonstrated by Brncic et al. (2007), paleoenvironmental data collected from the Atlantic equatorial zone and does not necessarily predict forest response in the central equatorial zone. In central Africa, several unique paleoenvironmental records spanning the Late Holocene from Cameroon (Ngomanda 2005, 2009), Congo-Brazzaville (Brncic et al. 2007, 2009), and the Central African Republic (Runge 2002; Neumer and Becker 2008) show how these factors influence local environment. For instance, Ngomanda et al. (2009) interprets complex shifts in the dominance of light-demanding and pioneer taxa from the pollen record at Nyabessan Swamp as "a rapid turnover in terms of gap formation and closure, suggesting brief and recurring disturbances of a decadal scale that have no modern analogue." These shifts in light-demanding and pioneer pollen taxa are also evident in palynological and geochemical reconstructions undertaken by Brncic et al. (2007, 2009), but they indicate that that dry conditions between 4,000 and 2,000 BP were regularly interrupted by periods of increased rainfall. Additional data from stable carbon isotope analyses of alluvial deposits in the Central African Republic support this interpretation that Late Holocene environments were characterized by regular fluctuations in the availability and timing of precipitation in the region (Runge 2002).

To complicate matters further, the Medieval Warm Period (MWP) and Little Ice Age (LIA) are climatic anomalies visible in a number of paleoenvironmental records worldwide and date to approximately 1,600–1,000 BP and 700–100 BP, respectively (de Monocal 2000). These events are less visible or sometimes omitted entirely in some palynological records (Maley and Brenac 1998; Bonnefille and Chalié 2000; Elenga et al. 2004), largely because of chronological inaccuracies introduced by fluctuations in the 14C calibration curve and the millennial-scale sampling strategy employed by many researchers (also see Ngomanda et al. 2005/2009, figure 1). However, reconstructions from lacustrine records in Atlantic equatorial Africa (Ngomanda et al. 2005, 2009) and east Africa (Barker et al. 2010; Verschuren et al. 2000; Vincens et al. 1999) show overall reductions in lake levels and increases in pollen belonging to light-demanding and pioneer vegetation formations. Ngomanda et al. (2005) conclude that "this major extension of secondary forest and inference of drought in western equatorial Africa between 1240 and 550 cal. BP coincides with the driest conditions in equatorial East Africa," but Brncic et al. (2007) come to a slightly different conclusion, noting that "semi-evergreen forest taxa persisted throughout the last 3,300 years with no sign of savanna expansion, despite extensive periods of reduced (and increased) moisture." Palynological records from the Ngotto Forest, Central African Republic, show persistent forest elements and complex successional formations through much of the MWP with dramatic shifts toward light-demanding and pioneer formations taking place in the last 800 years (Kiahtipes et al. 2011). Stable carbon isotopes derived from alluvial sediments in Ngotto corroborate this interpretation, showing no major shifts toward savanna vegetation until relatively recently (Neumer et al. 2008). Cumulatively, the results of these studies show a complex picture of vegetational change resulting from several climatic perturbations have characterized the last 4,500–5,000 years in west and west-central Africa. This does not eliminate the idea that climate contributed to Bantu migrations but weakens a direct cause and effect inference.

The Evolution and Emergence of the Forager-Bantu Dyad: A Hypothesis about the Rise of Social Currencies and Inequalities

If heterogeneous prehistoric and historic circumstances characterize forested regions of western and central Africa and traditional overarching explanations such as starch insufficiency do not explain the emergence of forager-Bantu interactions, it reopens the questions of how, when, and why these interrelationships emerged in the first place. If there is no nutritional-functional basis for these interrelationships, then how do we explain the emergence of purportedly similar cultural manifestations between foragers and Bantu populations? Some recent explanations for the emergence of these interrelationships focus on historic processes such as the introduction of European goods, impact of the slave trade, and economic change (Bahuchet and Guillaume 1979;

Kleiman 1999). The effects of these historic events undoubtedly impacted preexisting interrelationships among these populations (see Lupo 2011), but the evolution of the forager-Bantu dyad's must be viewed from within a longitudinal prehistoric trajectory.

Viewed from this context, we nominate metallurgy, a process known to be associated with the rise of social and political inequalities in Africa (Holl 2009) and elsewhere (Kim 2001), as a pivotal factor in the emergence of these interrelationships. The advent of metallurgy is often mentioned among the constellation of factors cementing early interrelationships of foragers and Bantu populations. Iron production has a separate trajectory that is not necessarily concordant with the spread of ceramics and domesticated foods. Some very early dates for ironworking are reported by Zangato and Holl (2010), and most of these appear to be clustered in savanna zones (Eggert 2010; Holl 2009), but the earliest dates for iron production in forested areas from Cameroon to Congo fall between 2,600 and 2,100 years BP (Eggert 1993; Essomba 1989; Holl 2009; Lanfranchi et al. 1998; Schwartz et al. 1990; Yandia 1996). The identity of the earliest ironworkers is not entirely clear. For example, evidence suggests some forested regions in west-central Africa were abandoned and later reoccupied by different waves of migrations by farming and ironworking populations rather than a homogenous spread of people (Oslisly 2001). Most of the recent discussions surrounding the spread of ironworking are focused on how it emerged, where it came from, and the trajectory of the spread (i.e., transmission of ideas or population movement and independent innovation) (Chirikure 2010; Craddock 2010; Eggert 2010; Killick 2004). This trend has inadvertently diverted attention away from some substantive questions regarding the impact of metallurgy on the sociopolitical structure and the interactions between populations (but see Dupré and Pincon 1997; de Barros 2000). In fact, very little consideration is given to how the introduction of a highly valued and desirable set of implements impacted the sociopolitical dynamics of population interaction, especially involving groups with different existing sociopolitical structures and economics. When discussed at all, the impact is often couched within the framework of how the availability of superior and more efficient technology tied forest foragers to settled Bantu farmers. While it is probable that some iron implements increased foraging efficiency, historically, iron implements had high value as items of wealth and prestige (see Lupo 2011). In the late nineteenth and early twentieth centuries, iron objects were used as a means of exchange, especially in the purchase of brides (see Guyer 1986 and the references therein). Another often cited impact on forests and foragers associated with the advent of metallurgy is habitat destruction resulting from fuel harvesting. The advent of metallurgy is viewed as exasperating environmental impacts initiated by slash-and-burn agriculture (see Holl 2000; Wilkie and Curran 1993). Both of these impacts resulted in a cavalcade scenario that further reduced foraging returns and increased forager dependency on domesticated crops. While it is clear that both processes can cause significant environmental impacts, the nature, scale, and intensity of these activities in the prehistoric past remains unclear, and researchers cannot assume a uniform impact in forested regions across central and western Africa (Bayon et al. 2012).

We argue that iron production and the advent of iron implements had a profound impact on the existing social landscape and amplified existing inequalities between prehistoric forest forager and Bantu populations (also see de Barros 2000). Historically, metallurgists exclusively controlled the supply, production, and distribution of iron implements. They also controlled the demand for these objects by restricting manufacturing knowledge and the process of production to select groups (de Barros 2000). In the ethnohistoric past, forest foragers, for example, had to depend on Bantu trading partners to obtain unmodified iron objects and finished products (see Bahuchet 1979; Johnston 1902; but see Mercader et al. 2000a). Further control of iron objects was manifested in the practice of Bantu partners lending hunting tools to foragers (Bahuchet 1979). Because ownership confers a partial share of the animal killed with the tool, Bantu partners can claim some of the carcass and retain ownership of the tool. Thus, in the historic and ethnographic period, at least, these interrelationships were characterized by the control of highly desirable, expensive, and prestigious items by metallurgists. Some iron items had utilitarian value, but others such as ornaments may have served as costly signals of social distinction among Bantu populations (Holl 2009). Foragers may have desired these items to serve utilitarian purposes and as costly signals of affiliation with Bantu trade partners (Lupo 2011). Affiliation with Bantu trade partners signaled potential access to a range of desirable items and other nontangible benefits (Lupo 2011).

A key question is when iron production and implements became linked to wealth and prestige in forested regions of west-central Africa. In west Africa, there is historical evidence documenting the importance of iron in communities in the Grassfield region (the so-called iron belt) of Cameroon from the seventeenth century (Warnier 2012). Historically, chiefdoms of the Ndop Plain in southern Cameroon were based on an iron industry (Warnier 2012), and de Barros (1997) argues that these chiefdoms emerged sometime before AD 1000. Evidence of regional trade networks linked via rivers emerged in central Africa by AD 500–1000 (Kleiman 1999). It is not yet clear whether these early trade networks extended into the interior forested portions of central Africa or whether iron was an extensive part of this trade. The fact that ironworking sites become very common in the archaeological record, especially after 1,000 years ago, may also indicate that increased importance metallurgy in these societies (see Holl 2009). More recently, Meister (2010) reports the inclusion of iron artifacts in purported grave features dating from 2,000 to 1,700 BP from a number of sites in southern Cameroon. The inclusion of these items in interments reflects their high value in these societies. The data are too limited to determine if there was differential representation of iron implements (as an indicator of social inequalities) in these grave features. Holl (2009) argues that the earliest iron objects were not utilitarian but were ornaments used in social arenas to advertise distinction. Utilitarian items do not appear until later in time when the access to iron is less restricted and more democratized. De Barros (1997, 2000) goes so far as to suggest that metallurgy is a catalyst of social differentiation because it requires a specialized and diversified workforce that he links to the emergence of centralized power in larger chiefdoms and kingdoms (also see Holl 2009).

Another question is whether there is any evidence for the prehistoric exchange or production of iron among prehistoric forest foragers. Data are very limited and largely come from rock shelters in the Ituri Forest, Democratic Republic of Congo (Mercader et al. 2000). Metallurgical evidence in the form of slag, iron objects, vitrified clay, and a few tuyere fragments from six archaeological sites suggest *in situ* iron manufacturing. Only one of these sites, Isak Baite Southwest, had extensive amounts of slag, evidence of smelting, and a forge, but most of this material was on the surface of the site. The chronology remains unclear, but Mercader et al. (2000a) estimated that the assemblage dated to the second millennium AD. In one of these sites, Matangai Turu, Mercader et al. (2001) recovered an 800-year-old burial containing a skeleton believed to be a hunter-gatherer based on the lack of dental caries. The burial contained lithic artifacts, plant and animal remains, ceramics, and one iron bar. Lithic artifacts comprise the majority of the assemblage, and organic remains from forest products also suggest that the occupants were forest foragers. Based on this evidence, Mercader et al. (2000a) argued that iron production was practiced on a small-scale by prehistoric forest foragers in the Ituri. Although some forest foragers may have produced iron in the past (also see Barham and Mitchell 2008), the scale and intensity of production are as yet unknown. Certainly, by the historic period, forest foragers did not know how to produce iron and had to rely on trade and Bantu partners to obtain implements.

The evidence of iron as a social currency and impetus for amplifying social inequalities among past foragers and Bantu populations is at this point limited. However, future archaeological study on the distribution, range, and timing of the spread of different kinds of iron artifacts may shed light on this question. For example, Kim (2001) recently contrasted prehistoric Korea and Danish scenarios characterized by the elite control over the distribution, spread, and demand of iron to non-elites. Elite's control over the supply and demand of iron can have appreciable effects on the types, distribution, deployment, and speed of the spread of different kinds of objects depending, on the overall goals of the elite (i.e., investment in prestige versus subsistence economies).

While there is no evidence in west-central Africa for an elite power structure during the early Iron Age, this does not eliminate the possibility that iron production and distribution were controlled by metallurgists. However, the current archaeological available dataset for west-central Africa is insufficient to evaluate these possibilities. However, if the production, access, and demand of iron implements were controlled by some segment of the population during the early Iron Age then one might expect a rarity of iron artifacts and persistence of stone tools well after the emergence of metallurgy. In fact, the "Stone-to-Metal Age" is characterized by the retention of stone artifacts (and ceramics) after the emergence of ironworking, which could reflect the continued use of traditional technologies by foragers or other populations or segments of society, if iron was controlled by a segment of society. The impact from the advent of iron production may well have had further implications for biomolecular patterns of population admixture identified by Quintana-Murci et al. (2008). Recall that Quintana-Murci et al. (2008) found evidence for recurrent and substantial maternal gene flow from foragers to Bantu populations commencing sometime after 40,000 years ago and diminishing sometime within the last few thousand years. If early iron production, access, and demand were controlled by a segment of society, the advent of ironworking may have destabilized the cost of obtaining a mate and inadvertently led to higher valuation of Bantu over forager women.

Conclusions

Clearly, the last 2,000–4,500 years has been a pivotal period in western and central Africa. Although studies in a variety of domains have greatly expanded what we know and have dispelled many assumptions from conventional interpretations, many questions remain. One problem facing current analysts concerns integrating results from intellectual domains that have different temporal and geographic scales of analyses. The large error ranges in time associated with biomolecular studies make them difficult to link to local archaeological sequences. Paleoenvironmental analyses sometimes offer greater resolution but often fail to distinguish distinct and important events within the last few thousand years. A good example is the impact of slash-and-burn agriculture and advent of metallurgy, which often get compressed into a single event in paleoenvironmental analysis (e.g., Rayon et al. 2012). Finer-grained chronologies from these domains in concert with additional archaeological studies in west and central Africa can resolve some of these difficulties. The focus on fine-grain events and prehistoric human responses as manifested in the archeological record makes these studies the lynchpin that will link these different domains. A second problem, however, concerns the lack of focus on how different recent prehistoric events such as the spread and introduction of different technologies influenced sociopolitical interactions among forest populations. Specifically, how does the introduction of high-value items that have utilitarian and prestige value impact the social interactions among these populations? This is a more difficult question to address but has greater intellectual traction.

References

Bahuchet, S. 1993. "History of the Inhabitants of the Central African Rainforest: Perspectives from Comparative Linguistics. In T*ropical Forests, People and Food: Biocultural Interactions and Applications to Development,* Man and the Biosphere Series, edited by C. M. Hladik, A. Hladik, O. F. Linares, H. Pagezy, A. Semple, and M. Hadley. Paris: UNESCO and Parthenon Publishing Group, 37–54.

Bahuchet, S., D. Mckey, and I. De Garine. 1991. "Wild Yams Revisited: Is Independence from Agriculture Possible for Rain Forest Hunter-gatherers?" *Human Ecology* 19:213–43.

Bahuchet, S., and H. Guillaume. 1979. "Relations Entre Chasseurs-Collecteurs Pygmées et Agriculteurs de la Forêt du Nord-Ouest du Bassin Congolais." In *Pygmeés de Centrafrique,* edited by S. Bahuchet. Paris: SELAF.

Bailey, R. C., and T. N. Headland. 1991. "The Tropical Rainforest: Is It a Productive Environment for Human Foragers?" *Human Ecology* 19(2):261–85.

Bailey, R. C., G. Head, M. Jenike, B. Owen, R. Reichman, and E. Zechenter. 1989. "Hunting and Gathering in Tropical Forests: Is It Possible? *American Anthropologist* 91:59–82.

Barham, L. S. 2001. "Central Africa and the Emergence of Regional Identity in the Middle Pleistocene." In *Human Roots: Africa and Asia in the Middle Pleistocene,* edited by L. Barham and K. Robson-Brown. Bristol: Western Academic and Specialist Press, 65–80.

———. 2002. "Backed Tools in Middle Pleistocene Central Africa and Their Evolutionary Significance." *Journal of Human Evolution* 43:585–603.

Barham, L. S., and P. Mitchell. 2008. *The First Africans: African Archaeology from the Earliest Toolmakers to the Most Recent Foragers.* Cambridge: Cambridge University Press.

Barker, G., H. Barton, M. Bird, P. Daly, I. Datan, A. Dykes, L. Farr, D. Gilbertson, B. Harrisson, C. Hunt, T. Higham, L. Kealhofer, J. Krigbaum, H. Lewis, S. Mclaren, V. Paz, A. Pike, P. Piper, B. Pyatt, R. Rabett, T. Reynolds, J. Rose, G. Rushworth, M. Stephens, C. Stringer, J. Thompson, and C. Turney. 2007. The 'Human Revolution' in Lowland Tropical Southeast Asia: The Antiquity and Behaviour of Anatomically Modern Humans At Niah Cave (Sarawak, Borneo). *Journal of Human Evolution* 52:243–61.

Barker, P., R. Telford, O. Merdaci, D. Williamson, M. Taieb, A. Vincens, and E. Gilbert. 2000. The Sensitivity of A Tanzanian Crater Lake To Catastrophic Tephra Input and Four Millennia of Climate Change. *The Holocene* 10:303–3.

Barton, H., T. Denham, K. Neumann, and M. Arroyo-Kalin. 2012. Long Term Perspectives On Human Occupation of Tropical Rainforests: An Introductory Overview. *Quaternary International* 249:1–3.

Barton, H. 2005. The Case For Rain Forest Foragers:The Starch Record At Niah Cave, Sarawak. *Asian Perspectives* 44(1):56–72.

Batini, C., J. Lopes, D. M. Behar, F. Calafell, L. B. Jorde, L. Van Der Veen, L. Quintana-Murci, G. Spedini, G. Destro-Bisol, and D. Comas. 2011. Insights

into the Demographic History of African Pygmies from Complete Mitochondrial Genomes. *Molecular Biology and Evolution* 28(2):1099–110.
Batini, C., V. Coia, C. Battaggia, J. Rocha, M. M. Pilkington, G. Spedini, D. Comas, G. Destro-Bisol, and F. Calafell. 2007. Phylogeography of the Human Mitochondrial L1c Haplogroup: Genetic Signatures of the Prehistory of Central Africa. *Molecular Phylogenetics and Evolution* 43:635–44.
Bayon, G., B. Dennielou, J. Etoubleau, E. Ponzevera, S. Toucanne, and S. Bermell. 2012. "Intensifying Weathering and Land Use in Iron Age Central Africa." *Science* 335 (March 9):1219–22.
Berniell-Lee, G., F. Calafell, E. Bosch, E. Heyer, L. Sica, P. Mouguiama-Daouda, L. Van Der Veen, J-M. Hombert, L. Quintana-Murci, and D. Comas. 2009. Genetic and Demographic Implications of the Bantu Expansion: Insights from Human Paternal Lineages. *Molecular Biology nd Evolution* 26(7):1581–9.
Bonnefille, R., and F. Chalié. 2000. Pollen-Inferred Precipitation Time-Series from Equatorial Mountains, Africa, the Last 40 Kyr BP. *Global and Planetary Change* 26:25–50.
Bouquiaux, L., and J. M. C. Thomas. 1980. Le Peuplement Obanguien: Hypotheses De Reconstruction De Mouvements Migratoires Dans La Région Oubanguienne D'après Des Données Linguistiques, Ethnolonguistiques Et De Tradition Orale. In *L'expansion Bantoue*. L. Bouquiaux, Ed. Pp. 807–24. Paris: Selaf.
Brncic, T. M., K. J. Willis, D. J. Harris, and R. Washington. 2007. Culture of Climate? The Relative Influences of Past Processes On the Composition of the Lowland Congo Rainforest. *Philosophical Transactions of the Royal Society B* 362:229–42.
Brncic, T. M., et al. 2009. Fire and Climate Change Impacts On Lowland Forest Composition in Northern Congo During the Last 2580 Years from Palaeoecological Analyses of A Seasonally Flooded Swamp. *The Holocene* 19(1):79–89.
Bruges Armas, J., Et Al. 2003. HLA Class I Variation in the West African Pygmies and Their Genetic Relationship With Other African Populations. *Tissue Antigens* 62(233–42).
Casey, J. 2005. Holocene Occupations of the Forest and Savanna. In *African Archaeology*. A. B. Stahl, Ed. Pp. 225–48: Blackwell Publishing Ltd.
Chirikure, S. 2010. On Evidence, Ideas and Fantasy: The Origins of Iron in Sub-Saharan Africa: Thoughts On É. Zangato & A. F. C. Holl's "On the Iron Front." *Journal of African Archaeology* 8(1):25–8.
Colinvaux, P. A., and M. B. Bush. 1991. The Rain Forest Ecosystems As A Resoruce For Hunting and Gathering. *American Anthropologist* 93:153–62.
Cornelissen, E. 2002. Human Responses ao Changing Environments in Central Africa Between 40,000 and 12,000 BP. *Journal of World Prehistory* 16:197–235.
———. 2003. On Microlithic Quartz Industries At the End of the Pleistocene in Central Africa: The Evidence from Shum Laka (NW Cameroun). *African Archaeological Review* 20(1):1–24.
Craddock, P. 2010. New Paradigms For Old Iron: Thoughts On É. Zangato & A. F. C. Holl's "On the Iron Front." *Journal of African Archaeology* 8(1):29–36.
Daniell, W. F., and R. G. Latham. 1848. On the Natives of Old Callebar, West Coast of Africa. *Journal of the Ethnological Society of London* (Royal Anthropological Institute of Great Britian and Ireland) 1(210–27).
De Barros, P. 2000. Iron Metallurgy: Sociocultural Context. In *Ancient African Metallurgy: The Sociocultural Context*. J. O. Vogel, Ed. Pp. 147–98. Walnut Creek, CA: Altamira Press.

Delègue, M., M. Fuhr, D. Schwartz, A. Mariotti, and R. Nasi. 2001. Recent Origin of A Large Part of the Forest Cover in the Gabon Coastal Area Based On Stable Carbon Isotope Data. *Oecologia* 129(1):106–13.

De Maret, P. 1994/1995. Pits, Pots, and the Far-West Streams. *Azania* 29/30:318–23.

———. 2005. from Pottery Groups To Ethnic Groups in Central Africa . In *African Archaeology*. A. B. Stahl, Ed. Pp. 420–40: Blackwell Publishing.

De Menocal, P., J. Ortiz, T. Guilderson, J. Adkins, M. Sarnthein, L. Baker, and M. Yarosinsky. 2000. Abrupt Onset and Termination of the Central African Humid Period; Rapid Climate Change Responses To Gradual Insolation Forcing. *Quaternary Science Review* 19:347–61.

Destro-Bisol, G., V. Coia, I. Boschi, F. Verginelli, A. Caglià, V. Pascali, G. Spedini, and F. Calafell. 2004a. The Analysis of Variation of Mtdna Hypervariable Region 1 Suggests That Eastern and Western Pygmies Diverged Before the Bantu Expansion. *The American Naturalist* 163(212–226).

Destro-Bisol, G. F. Donati, V. Coia, I. Boschi, F. Verginelli, A. Caglià, S. Tofanelli, G. Spedini, and C. Capelli. 2004b. Variation in Femal and Male Lineasges in Sub-Saharan Populations: The Importance of Sociocultural Factors. *Molecular Biology and Evolution* 21(9):1673–82.

Dounais, E. 1993. Perception and Use of Wild Yams by the Baka Hunter-Gatherers in South Cameroon. In *Tropical Forests, People and Food*. A. H. C. M. Hladik, O. F. Linares, H. Pagezy, A. Semple, and M. Hadley, Ed. Pp. 621–32. Man and the Biosphere. Paris: UNESCO.

———. 2001. The Management of Wild Yam Tubers By the Baka Pygmies in Southern Cameroon. *African Study Monographs, Supplementary* 26:135–56.

Dupré, M-C., and B. Pinçon. 1997. Métallurgie Et Politique En Afrique Centrale: Deux Mille Ans De Vestiges Sur Les Plateaux Batéké, Gabon, Congo, Zaire. Paris: Karthala.

Eggert, M. K. H. 1992. The Central African Rain Forest: Historical Speculation and Archaeological Facts. . *World Archaeology* 24(1):1–24.

———. 2002. Southern Cameroun and the Settlement of the Equatorial Rainforest: Early Ceramics from Fieldwork in 1997 and 1998–99. In *Tides of the Desert-Gezeiten Der Wüste*. J. 8, Ed. Pp. 507–22. Köln: Heinrich-Barth-Institut.

———. 2005. The Bantu Problem and African Archaeology. In *African Archaeology*. A. B. Stahl, Ed. Pp. 301–26: Blackwell Publishing Ltd.

———. 2010. Too Old? Remarks On New Evidence of Ironworking in North-Central Africa. *Journal of African Archaeology* 8(1):37–8.

Eggert, M. K. H., et al. 2006. Pits, Graves and Grains: Archaeological and Archaeobotanical Research in Southern Cameroun. *Journal of African Archaeology* 4(2):273–98.

Elenga, H., J. Maley, A. Vincens, and I. Farrera. 2004. Paleoenvironments, Paleoclimates and Landscape Developments in Atlantic Equatorial Africa: A Review of Key Sites Covering the Last 25 Kyrs. In *Developments in Paleoenvironmental Research: Past Climatic Variability Through Europe and Africa*. F. Gasse, C. E. Stickley, R. W. Battarbee, Eds. Pp. 46–61. Dordretch, The Netherlands: Springer.

Elenga, H., D. Schwartz, A. Vincens, and J. D. Bertaux. 1996. Diagramme Pollinique Holocéne Du Lac Kitina (Congo): Mise En Evidence De Changements Paleobotaniques Et Paleoclimatiques Dans Le Massif Forestier Du Mayombe.

Comptes Rendus De L'academie Des Sciences. Série II, Sciences De La Terre Et Des Planets 323:403–10.

Essomba, J. M. 1993. *Les Fer Dans Le Passé Des Sociétés Du Sud Cameroun*. Paris: L'Harmattan.

Gotilogue, S. 2000. État De Recherches Archeologiques En Republique Centrafricaine. Recent Research into the Stone Age of Northeastern Africa. *Studies in African Archaeology, Poznan Archaeological Museum* 7:239–57.

Greenberg, J. H. 1949. Studies in African Linguistic Classification: III. The Position of Bantu. *Southwestern Journal of Anthropology* 5:309–17.

———. 1955. *Studies in African Linguistic Classification*. New Haven: Compass.

Guthrie, M. 1962. Bantu Origins: A Tentative New Hypothesis. *Journal of African Languages* 1:9–21.

Guyer, J. I. 1986. Indigenous Currencies and the History of Marriage Payments: A Case Study from Cameroon (Monnaies Indigènes Et Histoire De La Dot Au Cameroun). *Cahiers d'Études Africaines* 26(104):577–610.

Hammer, M. F., A. E. Woerner, F. L. Mendez, J. C. Watkins, and J. D. Wall. 2011. Genetic Evidence For Archaic Admixture in Africa. *Proceedings of the National Academy of Science* 108(37):15123–8.

Hassan, F. 2002. Introduction. In *Droughts, Food and Culture: Ecological Change and Food Security in Africa's Later Prehistory*. F. Hassan, Ed. Pp. 1–10. New York: Kluwer Academic/Plenum Publishers.

Hewlett, B. 1990. *Foragers and Rural Development*. Republique Centrafricaine, Projet ECOFAC- Composante RCA, Ngotto Reserve.

———. 1991. *Intimate Fathers: The Nature and Context of Aka Pygmy Paternal Infant Care*. Ann Arbor: University of Michigan Press.

Hill, R., and A. Baird. 2003. Kuku-Yalanji Rainforest Aboriginal People and Carbohydrate Resource Management in the Wet Tropics of Queensland, Australia. *Human Ecology* 31(1):27–52.

Hladik, A., et al. 1984. Les Plantes À Tubercules De La Forêt Dense d'Afrique Centrale. *Revue D' Ecologie (Terre Et Vie)* 39:249–90.

Hladik, A., and E. Dounias. 1993. Wild Yams of the African Forest as Potential Food Resources. In *Tropical Forests, People and Food*. A. H. C. M. Hladik, O. F. Linares, H. Pagezy, A. Semple, and M. Hadley, Ed. Pp. 163–76. Man and the Biosphere, Vol. 13. Paris: UNESCO.

Holden, C. J. 2002. Bantu Language Tree Reflects the Spread of Farming Across Sub-Saharan Africa: A Maximum Parsimony Analysis. *Proceedings of the Royal Society: Biological Sciences* 269 (1493):793–9.

Holl, A. F. C. 2000. Metals and Precolonial African Society. In *Ancient African Metallurgy*. J. Vogel, Ed. Walnut Creek: Altamira Press.

———. 2009. Early West African Metallurgies: New Data and Old Orthodoxy. *Journal of World Prehistory* 22:415–38.

Johnston, Sir H. H. 1902. *The Uganda Protectorate: An Attempt To Give Some Description of the Physical Geography, Botany, Zoology, Anthropology, Languages and History of the Territories Under British Protection in East Central Africa, Between the Congo Free State and the Rift Valley and between the First Degree of South Latitude and the Fifth Degree of North Latitude*. London: Hutchinson & Co.

Joiris, D. V. 1998. *La Chasse, La Chance, Le Chant: Aspects Du System Rituel Des Baka Du Cameroun*, Université Libre De Bruxelles.

——V. 2003. The Framework of Central African Hunter-Gatherers and Neighboring Societies. *African Study Monographs, Supplementary* 28:55–79.

Kadomura, H., and J. Kiyonaga. 1995. Origins of Grassfields Landscape in the West Cameroun Highlands. In *Savannization Process in Tropical Africa II.* H. Kadomura, Ed. Pp. 47–85. Tokoyo: Metropolitan University.

Kahlbeber, S., K. Bostoen, and K. Neumann. 2009. Early Plant Cultivation in the Central African Rainforest: 1st Millenium BC Pearl Millet from Southern Cameroun. *Journal of African Archaeology* 7(2):253–72.

Kiahtipes, C. A., K. D. Lupo, D. N. Schmitt, J-P Ndanga, J. G. Jones, and R. Lee. 2011. Prehistory and the Present: Palaeoenvironments in the Northern Congo Basin. *Before Farming.*

Killick, D. 2004. Review Essay: What Do We Know About African Iron Working. *Journal of African Archaeology* 2(1):97–112.

Kim, J. 2001. Elite Strategies and the Spread of Technological Innovations: the Spread of Iron in the Bronze Age Societies of Denmark and Southern Korea *Journal of Anthropological Archaeology* 20:442–78.

Kitanishi, K. 1995. Seasonal Changes in the Subsistence Activities and Food Intake of Aka Hunter-Gatherers in Northeastern Congo. *African Study Monographs* 16(2):73–118.

Kleiman, K. 1999. Hunter-Gatherer Participation in Rainforest Trade-Systems: A Comparative History of Forest Vs Ecotone Societies in Gabon and Congo, C. 1000-1800 AD. In Central African Hunter-Gatherers in a Mulitdisciplinary Perspective: Challenging Elusiveness. K. Biesbrouck, S. Elders, and G. Rossel, Eds. Pp. 89–104. Leiden: University of Leiden Press.

Lanfranchi, R., J-P Ndanga, and H. Zana. 1998. New Carbon 14C Datings of Iron Metallurgy in the Central African Dense Forest. *Yale F & ES Bulletin*:41–50.

Latinas, K. 2000. The Development of Subsistence System Models For Island Southeast Asia and Near Oceania: The Nature and Role of Arboriculture and Arboral Based Economies. *World Archaeology* 32(1):41–67.

Lavachary, P. 1996. Shum Laka Rock Shelter Late Holocene Deposits: from Stone To Metal (North Western Cameroun). In *Aspects of African Archaeology: Papers from the 10th Congress of the Panafrican Association For Prehistory and Related Studies.* G. Pwiti and R. Soper, Eds. Lusaka: University of Zimbabwe Publications office.

Lavachary, P. 2001. The Holocene Archaeological Sequence of Shum Laka Rock Shelter (Grassfields, Western Cameroun). *African Archaeological Review* 18(4):215–47.

——. 1996. Aspects of African Archaeolgy. 10th Congress of the Panafrican Association For Prehistory and Related Studies, Harare, 1996, Pp. 265–74. University of Zimbabwe.

Lupo, K. 2011. Implications of Bofi & Aka Ethnoarchaeology in the Congo Basin For Understanding Late Holocene Technological Change. *Before Farming* 4: Online.

Lupo, K. D., and D. N. Schmitt. 2005. Small Prey Hunting Technology and Zooarchaeological Measures of Taxonomic Diversity and Abundance: Ethnoarchaeological Evidence from Central African Forest Foragers, *Journal of Anthropological Archaeology*, 24(4):335–53.

——. 2004. Meat-Sharing and the Archaeological Record: A Preliminary Test of the Show-off Hypothesis Among Central African Bofi Foragers. In *Hunters and Gatherers in Theory and Archaeology*, G. Crothers (Ed), Pp. 241–60. Center

For Archaeological Investigations Occasional Paper No. 31. Carbondale: Southern Illinois University.
Maley, J. 2001. The Impact of Arid Phases On the African Rain Forest Through Geological History. In *African Rain Forest Ecology and Conservation*, W. Weber, L. J. T. White, A. Vedder, and Naughton-Treves, L., Ed. Pp. 68–87. New Haven Yale University Press.
Maley, J., and P. Brenac. 1998. Vegetation Dynamics, Palaeoenvironments and Climatic Changes in the Forests of Western Cameroun During the Last 28,000 Years BP. *Review of Palaeobotany and Palynology* 99:157–87.
Maley, J., and A. Chepstow-Lusty. 2001. *Elais Guineensis* Jacq. (Oil Palm) Fluctuations in Central Africa During the Late Holocene: Climate Or Human Driving Forces For This Pioneering Species? *Vegetation History and Archaeobotany* 2001:117–20.
Mbida, C. M., W. Van Neer, H. Doutrelepont, and L. Vrydaghs. 2000. Evidence of Banana Cultivation and Animal Husbandry During the First Millennium BC in the Forest of Southern Cameroon. *Journal of Archaeological Science* 27(151–62).
Meister, C. 2010. Remarks On Early Iron Age Burial Sites from Southern Cameroon. *African Archaeological Review* 27:237–49.
Mercader, J. 2002. Forest People: The Role of African Rainforests in Human Evolution and Dispersal. *Evolutionary Anthropology* 11(117–24).
———. 2003. Foragers of the Congo: The Early Settlement of the Ituri Forest. In *Under the Canopy*. J. Mercader, Ed. Pp. 93–116. New Brunswick: Rutgers University Press.
Mercader, J., and A. S. Brooks. 2001. Across Forests and Savannas: Late Stone Age Assemblages from the Ituri and Semliki, Democratic Republic of Congo. *Journal of Anthropological Research* 57(2):197–217.
Mercader, J., R. Salvador, and P. Gómez-Ramos. 2000a. Shared Technologies: Forager-Farmer Interaction and Ancient Iron Metallurgy in the Ituri Rainforest, Democratic Republic of Congo. *Azania* 35(1):107–22.
Mercader, J., M. Garci-Heras, and I. Gonzalez-Alvarez. 2000b. Ceramic Tradition in the African Forest: Characterisation Analysis of Ancient and Modern Pottery from Ituri, D. R. Congo. *Journal of Archaeological Science* 27(2):163–82.
Mignouna, H. D., and A. Dansi. 2003. Yam (*Dioscorea* Spp.) Domestication By the Nago and Fon Ethnic Groups in Benin. *Genetic Resources and Crop Evolution* 50:519–28.
Neumann, K., and E. Hildebrand. 2009. Early Bananas in Africa: The State of the Art. *Ethnobotany Research & Applications* 7:353–62.
Neumer, M., E. Becker, and J. Runge. 2008. Palaeoenvironmental Studies in the Ngotto Forest: Alluvial Sediments As Indicators of Recent and Holocene Landscape Evolution in the Central African Republic. *Palaeocology of Africa* 28:121–37.
Ngomanda, A., A. Chepstow-Lusty, M. Makaya, P. Schevin, J. Maley, M. Fontugne, R. Oslisly, N. Rabenkogo, and D. Jolly. 2005. Vegetation Changes During the Past 1300 Years in Western Equatorial Africa: A High-Resolution Pollen Record from Lake Kamalete, Lope Reserve, Central Gabon. *The Holocene* 15(7):1021–31.
Nguetsop, V. F., M. Servant, and S. Servant-Vildary. 1998. Paléolimnologie Et Paléoclimatologie De l'Ouest Cameroun Au Cours Des 5000 Derniéres Années, 'A Partir De L'étude Des Diatomées Du Lac Ossa. *Comptes Rendus De l'Academie Des Sciences*, Paris, Série Ila 327:39–45.

Nguetsop, V. F, S. Servant-Vildary, and M. Servant. 2004. Late Holocene Climatic Changes in West Africa, A High Resolution Diatom Record from Equatorial Cameroun. *Quaternary Science Review* 23:591–609.
Oliver, R. 1966. The Problem of the Bantu Expansion. *Journal of African History* 7:361–76.
Oslisly, R. 1999. Contribution De L'anthracologie À L'étude De La Relation Homme Milieu Au Cours De l'Holocène Dans La Vallée De l'Ogooué Au Gabon. In *Wood To Survive, Annales Sciences Economiques,* Vol. 25. Musée Royal De l'Afrique Centrale, Tervuren, Belgique.
Oslisly, R. 2001. The History of Human Settlement in the Middle Ogooué Valley (Gabon): Implications For the Environment. In *African Rain Forest Ecology and Conservation.* W. Weber, L. J. T. White, A. Vedder, and L. Naughton-Treves, Eds. Pp. 101–18. New Haven: Yale University Press.
Patin, E., G. Laval, L. B. Barreiro, A. Salas, O. Semino, S. Santachiara-Benerecetti, K. K. Kidd, J. R. Kidd, L. Vand Der Veen, J-M. Hombert, A. Gessain, A. froment, S. Bahuchet, E. Heyer, and L. Quintana-Murci. 2009. Inferring the Demographic History of African Farmers and Pygmy Hunter-Gatherers Using A Multilocus Resequencing Data Set. *PLOS Genetics* 5(4): E1000448.
De Quatrefages, A. 1895. *The Pygmies.* London: Macmillan and Co.
Quintana-Murci, L., H. Quach, C. Harmant, F. Luca, B. Massonnet, E. Patin, L. Sica, P. Mouguiama-Daouda, D. Comas, S. Tzur, O. Balanovsky, K. K. Kidd, J. R. Kidd, L. Van Der Veen, J-M. Hombert, A. Gessain, P. Verdu, A. froment, S. Bahuchet, E. Heyer, J. Dausset, A. Salas, and D. M. Behar. 2008. Maternal Traces of Deep Common Ancestry and Asymmetric Gene Flow Between Pygmy Hunter-Gatherers and Bantu-Speaking Farmers. *Proceedings of the National Academy of Science* 105(5):1596–1601.
Reynaud-Farrera, I., J. Maley, and D. Wirrman. 1996. Végétation Et Climat Dans Les Forêts Du Sud-Ouest Cameroun Depuis 4770 ans BP: Analyse Pollinique Des Sédiments Du Lac Ossa. *Comptes Rendus De l'Academie Des Sciences, Paris, Série Ila: Sciences De La Terre Et Des Planètes* 322:749–58.
Rossel, G. 1999. Crop Names and the History of Hunter-Gatherers in Northern Congo. In *Central African Hunter-Gatherers in Multidisciplinary Perspectives: Challenging Elusiveness.* K. Biesbrouck, S. Elder, and G. Rossel, Eds. Pp. 105–16. Leiden: Research School For Asian, African, and Amerindian Studies Universiteit Leiden.
Runge, J. 2002. Holocene Landscape History and Paleohydrology Evidenced By Stable Carbon Isotope (Delta C-13) Analysis of Alluvial Sediments in the Mbari Valley (5 Degrees N/23 Degrees E), Central African Republic. *Catena* 48(1–2):67–87.
Salzmann, U., and P. Hoelzmann. 2005. The Dahomey Gap: An Abrupt Climatically Induced Rain Forest Fragmentation in West Africa During the Late Holocene. *The Holocene* 15(2):190–9.
Sato, H. 2001. The Potential of Edible Wild Yams and Yam-Like Plants As A Staple Food Resource in the African Tropical Rain Forest. *African Study Monographs, Supplementary* 26:123–34.
Schebesta, P. 1936. *My Pygmy and Negro Hosts.* London: Hutchinson and Co.
Schmitt, D. N., and K. Lupo. 2008. Do Faunal Remains Reflect Socioeconomic Status? An Ethnoarchaeological Study of Central African Farmers in the Northern Congo Basin. *Journal of Anthropological Archeology* 27:315–25.

Schwartz, D., H. De Foresta, R. Dechamps, and R. Lanfranchi. 1990. Découverte D'unpremier Site De l'Age De Fer Ancien (2,100 B.P.) Dans Le Mayumbe Congolais. Implications Paléobotaniques Et Pédologiques. *Comptes Rendus De L'academie Des Sciences. Série II, Sciences De La Terre Et Des Plants* 310(2):1293–8.

Sowunmi, M. A. 1999. The Significance of the Oil Palm (*Elaeis Guineenis Jacq.*) in the Late Holocene Environments of West and West Central Africa: A Further Consideration. *Vegetation History and Archaeobotany* 8:199–210.

Stahl, A. B. 1993. Intensification in West African Late Stone Age: A View from Central Ghana. In *The Archaeology of Africa: Foods, Metals and Towns.* T. Shaw, P. Sinclair, B. andah, and A. Okpoko, Eds. Pp. 261–73. London: Rutledge.

Taylor, N. 2011. The Origins of Hunting & Gathering in the Congo Basin: A Perspective On the Middle Stone Age Lupemban Industry. *Before Farming* 1: Online.

Turnbull, C. M. 1962. *The Forest People: A Study of the Pygmies of the Congo.* New York: Doubleday & Co., Inc.

Vansina, J. 1994/1995. A Slow Revolution: Farming in Subequatorial Africa. *Azania* 29/30:1–14.

Vansina, J. 1990. *Paths in the Rainforest.* Milwaukee: University of Wisconsin Press.

Verdu, P., F. Austerlitz, A. Estoup, R. Vitalis, M. Georges, S. Thery, A. froment, S. Lebomin, A. Gessain, J-M Hombert, L. Van Der Veen, L. Quintana-Murci, S. Bahuchet, and E. Heyer. 2009. Origins and Genetic Diversity of Pygmy Hunter-Gatherers from Western Central Africa. *Current Biology* 19:312–8.

Verschuren, D., K. R. Laird, and B. Cumming. 2000. Rainfall and Drought in Equatorial East Africa During the Past 1,100 Years. *Nature* 403:410–4.

Warnier, J-P. 2012. *Cameroun Grassfields Civilzation.* Bamenda, Cameroon: Langaa Research & Publishing Common Initiative Group.

Yandia, F. 1996. État De Recherches Archéologiques Et Notamment De La Métallurgie Du Fer En République Centrafriciane. In *Aspects of African Archaeology: Papers From the 10th Congress of the Panafrican Association For Prehistory and Related Studies.* G. Pwiti and R. Soper, Eds. Pp. 509–14. Harare: University of Zimbabwe Publications.

Yasuoka, H. 2006. Long-Term Foraging Expedition (Molongo) Among the Baka Hunter-Gatherers in the Northwestern Congo Basin, With Special Reference To the "Wild Yam Question." *Human Ecology* 34(2):275–96.

———. 2009. Concentrated Distribution of Wild Yam Patches: Historical Ecology and the Subsistence of African Rainforest Hunter-Gatherers. *Human Ecology.*

Zangato, E., and A. R. Holl. 2010. On the Iron Front: New Evidence from North Central Africa. *Journal of African Archaeology* 8(1):7–23.

Zangato, E., and J. F. Saliege. 1998. Radiocarbon Dates of Sites in North-West Central African Republic: The Case of Ndio. *West African Journal of Archaeology* 28(1):19–37.

4

"Do Pygmies Have a History?" Revisited: The Autochthonous Tradition in the History of Equatorial Africa[1]

Robert E. Moïse

Introduction

In his groundbreaking work *Paths in the Rainforests: Toward a History of Political Tradition in Equatorial Africa,* the eminent historian Jan Vansina presents an account of the history of the forest region of equatorial Africa since the arrival of Bantu-speaking peoples there over three millennia ago (1990).[2,3] In particular, he addresses the role played by Bantu political institutions in shaping equatorial history and refers to this complex of institutions as the "equatorial tradition"—a political tradition of grand scale on a par with other traditions in world history, such as the Christian tradition, the Confucian tradition, and so on. Although his work is the first and most comprehensive attempt to articulate the broad history of the region over the *longue durée,* the equatorial peoples commonly referred to as "Pygmies" play only a minor role in it.[4] Since the writing of a regional history that fails to consider the peoples commonly thought to be its original inhabitants could be seen as problematic, he provides the reason for this lacuna. He considers the existing literature on Pygmies to be of little utility for historical purposes because of the failure of scholars to study their relations with other equatorial peoples and the fact that they have been studied "for their presumed potential to elucidate life at the dawn of humanity, not for their own sake" (Vansina 1990, 29).

The tendency to regard Pygmies and other peoples practicing hunting-gathering economies as windows to the remote past is not simply the habit of a particular breed of scholar. It runs very deep in Western culture, and the figure of the ancient "hunter-gatherer" practicing a culture-of-nature has figured prominently as a foil in the creation of modern identities. Thus, contemporary "hunter-gatherers" have acted as symbolic mediators between a present firmly grounded in culture and a distant past thought to be firmly grounded in nature. In the case of Pygmies, this association with the remote past is so ubiquitous that it has had profound consequences on the course taken by contemporary scholarship, including attempts to develop an understanding of their history.

First, historians have tended to consider Pygmies as being the province of those who study *prehistory* rather than history, thereby relegating them to the discipline of anthropology (Klieman 2003). Second, the major interest in Pygmies expressed in anthropology has been by those wanting to elucidate the deep past. In the contemporary era, this evolutionary emphasis has been institutionalized in a specialist subfield arising in the 1960s, "hunter-gatherer studies" (Lee and Devore 1968). Given the strong tendency toward specialization in professional scholarship, scholars wishing to study contemporary Pygmy groups have been channeled into this subfield. In addition to the challenges posed by such specialization, the subfield's analytical models have tended to discourage inquiry into the *actual* history of Pygmies as a set of peoples interacting with other equatorial peoples over time (Klieman 2003).

One of the major obstacles presented by the classic paradigm of hunter-gatherer studies is that it has presumed this history is already known and, therefore, unproblematic. That is, the fact that, within the era of ethnographic observation, Pygmy peoples have been practicing forms of economy based on hunting and gathering is taken as a sign that, through some form of historical isolation, Pygmy cultural practices have remained unchanged since ancient times. Another important obstacle is the notion that the key source of cultural continuity within Pygmy history is the mode of production itself: that which has facilitated such profound continuity over large spans of time has been a faithful adherence to a "way of life" based on a hunting-gathering mode of production. As a result, all the other spheres of culture that scholars seeking answers to historical questions would normally explore—politics, ideology, individual agency, etc.—are relegated to the shadows, as economy is presumed to be the major driver of history.

Since the initial formulation of the classic hunter-gatherer studies paradigm, its notion of social isolation and highly economistic models have been the object of various critiques (Bahuchet and Guillaume 1982; Denbow 1984; Dentan 1988; Gordon 1984; Mosko 1987; Myers 1988c; Rupp 2003; Wilmsen 1983, 1989b). In addition, there have been various efforts to address historical issues among Pygmy peoples. In the discipline of history, Klieman's seminal study (2003) represents the first attempt by a historian to reconstruct the history of Pygmies over the *longue durée*, and it underlines the fundamental importance of viewing their interaction with other equatorial peoples in terms of the conceptual frameworks created by immigrant ideologies. In the discipline of anthropology, it is Bahuchet who has made the most sustained effort to address historical issues (1993a, 1993c, 2012, but also see Guillaume 2003; Seitz 1993; Waehle 1986). Of particular relevance to the present discussion is his notion of the "*paradoxe des Pygmées*" (Pygmy paradox), which he uses to describe the fundamental continuities found in Pygmy cultural forms, despite their widespread borrowing of the languages of others (Bahuchet 1993c).

Due to limitations of space, providing a detailed account of Pygmy history over the *longue durée* is beyond the scope of this paper. Yet I do explore some of the outstanding themes of this history in an effort to build on the work of Vansina, Klieman, Bahuchet, and others. In particular, I concentrate on four issues. First, do Pygmies possess a distinct set of principles, practices, and institutions in the political domain that could be considered to constitute a political tradition comparable to that outlined by Vansina for equatorial Bantu? If so, what are its core elements, and how does it compare to the tradition he describes?[5] Second, what is the general trajectory of relations between Pygmies and other equatorial peoples over the *longue durée*, and how did their lower status in equatorial societies come about? Have they always been victims of a timeless "discrimination," as contemporary human rights discourses presume (Lewis 2000; Jackson 2003; Woodburn 1997), or was there, as Vansina suggests, a historical evolution from an attitude of ancient respect for autochthons to one of later disdain (1990)? If such an evolution occurred, what are the factors that may have produced it? Third, how do Pygmies make history, and what kind of a history have they made? Fourth, if historical continuities in Pygmy cultural forms are not the result of social isolation, how have they been produced?

I begin the discussion with a background section that outlines some key elements of Pygmy political culture, which I consider to be the building blocks of what I refer to as the "autochthonous tradition." In addition, I compare its features to those of the equatorial tradition, as described by Vansina. In the second section, I describe the general trajectory of autochthon/immigrant relations over the *longue durée* of precolonial history and consider the problem of how Pygmies came to occupy a low status within equatorial societies. The question of how Pygmies make history is the subject of the third section and I present case material from the last four centuries to provide a concrete sense of these processes. Finally, these historical data are reviewed in a discussion of what insights they can provide for the various issues raised in this introduction.

The data on which the paper is based derive from two sources. The first is forty-eight months of field research in the Central African Republic (CAR) and Cameroon, during which I collected ethnographic and oral historical data from Bayaka (CAR) and Baka (Cameroon) Pygmy communities as well as historical data from colonial archives. The second is secondary sources, which include scholarly works, both ethnographic and historical, and early accounts—of explorers, travelers, colonial administrators, missionaries, etc.—that describe Pygmy groups throughout the equatorial region from the dawn of the seventeenth century onward.

Background: Equatorial Traditions

In developing his model of equatorial history, Vansina employs the notion of historical "tradition" in a particular way. He considers it to involve elements of both continuity and change: continuity is manifested in the long-term reproduction of certain fundamental cultural principles, "basic choices, which, once made, are never again put into question," while change is manifested in the continual transformation of elements of the tradition as it moves through time (Vansina 1990, 258).

For the equatorial tradition, he identifies three key elements that can be regarded as its cornerstones. The first is a form of social organization based on what he terms the "house"—a grouping of people gathered around a man of influence, or "big man," whose followers are linked to him through ties of kinship, clientship, or other forms of dependency (Vansina 1990, also see Grinker 1994).[6] The second is a dynamic of political competition between big men, who vied with one another for followers, alliance relationships, and, in certain historical circumstances, access to regional opportunities (Vansina 1990). Third, there was a very strong penchant for political autonomy, which was generalized throughout the social fabric (Vansina 1990). As Ngoa notes, in precolonial equatorial societies "dignity consisted in total liberty of action" (quoted in Guyer 1986, 605). In much of the equatorial region, this profound emphasis on autonomy impeded the development of centralization and produced a political order based on the autonomy of local units (Copet-Rougier 1987, 1997; Geschiere 1982; Kalck 1971; Laburthe-Tolra 1981; Vansina 1990). In addition, in those areas where centralization emerged, the emphasis on autonomy could produce a political order like the kingdom of the Kongo, in which the administrative structures of the state were obliged to operate in conjunction with the local clan organization (Balandier 1968).

* * *

Based on the available historical and ethnographic data for the various ethnic communities identified as "Pygmies," are there any fundamental structures or principles that could be considered as cornerstones of an autochthonous tradition?

In the realm of social organization, one finds a basic unit like Vansina's house, although it does not seem to have functioned in the same way as a "magnet" for dependents. It consists of a small group of coresident households, which is usually referred to in the ethnographic literature as a "camp" (Bruel 1911; Bahuchet 1979b, 1985; Pedersen and Waehle 1986), although it has also been termed "segment" (Dodd 1984) or "sub-band" (Ichikawa 1978). The basis for membership in such groups is a matter of debate,[7] but as its core adult members usually refer to one another as "siblings," I consider their unity to come primarily from the shared experience of growing up together (Moïse 1992). In addition, I refer to it as a "segment" because Pygmy residential groups ("camps") can include several of them (Terashima 1985). The segment acts as a unit of residence (its members normally reside together), of intimate exchange (its members share meals, domestic labor, etc.), and of politics (it acts as a unit of fission when residential groups split up).

In the realm of political culture, the penchant for autonomy in the Bantu political tradition finds a direct correlate in the autochthonous tradition, although Pygmies give it much greater emphasis. As a Bayaka Pygmy man from the Lobaye region of the Central African Republic (CAR) articulated the principle of autonomy: "(n)ever in life can you force someone to do something that is beyond his or her abilities, against his or her will, or contrary to his or her well-being" (Mossina Dibaba, personal communication 1991). A profound emphasis on autonomy within Pygmy political culture was noted by various early writers (Battel [1625] in Ravenstein 1901; Burrows 1898; Douet 1914; Regnault 1911), as it has been by later ethnographers working with Pygmy groups across the equatorial region (Bahuchet 1985; Dodd 1984; Hewlett 1991; Moïse 1992; Putnam 1948; Turnbull 1965b; Vallois and Marquer 1976).

The principle of autonomy has fundamental implications for Pygmy political life. First, because the autonomy of the individual is regarded as inviolable, almost sacred, it cannot be transgressed; this, in turn, prevents forms of hierarchy (based on age, political authority, etc.) from ever becoming institutionalized as relations of coercive power (Moïse 2003).[8] As a result, one element of the Bantu equatorial tradition identified by Vansina as key—political competition between big men—is conspicuously absent in the autochthonous tradition.

The second consequence of this fundamental commitment to autonomy is that it provides an overarching logic to political relations, regulating interactions at all levels of organization. Within Pygmy communities, those with authority (leaders, elders, etc.) have the right to give advice, which is listened to respectfully by others, but the decision whether to follow it always rests with the individual; in the face of directives from those with authority, individuals maintain a universal right of refusal (Dodd 1984; Moïse 1992). By the same token, the social unit of the segment always retains its autonomy as a unit of decision making for residential arrangements: no matter how large residential aggregations may become, it is always up to the members of a segment to decide where, and with whom, they will live. In the wider region, the principle of autonomy also plays a key role as a response to historical constraints: ever since the rise of their social inferiority, Pygmies appear to have distanced themselves from violence, domination, and mistreatment by using the forest as a refuge, a space of collective autonomy (Bertaut 1943; Deschamps 1962; Hewlett 1991; Turnbull 1961, 1965b; van de Sandt 1996).

A second principle of Pygmy political culture that can be regarded as a cornerstone of the autochthonous tradition is what I refer to as "entrepreneurialism." It can take various forms, including (1) the pursuit of a career as a "master of knowledge," (2) responses to pluralistic environments involving a marked openness to creating alliance relations across social boundaries, (3) the forging of alliance relationships to gain access to resources controlled by others, and (4) strategies employed to take advantage of historical opportunities. The last type is of particular relevance to the material presented here. When historical opportunities arise, individuals seek access to the resources they offer by (1) migrating to locations where they can access them, (2) adopting whatever productive modes are necessary, and (3) forging alliance relations with whatever outsiders are useful. It is this form of entrepreneurialism that provides an avenue for realizing the "profound aspirations" that Althabe attributes to certain Pygmy individuals (1965).

Yet historical opportunities can also produce general responses within a region so that the members of numerous residential groups may migrate to take advantage of them. In such cases, a generalized entrepreneurialism can occur, which I refer to as a "regional social movement" (Moïse n.d.).[9] Yet even though such movements may produce large-scale aggregations, they fail to lead to the development of more centralized, hierarchical forms of organization because collective action is undertaken in such a way that the units of social action—leaders, elders, and members of residential groups—always retain their autonomy as independent actors.

Pre-colonial Historical Frameworks: Autochthon/Immigrant Relations over the Longue Durée

To provide a wider context for the historical issues addressed in this chapter, I begin by outlining the general trajectory of autochthon/immigrant relations over the *longue durée* of precolonial history. The period considered extends from the entry of Bantu-speaking peoples into the equatorial region—dated variably between 4000 and 1500 BCE—to the arrival of Europeans in the late nineteenth century. I divide this period into two historical phases and consider the transition between them to have been produced by a complex of historical developments, which also led to the emergence of Pygmy social inferiority.

The Regime of Complementarity: Early Relations between Autochthons and Immigrants, Beginnings to ca. 500 CE

The first peoples to migrate into the equatorial forest from the north were speakers of Bantu languages, and their movement from their homeland in the Nigeria/Cameroon borderlands across most of the continent south of the equator is commonly referred to as the "Bantu expansion" (Eggert 2005; Ehret 2001; Guthrie 1967–1971; Lupo in this volume; Oliver 1966; Vansina

1984, 1990, 1994/95, 1995). The expansion out of the homeland is generally considered to have begun between three to five thousand years ago, but there is also evidence for earlier movements along the coast (Clist 1999; Klieman 2003) and more generally (Montano et al 2011).[10] Vansina posits that the movement into the equatorial forest started between 2000 BCE and 1500 BCE (1986, 1990), while Klieman considers its initial stages to have begun by 4000 BCE (2003; also see Montano et al 2011). Whatever its precise timing, the expansion was clearly a slow movement of people across the landscape unfolding over the millennia.[11]

As Bantu peoples first encountered Pygmy autochthons, it is quite possible they had violent clashes. Yet there are several factors that are likely to have encouraged the two groups to establish peaceful relations of alliance and exchange. First, security concerns within a decentralized political environment would have provided strong incentives for the creation of alliance relations (Moïse 2003). In such environments, peace is not a given, but a human achievement: it is made possible by the creation and maintenance of relationships of alliance and exchange—to exchange with one's neighbors is to live peacefully with them, and to live peacefully with them is to exchange with them (Copet-Rougier 1997; Mauss 1967; Myers 1986). Second, Bantu ideologies are likely to have given immigrants strong incentives to establish relations with autochthons. As Klieman argues, early Bantu peoples, who would have been operating out of wider Niger-Congo belief systems, would have cast autochthons as "first-comers" who, as "owners of the land," would have enjoyed a direct relationship with the mystical forces animating it and possessed ritual knowledge to gain access to such forces (2003). Because precolonial Bantu political authority rested on mystical foundations (de Heutsch 1962, 1972; Wrigley 1992), leaders of "late-comer" immigrant societies would have sought to integrate autochthonous peoples into a common social formation in which autochthons could perform essential ritual functions (see Kopytoff 1987). Third, even though early Bantu were well-acquainted with forest environments (Klieman 2003), they were immigrants to the region and must have looked to autochthons as sources of local knowledge. Fourth, the entrepreneurial approach to pluralistic environments exhibited by Pygmies in later eras may well have been in place, which would have encouraged their participation in such relations. Whichever of these factors may have been involved, it is clear from the later history that autochthons and immigrants established relations of alliance and exchange from an early time.[12] Indeed, archeological evidence suggests that exchange with neighboring hunter-gatherer groups was an element of Bantu strategies from the earliest phases of the expansion (Clist 1999; Klieman 1997, 2003).

What elements might such exchange relations have involved? Early Bantu economy and material culture certainly included items that would have been novel, and probably appealing, to autochthons: ceramics, polished stone tools, and some cultigens (Klieman 2003; Vansina 1990). For their part, autochthons could have contributed items such as meat and honey, as they did to their allies in later eras, or various other forest products, as they later did to regional trade networks.[13] However, a variety of evidence suggests that a major contribution autochthons made to these relations, especially in early eras, was knowledge.

Equatorial oral traditions are almost unanimous in representing Pygmies as the original inhabitants of the region who guided immigrants to their present locations (Bahuchet and Guillaume 1982; Delobeau 1977; Deschamps 1962; Koch 1968; Seitz 1993; Vansina 1990). As a Gabonese man told Deschamps, "the Babongo (Pygmies) . . . acted as our compass" (1962:25, translation mine; Klieman 2003; Vansina 1990). In this way, Pygmies provided the necessary environmental knowledge to facilitate immigrant settlement. In addition, Pygmies are often represented in these traditions as "civilizers"—those who bequeathed to immigrants knowledge of the fundamentals of culture (Avelot 1905; Bahuchet and Guillaume 1982; Delobeau 1977; Klieman 2003; Seitz 1993). Although some of these representations stretch the limits of historical credibility, they may be immortalizations in myth of the fundamental contribution that autochthons made to the early social order as bearers of essential knowledge.[14]

An additional body of evidence suggesting a key role for autochthonous knowledge in early relations comes from the social institutions that developed around indigenous "masters" of knowledge. Within the era of ethnography, the primary positions for such individuals have been that of *nganga*—healer, diviner, and ritual specialist—and *tuma*, master of the hunt, especially for large game such as elephant. Because Pygmy *tuma* and *manganga* (plural form) provide services to a range of people in their regions, they act as important nodes in regional political economies, and practitioners who are accomplished can gain considerable renown, taking their place in local pantheons alongside such "notables" as village chiefs, local big men, and so on. The fact that the positions of *nganga* and *tuma* are found throughout the equatorial region, are everywhere associated with similar practices, and the terms for them date to the earliest periods of the Bantu expansion (Klieman 2003; Vansina 1990), all suggest these institutions are ancient.[15] In this way, it appears that indigenous masters of knowledge played a critical role in regional society from the beginning of immigrant-autochthon interactions and that Bantu speakers applied their terms to them wherever they were encountered.[16, 17]

* * *

Given the significant contribution that autochthons appear to have made to early relations with immigrants, the question arises, what was the nature of political relations between them? Bantu notions of the mystical foundations of political authority, as well as various material needs, would likely have encouraged the development of regional networks of Pygmy *nganga*, *tuma*, and possibly other cult masters, if these were not already in place. This, in turn, would have created a generalized landscape of respect for autochthons and, as the material and political conditions for the emergence of later inequalities were not yet in place, the atmosphere of respect in the ritual domain may well have extended to the political domain. If this were the case, then relations between autochthons and immigrants are likely to have been characterized by a relative parity—possibly even enough for intermarriage between the two groups. Such a scenario is supported by both linguistic and genetic evidence.

In his studies of Pygmy languages in the area of Cameroon/CAR, Bahuchet has identified a sizable substratum of vocabulary common to both Aka and Baka languages (approximately 20 percent), even though they belong to different language families—Bantu and Ubangian, respectively (1993a). This he considers to be evidence of an earlier historical phase in which Aka and Baka both spoke a shared "proto" language, which he refers to as *Baakaa (Bahuchet 1993a, 1993c, 2012). In addition, he notes considerable linguistic "flow" between *Baakaa and the languages of various immigrant groups, moving in both directions: out of a corpus of 3,000 words, he notes about 35 percent are words of Pygmy origin that were borrowed by immigrant groups, while another 35–40 percent are words of immigrant groups borrowed by Pygmies (Bahuchet 1993c). Also of interest is the fact that various terms associated with marital relations are shared between Pygmy and non-Pygmy languages. He notes:

> 29% of the vocabulary dealing with social relations, particularly terms concerning marriage, are common to both languages, suggesting a history of marriage and exchange between the two populations (Pygmies and farmers). (Bahuchet 1993a, 40)

Although Klieman's interpretation of the history of this speech community differs from that of Bahuchet, she echoes his analysis in viewing this early proto-period as one of extensive cultural flow between Pygmy and Bantu groups (2003).[18] She states:

> We must speak of a long period of close interactions between agriculturalists' and forest specialists' (Pygmies') linguistic ancestors—one that eventually led not only to the sharing of languages, but probably also cultures, technologies, and genes. (Klieman 2003, 110)

This model of a long period of intimate interaction between immigrants and autochthons, involving considerable sharing and flow, is also supported by the genetic evidence. Various scholars have noted that "admixture"—the presence of genetic elements from non-Pygmy populations—is a feature of all contemporary Pygmy groups (Cavalli-Sforza et al 1969, 1986; Patin et al 2009; Tishkoff et al 2009; Verdu in this volume; Wijsman 1986). In addition, it is significantly higher among Pygmies in the western part of the region (Cavalli-Sforza 1986; Verdu et al 2009), which is the area of early Bantu migration and settlement. Furthermore, there has been significantly more gene flow from non-Pygmy populations to Pygmy populations than the reverse (Cavalli-Sforza 1986; Verdu et al 2009); under current conditions of patrilocality and patrilineality, which are very widespread in the region, this would suggest a flow of immigrant women into marriages with autochthon men.[19] At the very least, such a genetic profile is consistent with an early practice of intermarriage between immigrants and autochthons. Yet it might also be evidence of the giving of immigrant wives to autochthon husbands during the early phases of the expansion, a scenario Cavalli-Sforza considers a likely explanation of the high degree of non-Pygmy to Pygmy gene flow (1986).[20] Taken together, the considerable contribution autochthons appear to have made to early social formations, the evidence for extensive linguistic borrowing and widespread gene flow, and the nearly universal presence of alliance relations between Pygmies and other equatorial peoples, all support a model of the era of complementarity as a period of long-term, intimate interaction with considerable cultural flow in a political environment of relative parity.

From Complementarity to Inequality: The Rise of Pygmy Social Inferiority

In keeping with the economistic emphasis of the classic paradigm of hunter-gatherer studies, many scholars have presumed an economic explanation for the lower social status conferred on Pygmies in equatorial societies: once Bantu, and other immigrant groups, were established in the region, the "competitive edge" that food-production provided them allowed them to occupy a dominant position vis-à-vis the hunter-gatherer autochthons among whom they had settled (Diamond 1999).[21] Vansina takes a more nuanced position, but the explanation he provides for Pygmy social inferiority remains an economic one: although Bantu would not have enjoyed economic dominance during the early phases of settlement, once they adopted banana cultivation in the first millennium CE it allowed them to produce agricultural surpluses for trade, which drew hunter-gatherers into a relation of permanent dependency (1990). Yet aside from the potential role played by economic factors, what other factors might have been involved?

As noted in the last section, Vansina considers the principle of autonomy to be a fundamental element of Bantu political traditions. In a discussion of nineteenth-century Beti of southern Cameroon, Guyer echoes this notion of political autonomy but also extends it: "(t)o be *nti* (a "free" man, as opposed to a slave) . . . a man needed to cultivate a manner and a social position which conveyed freedom from constraint *and from social indebtedness*" (1986, 595, emphasis mine). Thus, the "autonomy" that was sought by other equatorial peoples in the late precolonial era was not simply a lack of political constraint, but also an independence in a larger social sense, which required a certain degree of economic independence as well. In this way, to be indebted to another for such basic things as staple foods disqualified one as a man, at least as an "admirable man" (Fallers 1973). Thus, by the later phases of the precolonial era, agricultural production had become so fundamental to the provisioning of the household among many equatorial peoples that to lack it would immediately place one in a position of indebtedness to others—a condition considered morally objectionable.

Prior to the introduction of banana cultivation, however, the economy would have depended as much on gathered starch foods as cultivated ones (Vansina 1990, 1994/95). As Vansina notes, the only starch early Bantu cultivated was the yam, whose yield is one-tenth that of the banana (1990). Thus, if it was agricultural production that caused a fundamental shift in attitudes toward Pygmies, this is likely to have happened only after the adoption of banana cultivation had transformed Bantu economies into fully agricultural ones. In addition, if banana cultivation was the causal factor, it is likely that it would have been because of Bantu attitudes toward indebtedness, rather than surplus production transforming a hitherto voluntary exchange relationship into an obligatory one. For, as was argued in the last section, there were various factors that were likely to have made exchange relationships between autochthons and immigrants very attractive, if not obligatory, from the moment Bantu arrived in the forest.

In his account of equatorial history, Vansina describes another set of events that are likely to have played an even greater role in the rise of Pygmy inferiority (1990). Once Bantu had taken up banana cultivation and iron production in the centuries surrounding the beginning of the Common Era, this initiated a historical trajectory that increased opportunities for economic specialization and produced demographic growth, the emergence of new forms of wealth, an increase in regional trade, and an associated rise in political competition. Over time, a big-man political culture, fueled by competition between groups, produced an expansion of scale resulting in chiefdoms, principalities, and kingdoms in some areas. In conjunction with this evolution, or perhaps later, decentralized areas witnessed the emergence of regional hierarchies.

Placing autochthons within this historical framework, one can develop an alternative scenario for their decline in status. Once banana cultivation and iron production became well-established in the region, new forms of specialization emerged that created divisions between immigrants and autochthons: Bantu groups became occupied with the spheres of trading wealth, political power, and warfare, while autochthonous groups remained focused on the material resources and mystical powers issuing from the forest. Then, as the effects of these innovations began to be felt within the region, increasing levels of political, economic, and military power were required to hold one's own. Yet because the political culture of autochthons mitigated against the accumulation of political power, and they were unable—or unwilling—to respond in economic and military domains, they failed to keep pace with this escalation. As they came to lack those things that conferred value on Bantu—wealth, power, military force, etc.—they came to be regarded as social inferiors, dependents, "men of no worth" (Laburthe-Tolra 1981). As such, they became unfit as spouses and deserving only of a position of subordinate clientship. In this new regime of social inferiority, autochthons came to depend on their alliance partners not only for agricultural food but for the protection and influence necessitated by the new regional environment.[22]

Although it is certainly possible that the adoption of banana cultivation or iron production on the part of Bantu was sufficient to engender perceptions of autochthons as socially inferior, once regional society had been transformed into a truly hierarchical landscape, it seems likely that a low social status for Pygmies was a *fait accompli*. According to the chronology provided by Vansina, this would place its emergence somewhere between the middle of the first millennium CE and the middle of the second (1990).[23] Although such processes undoubtedly occurred at different times in different places (Klieman 2003), the general time frame of 500–1500 CE can be regarded as a period of transition in which the former regime of complementarity gradually gave way to a regime of inequality. If this is the case, it is important to note the relative extent of these two phases of precolonial history. Regardless of the precise dates one employs for Bantu expansion into the forest and the emergence of the regime of inequality, the era of complementarity comprises well over half of the period since the arrival of Bantu in the region.[24] Although scholars have paid much less attention to this early era, from an autochthonous perspective, it is the most empowered phase of their history, comprising the long span of time before their marginalization by the immigrant political economy began. Indeed, it is this era which has been enshrined in their oral traditions as a "golden age."[25]

How Pygmies Make History: The Autochthonous Tradition in Action

Now that the broader historical frameworks for the precolonial era have been outlined, I turn to political process to explore how Pygmies have made history under the conditions created by these frameworks. Unfortunately, the data for the early era of complementarity and the early phases of the era of inequality are very general. However, there is documentary evidence from the last four centuries, as well as more detailed accounts from the era of ethnography, that can elucidate the subject. Although these data reflect more recent historical conditions, their level of detail allows for the development of a more fine-grained understanding of Pygmy political process. Because alliances between Pygmy extended-family groups and Bantu extended-family groups were the most common type of alliance relationship by the end of the precolonial era, I begin by discussing how Pygmies negotiated them.

Preserving Autonomy in Alliance Relationships with Local Bantu

According to the logic of the regime of inequality, the role played by Pygmies in alliance relationships with local Bantu was that of "subordinate client."

But despite their inferior position, they developed strategies for minimizing its inconveniences while still gaining access to the resources their alliance partners controlled. The first such strategy is what could be called "contextual clientship." Accounts of their very different behavior in Pygmy and Bantu social contexts make clear how it operated. For the area of east Cameroon, Bertaut, an official with the colonial administration who had regular contacts with Baka Pygmies in the 1930s, notes:

> Seen in public, they (Baka) seem disoriented, fearful; they avoid eye contact, and only observe their surroundings surreptitiously. Alone in the forest, they abandon their fearful air, become cheerful . . . (and) communicative. (1943, 92, translation mine)

Or, for Aka (Bayaka) in southwestern CAR, Hewlett states:

> In the forest, Aka sing, dance, play, and are very active and conversant. In the (Bantu) village, their demeanor changes dramatically—they walk slowly, say little, seldom smile, and try to avoid eye contact with others. (1991, 29)

Thus, Pygmies adopted a deferent demeanor appropriate to social subordinates whenever they operated in the spaces of other equatorial peoples.[26] Yet the fact that this demeanor was quickly abandoned whenever they returned to their own milieu indicates that such deference was a choice. In this way, they made their position as subordinate clients into something that was contextual: it only operated in Bantu spaces (see Turnbull 1965b).

A second strategy Pygmies employed to negotiate their position within the alliance relationship was minimizing the amount of time they operated in such contexts of deference and subordination. This was achieved largely by putting what could be called "buffer zones" between their settlements and those of their alliance partners. Whereas Bantu individuals who occupied a client status in the precolonial era normally resided in their patron's village (Geschiere 1982; Vansina 1990), Pygmy clients seem to have usually made their settlements at a considerable distance from those of their patron (Moïse 2003; Mounteney-Jephson 1891). For the Lobaye region of CAR, Bahuchet notes that the Bayaka annual cycle in the precolonial era consisted of residence in settlements deep in a forest territory throughout the entire year except for a one- to two-month period at the beginning of the dry season, during which they could migrate to the riverside villages of their Ngando (Bantu) alliance partners to help with the clearing of forest areas for gardens; Ngando would also reside near the forest settlements of their Bayaka allies periodically for assistance with large, collective net hunts (1986). Precolonial data for Baka in Cameroon suggest a similar pattern (Dodd, 1986; Moïse 2003). In this way, contact with one's allies was limited to seasonal productive activities, visits for exchange purposes, and important events such as funerals, elephant kills, and the like. The strategy informing this pattern was to tolerate inferiority when necessary, but to escape it whenever possible.

A third strategy Pygmies used to negotiate the alliance relationship is what Leonhardt refers to as the "exit option": abandoning one's alliance partner and finding another (n.d.). Accounts from the dawn of the colonial era attest to a frequent reliance on this strategy (Bruel 1911; Burrows 1898; Trilles 1903). In addition, Regnault notes that early colonial administrators along the Sangha River were the object of recurrent pleadings by local Bantu to help them win back their Pygmy allies who had abandoned them (1911). Of course, the exit option would have worked only if Pygmies had a certain appeal as alliance partners. Yet there are various data to suggest that they did, at least at certain times and places: there are accounts of violent conflicts between Bantu residential groups over the "theft" of one group's Pygmy allies by another (Putnam 1948); new Bantu partners could be required to pay compensation to old partners when Pygmy residential groups switched their allegiances (Seitz 1993), and, as Trilles states for the area of the Dja River at the turn of the twentieth century, local Pygmies are "sure to be given a good reception wherever they go" (1903, 431, translation mine).[27]

Entrepreneurial Pursuits in the Wider Region

Although the regime of inequality introduced constraints into Pygmy lives, including the transformation of the alliance relationship into a vertical relation, it also provided them with opportunities in which things such as their forest skills or their unique ontology came to acquire social value in regional political economies. In such cases, Pygmies took advantage of these opportunities to create alliance relations with new sets of partners who could offer better access to resources or better treatment. To examine such pursuits, I first consider an opportunity made possible by the emergence of kingdoms in the second millennium CE: alliance relations with precolonial royals.

Alliances with Royals

In many precolonial equatorial states, Pygmies played a key role in the life of the kingdom. In oral traditions, they were invariably represented as "first comers" who had a fundamental part in the founding of the polity (Klieman 2003; Seitz 1993; Vansina 1978). As such, they were considered intercessors to the spirits of the land and guardians of the sacred powers of the kingdom. Given this symbolic significance in royal ideology, their participation in rituals of enthronement, annual rituals of renewal, and other rites was essential (Abega n.d.; Laman 1953; Seitz 1993; Vansina 1978, 1990).

For their part, royals seem to have done everything possible to treat Pygmies well and retain their loyalty. After a visit to the Mangbetu court c. 1870, Schweinfurth described how King Munza feasted his Akka Pygmy clients on flasks of beer, plantain wine, ears of corn, and other delicacies (1874). Similarly, Denolf describes Bacwa Pygmies of the Kuba kingdom as "the people of the Lord (who) enjoyed his personal protection" (quoted in Seitz 1993, 228, translation mine), and, in the kingdom of Rwanda, Tutsi rulers gave Twa Pygmies rights to forest territories, rights to collect tribute for travel within them, and extreme jural leniency (Seitz 1993).

From the Pygmy perspective, alliances with royals offered numerous advantages over those with local Bantu: they were regarded with respect, they received very benevolent treatment, resources were abundant, and protection was clearly the "best in the land." Once centralization created opportunities for alliances with royals, Pygmies appear to have taken full advantage of them, as shown by the ubiquity of their ties to ruling groups throughout the region—in the kingdoms of Kuba, Tikar, Loango, Rwanda, Urundi, Bushi, Butembo, Mangbetu, and possibly others (Abega n.d.; Battel [1625] in Ravenstein 1901; Klieman 2003; Schweinfurth 1874; Seitz 1993; Vansina 1978). In addition, the entrepreneurial initiatives undertaken by Pygmies are illustrated by the considerable breadth of exchange relations they engaged in with royals: they provided a range of forest products, various ritual services, and military assistance; they acted as personal assistants in a range of capacities; and they enriched the life of the courts with their aesthetic skills—acting as singers, dancers, bards, and comedians to entertain the courts and their visitors (Lewis and Knight 1995; Schweinfurth 1874; Seitz 1993). Furthermore, there is documentary evidence for various cases—from the dawn of the seventeenth century to the late nineteenth century—of Pygmy entrepreneurialism in relation to royals: Barimba Pygmies in the Loango kingdom bypassed local Bantu to establish alliance and trading relations with provincial rulers (Battel [1625], in Ravenstein 1901; Klieman 2003); Bacwa Pygmies in the Kuba kingdom treated local chiefs and functionaries "with profound disdain" while maintaining intimate ties to the royal court (Seitz 1993; Vansina 1978); Schweinfurth describes a "small colony" of Pygmies living near the court of King Munza (1874); and Twa entrepreneurs established themselves as important figures in the Rwandan royal courts, with some becoming ennobled, others attempting to influence transfers of power, and still others staging revolts over perceived mistreatment (Kagabo and Mudandagizi 1974; Lewis 2000; Seitz 1993).

I now consider a contemporary case that allows for a more fine-grained understanding of the political processes involved in such expressions of entrepreneurialism.

Seeking Autonomy in the Postcolonial Era: Bayaka of the Lobaye

In keeping with the CAR's post-Independence emphasis on economic development, the decades of the 1960s and 1970s introduced into the Lobaye region a significant increase in cash-crop production as well as the establishment of logging operations, which attracted large numbers of "foreigners" (nonlocals) to the area (Bahuchet and Guillaume 1982). The rise of cash-crop production was welcomed by local Ngando villagers, as it greatly increased their access to modern goods and services. However, it placed additional burdens on local Bayaka, who still maintained alliance relationships with their traditional Ngando partners and were pressed into providing a considerable amount of the labor that was required. In addition, farms of cash-crops were always located near the Ngando villages, an environment local Bayaka generally associate with "trouble"—accusations of theft, disputes, mistreatment, etc. (BBC 1996; Biesbrouck 1999; Turnbull 1965b; Van de Sandt 1996). On the other hand, the introduction of outsiders into the area created a wider market for game meat as well as a large number of mobile traders who supplied it by trading agricultural foods and petty commodities for meat with Bayaka in their forest camps (Bahuchet 1985).

In the course of the 1970s and early 1980s, Bayaka in the area responded to these transformations by developing a new strategy vis-à-vis the Bantu world. First, they took up agriculture, creating farms of staple foods next to their base camps deep in the forest—usually one to three days walk from the villages of their traditional allies (Guille-Escuret 1998).[28] Second, they broadened their exchange networks to include the traders circulating in the region, who were appealing as exchange partners because their forest-based exchanges allowed Bayaka to procure the goods they desired without having to expose themselves to the social complications of the village milieu. Thus, the new strategy was an attempt to increase their autonomy from their traditional allies and make themselves less susceptible to the machinations of the Bantu world. At the same time, relations with their allies were not severed entirely—they continued, albeit in an attenuated form. This allowed Bayaka to continue to rely on them for "protective" functions, such as assistance in the resolution of disputes.

Guille-Escuret, who refers to the activities of Bayaka in the Lobaye as an "agricultural revolution" (*révolution agricole*), provides some detail on the emergence of agricultural production: in the course of the 1970s, there were initiatives by residential groups in different areas acting independently of one another, which was followed by a general diffusion throughout the region (1998). In this process, he notes that agricultural practices spread fairly easily among communities inhabiting the same forest territories, but that it also spread throughout the region in a period of less than fifteen years (Guille-Escuret 1998). Here, the range of ties linking groups within a region—of kinship, friendship, ritual, etc.—likely served as pathways for the diffusion of the new strategy (Hewlett et al 1982; Joiris 1996; Kisliuk 1998).

In this way, the strategy was adopted independently by each residential group, but a regional social movement emerged over the pathways linking groups to one another. I consider this phenomenon a "regional social movement" because there was widespread dissatisfaction among local Bayaka with their traditional alliance relationships, followed by the adoption of a new strategy in response—a change in practice that became generalized throughout the region. This movement was not produced by organized, coordinated action, but rather through "waves" of independent decisions by individuals and residential groups.

Discussion

Although limitations of space have precluded a detailed examination of the historical data, the material just presented can provide an answer to the question, "Do Pygmies have a history?" Clearly, they do. In this section, I explore the key themes of this history: the broad contours of precolonial history and how Pygmies responded to the constraints and opportunities they presented, how they have made use of the elements of the autochthonous tradition to make their own history, and what insights all this offers into the issues raised in the introduction.

Historical Developments, Pygmy Responses

The first phase of Bantu settlement in the equatorial region appears to have unfolded under a long-term regime of relative complementarity, characterized by common social formations of autochthons and immigrants, involving free flows of cultural elements across social boundaries. Some of the factors responsible for this regime's success were likely: the first-comer/late-comer ideological complex, which guaranteed Pygmies a place in the social order as autochthonous masters of knowledge, a relative openness on the part of Pygmies to creating relations across social boundaries, and a basic compatibility between immigrant and autochthonous cultural practice.

The second phase of this history witnessed an escalation of "force" in the economic, political, and military domains of Bantu life, which eventually gave rise to a hierarchical social order and left Pygmies in the position of subordinate clients. Once this regime of inferiority emerged, Pygmies seem to have disengaged from the larger social order and retreated into refuge spaces, while developing new strategies for preserving autonomy in the alliance relationship: contextual clientship, buffer zones, the exit option, and so on. Then, as new opportunities arose, they sought out better access to resources, more guarantees of autonomy, and partners who provided more parity in the alliance relationship. Although the only precolonial opportunity discussed here was creating alliances with royals, another important one was acting as independent producers of forest products for regional trade (Grenfell in Johnston 1969; Harms 1981; Klieman 2003; Moïse 2011; Stanley 1885; Stanley 1890; Trilles 1903; Ward [1890] in Schultz 1986).[29] Both these strategies have continued into later eras, with new alliance partners ranging from colonial administrators and company agents, to development and conservation projects, logging companies, missionaries, researchers, and so on (Capagory 1933; Douet 1914; Hewlett 2009; Moïse 2011; Pedersen and Waehle 1986; Parke 1891; Putnam 1954; Putnam 1948; Schebesta 1957; Schweinfurth 1874).

At the same time, it is important to recognize that there may have been multiple responses to the regime of social inferiority once it emerged. That is, once banana cultivation, iron production, and their associated escalations had become widespread, some autochthonous groups may have adopted them, remained in close association with immigrant groups, and "become Bantu," while others chose a path of autonomy, leading to a separate evolution and later membership in the Pygmy social category. Although there is no direct evidence for these processes, one can assume that the contemporary ethnic communities identified as "Pygmies" are a product of the latter historical trajectory, while some contemporary ethnic communities identified today as "Bantu" may contain elements of early autochthonous populations who chose the former (see Birmingham 1977 and Vansina 1986).

Responses to Inequality: Roots of a Separate Evolution

For those Pygmy groups who did pursue a strategy of autonomy once the regime of inequality emerged, what consequences would it have had for their subsequent history? In her analysis of the linguistic data, Klieman notes that contemporary Pygmy groups in Gabon and Republic of Congo (ROC) who share languages with their neighbors exhibit a marked linguistic "conservatism," retaining archaic phonologies and patterns of pronunciation employed in earlier historical periods (2003). Similarly, she describes how Babongo Pygmies in ROC failed to adopt linguistic innovations introduced by the expanding Tio kingdom in the fifteenth and sixteenth centuries (Klieman 2003; Vansina 1990). Such forms of linguistic conservatism she attributes primarily to Pygmy efforts to retreat into refuge spaces during the violent transformations of the Atlantic era (Klieman 2003). Yet regardless of its precise timing, such conservatism suggests a long-term trend toward a historical evolution that is distinct, and separate, from that of other equatorial peoples. Such a trend exhibits a basic similarity to the historical sequence described by Bahuchet, in which later historical eras were characterized by social distance and a separate linguistic evolution (1991).

In this way, the gradual rise of inequality seems to have caused a disengagement of many Pygmy groups from prior cooperative relations, a retreat into spaces of collective autonomy, and a subsequent separate evolution. All these processes, of course, would have exaggerated even further the cultural differences between them and other equatorial peoples. Thus, it is quite possible that a significant part of the difference observed in recent eras is a product of the separate evolution introduced once Pygmy groups began to experience significant inequalities and responded by pursuing a strategy of autonomy.[30]

Pygmies as Authors of Their Own History

In contrast to the image of isolated "hunter-gatherers," whose agency is limited by their deference to a cultural tradition generated by a mode of production, Pygmies have been shown to be quite entrepreneurial. They are adept at crossing boundaries of all kinds to gain access to regional resources: they engage in various modes of production, they forge alliance relations with all sorts of outsiders, and they develop skills in negotiating all kinds of environments. In addition, it is quite possible that they have always displayed an inclusiveness and openness to others that extends across boundaries of social difference.

Yet this strong commitment to an entrepreneurial approach to living with others has always been counterbalanced by an equally strong commitment to the principle of autonomy. Throughout their history they have always charted their own historical course, in keeping with their own values and principles. They chose to interact intimately enough with early Bantu that everywhere they came to share their languages and genes, yet they chose not to adopt their agricultural and ironworking technologies—preferring instead to trade forest products for cultigens and iron implements (Bahuchet 1993a). Similarly, they chose not to participate in the kinds of escalations in economic, political, and military domains that other equatorial peoples came to embrace in the later phases of the precolonial era. In addition, their approach to states—from the precolonial era to the present—has been to minimize their incorporation into administrative structures while creating mutually beneficial alliance relationships with those in power (Moïse 2003).

The principle of autonomy has also acted as a counterbalance to entrepreneurialism at the more fine-grained level of social action. That is, it is the commitment to autonomy that has prevented their entrepreneurial pursuits from gaining sufficient momentum to produce lasting change in their basic institutions or forms of organization. Thus, entrepreneurial pursuits always unfold in such a way that their authors retain their autonomy as independent actors: members of residential groups are never forced to comply with the initiatives of their leaders, residential groups involved in regional social movements never forfeit their independence as decision makers, and so on. As the basic units of social action remain eternally independent, the actions of one do not impose themselves on the actions of another. Large-scale actions may be produced, but always through waves of independent decisions that are never allowed to impinge on the actions of others.

In similar fashion, the principle of autonomy has mitigated against the development of more "complex" forms of political organization over time. Although regional social movements can result in large aggregations, and there is evidence for some large-scale coordination of local units at the end of the precolonial era (Mounteney-Jephson 1891; Stanley 1890), these do not become formalized as enduring political units, nor does decision making become centralized.[31] This is because the lower-level units that comprise them retain their political independence while in close proximity and, once they separate, the large-scale formation simply disappears; such processes have been noted for other equatorial peoples as well (Copet-Rougier 1997; Kalck 1972; Laburthe-Tolra 1981; Vansina 1990).

Because the strong commitment to autonomy mitigates against the emergence of institutionalized, hierarchical forms of organization, it is able to guarantee the independence of both individual actors and social groups as they respond to the ebb and flow of historical events within a region over time. In this way, it provides the key to how Pygmies can create regional social movements to respond to historical opportunities without producing enduring hierarchical organization and how they can engage in relations with all sorts of peoples but still retain a distinct identity and unique cultural forms. In short, it is the key to how Pygmies make history. In addition, it contributes to the solution of Bahuchet's "Pygmy paradox," in that it shows how Pygmies have produced fundamental continuity in their cultural forms despite millennia of intimate interaction with outsiders that resulted in the sharing of languages and genes. Furthermore, the principle of autonomy allows one to account for fundamental, long-term continuities in Pygmy cultural forms without resorting to the notion of social isolation—one of the basic premises of the classic hunter-gatherer studies paradigm that has proven untenable not only in equatorial Africa but in many other places as well.

Conclusion

The material presented in this paper shows clearly that there is not one political tradition in equatorial Africa, but two: that described by Vansina for Bantu peoples and that of the various autochthonous communities who came to be identified as "Pygmies." Although Vansina (1986) raised the question, "Do Pygmies have a history?" over two decades ago, while preparing his volume on the equatorial tradition, he considered the available historical and ethnographic materials to be insufficient to allow him to include them in his account. However, the material presented here, along with the important historical work that it has drawn on, demonstrate that writing a history of the autochthonous tradition in equatorial Africa is indeed possible. The present paper represents an initial effort toward that goal, one that can signal a direction for further writing and research.

Acknowledgments

Research for this chapter was made possible by the support of the Social Science Research Council's International Dissertation Research Fellowship (funds provided by the Andrew W. Mellon Foundation), the Wenner-Gren Foundation, and the James A. Swan Fund of the Pitt Rivers Museum. My sincere thanks to these institutions for their generous support.

Notes

1. The first part of the title of this paper is taken from an article by the well-known historian of Africa, Jan Vansina, which represents the first attempt by a historian to examine the role played by the peoples commonly referred to as "Pygmies" in the history of equatorial Africa (1986).

2. The equatorial forest region comprises an area that stretches from the Atlantic Coast in the west to the Great Lakes region in the east, a distance of approximately 2,500 kilometers (1,553 miles). It includes territories within the contemporary nations of: Cameroon, Gabon, Equatorial Guinea, Central African Republic, Republic of the Congo, Democratic Republic of the Congo, Rwanda, Burundi, and Uganda.
3. The languages of the non-Pygmy peoples of the equatorial region belong to three linguistic families: Bantu, Ubangian, and Sudanic. Speakers of languages in these families all have their origins in the lands to the north—from eastern Nigeria to the eastern Sudan. Because Bantu-speaking peoples were the first to come to the region, were by far the most numerous, and had the most historical impact, I concentrate on them in this paper. But as they exhibit many cultural similarities to Ubangian and Sudanic speakers, including the nature of their relations with Pygmy peoples, much of what is said about them is relevant to these other groups as well.
4. In this paper, I use "Pygmies" as a general term to refer to members of the various ethnic communities living under the regime of the Pygmy identity in equatorial Africa. The advantage of this term is that most readers know who is being referred to, but the disadvantage is that it is a term created by outsiders that can be used pejoratively in certain contexts. Alternative terms commonly used in the anthropological literature are "hunter-gatherers" and "foragers," but I consider these unsuitable for reasons that will become clear. In efforts to develop general terms based on African root words that avoid negative connotations, Klieman (2003) has proposed "Batwa" and Bahuchet (1993a) employs "Baaka." Yet neither of these terms have gained general acceptance so that most readers would know who is being referred to. Aside from its potential negative connotations, the term "Pygmy" offers certain rhetorical advantages. As Waehle argues: "(t) his term is not only internationally known, but probably also the most elastic of the existing terms for designating which groups we are talking about . . . ." (1999). For practical purposes, then, I employ "Pygmy" as a general term, but make use of ethnic terms—Baka, Mbuti, etc.—whenever possible.
5. Over and above the large body of ethnographic literature produced on Pygmies during the last century, Klieman (2003) could be considered the first scholar to broach the subject of an autochthonous tradition in the history of equatorial Africa. Yet here I include additional data on the political dimensions of this tradition which allow for a more complete account of it.
6. In terms of higher levels of organization, Vansina posits that houses grouped themselves into "villages," residential units which could include from one to many houses, while villages within a particular geographic area, which he refers to as a "district," were allied with one another for mutual support and defense (1990).
7. Scholars have proposed various factors as the basis for membership in segments and the larger residential groups they can comprise: Turnbull (1965a, 1968) emphasizes shared forest territories; various writers consider the core of such groups to be based on patrifiliation (Ichikawa 1978; Pedersen and Waehle 1988; Putnam 1948); Terashima (1985) presents a model of three types of residential groups, including patrilocal groups as well as larger units

drawn to regional resources; and Hill et al (2011) consider hunter-gatherer residential groups to be generally bilateral.

8. The only context in Pygmy political life in which something akin to power relations develops is the institution of "bride service"—a substitute for, or complement to, bridewealth. The period of bride service usually lasts a few years and, during this time, the new son-in-law lives with his affinal relatives and is obliged to make gifts of food (meat, honey, etc.) to them. At this time, he can be very much "under the thumb" of his wife's relatives, and they can place arduous demands on his labor that he must accept without comment—unless he is willing to forfeit his wife to avoid them (see Dodd 1984).
9. I consider such events as "social movements" because large-scale, collective action is undertaken with the goal of transforming the condition of peoples' lives, and they are "regional" because the same set of actions is undertaken by numerous communities within a geographic region. At the same time, the fit between Pygmy social movements and the kinds of political mobilizations commonly referred to as "social movements" in popular and scholarly discourse is quite loose. Pygmy social movements bear some resemblances to the common version in that they are popular movements of subalterns aimed at addressing, and hopefully overcoming, some of the disadvantages they experience at the hands of those who control the local political environment. However, they also exhibit fundamental differences: they are rural, not urban; they are diffuse, not organized or centralized; they do not take place within the wider context of formal politics; they do not employ modern media for the purposes of recruitment, organizing, or disseminating their ideas, and so on.
10. Conventional models of the Bantu expansion emphasize the role of banana cultivation as a driver for the spread of Bantu speakers throughout the continent. However, there is increasing evidence for movements of Bantu peoples within the equatorial region long before the coming of the banana. In the field of archaeology, there is evidence for early settlements containing polished stone tools and ceramics in the Bantu homeland by the turn of the fourth millennium BCE (Lavachery 1997–98), along the Gabonese coast during the third and fourth millennium BCE (Clist 1999; Klieman 2003), and up the Ogooué River by the turn of the second millennium BCE (Oslisly 1996). In the field of linguistics, Klieman's reinterpretation of the data using methods of glottochronology provides dates for the initial divergence of speech communities from the ancestral proto-Bantu community that are consistent with this time frame—beginning as early as 4000 BCE (2003). In the field of genetics, Montano et al analyzed Y-chromosome variation in populations in Nigeria, Cameroon, Gabon, and Congo (DRC) and found evidence for an initial expansion as early as 6000 BCE as well as an early entry of Bantu speakers into the forest region (2011). All this evidence argues against a model of the Bantu expansion as a single, large-scale population migration set in motion by the development of banana cultivation (Montano et al 2011; Ehret 2001; Vansina 1995).
11. Two forms of movement were involved in the expansion: a slow "drift" of local groups across the landscape and rapid migration on the region's waterways. The latter was made possible because proto-Bantu were

well-versed in water transportation, including maritime transport (Klieman 1997; Vansina 1990).

12. How early alliance relations may have started is unknown, but one possible scenario is provided by a pattern that had become widespread by the later phases of the precolonial era: as a group entered a new area, it had violent clashes with the local inhabitants, but through the forging of formal alliances—beginning with ritual blood pacts and usually proceeding to intermarriages—relations of hostility between strangers were transformed into relations of peaceful exchange between allies (Copet-Rougier 1997; Harms 1981).
13. For the area of coastal Gabon, archaeological evidence suggests that material exchange between early immigrants and autochthons included the exchange of ceramics for basalt and dolerite, which early Bantu used in the manufacture of stone tools (Klieman 2003).
14. Klieman considers the autochthon-as-civilizer theme in Bantu oral traditions not to be a case of historical memory, but an artifact of the move by Bantu, within the context of rising centralization, to legitimate their usurpation of the role of autochthons as intercessors to the spirits of the land (2003; also see Schadeberg 1999). In my view, these two interpretations are not mutually exclusive.
15. The root of *nganga* (*-ganga), which signifies "religious expert," enjoys "practically universal distribution" in the Bantu-speaking world and is considered to be proto-Bantu, the ancestral language of Bantu speakers (Vansina 1990, 298), while the term *tuma* also has very wide distribution, and early origins, in equatorial Bantu languages (Klieman 2003).
16. Linguistic terminology provides an additional piece of evidence for the importance of autochthonous knowledge in early relations. The term for "non-Pygmy" person found in several Pygmy languages in the western part of the region is *-míló (*-biló, plural), whose "literal definition . . . is simply 'the uninitiated'" (Klieman 2003, 111). This contrast between initiated autochthons and uninitiated immigrants is also manifest in ethnic perceptions among Aka (Bayaka) Pygmies living in association with Ngbaka-Ma'bo agriculturalists in southwestern CAR. Discussing how Aka perceive their Ngbaka-Ma'bo allies, Bahuchet states, "because they are not initiated . . . (they) are dependent on the Pygmies for the provision of forest products . . . (n)on-productive, they are like children" (in Klieman 2003, 84). Although contemporary scholars have interpreted Pygmy/Bantu relations through a lens of economic dependency (of Pygmies on Bantu), it appears that a ritual dependency (of Bantu on Pygmies) likely characterized these relations from the start.
17. Here, I do not intend to suggest that the only persons who occupied the positions of *nganga* and *tuma* would have been autochthons. In the era of ethnography, *manganga* have been found among various equatorial peoples (Koch 1968), while *tuma* have been primarily Pygmies. In the remote past, however, it is not unlikely that other equatorial peoples also became initiated as *tuma*.
18. Although Bahuchet cautions that his data on the speech community ancestral to contemporary Aka and Baka should not be projected back in time more than one millennium (1993a), Klieman considers this speech community

to have included other Bantu groups and to date from the latter half of the second millennium BCE (2003). My goal here is not to advance an argument about the dating of this particular community, but to suggest that it can be considered representative of the kinds of relations that appear to have obtained during the early era of complementarity.

19. Despite the strong prevalence of patrilocality and patrilineality in the final phases of the precolonial era, it is an open question whether that was the case in its early phases. For his part, Vansina (1990) considers early Bantu to have been bilateral, so that offspring would have been just as likely to be incorporated into their mother's as their father's kin group. Yet even if early Bantu were bilateral, it does not negate the possibility that a significant cause of admixture was their giving of wives to autochthon husbands.
20. Verdu seeks to explain the flow of genes from non-Pygmy to Pygmy populations through the only kind of intermarriage that occurs today with any frequency—that of Pygmy women to Bantu men (this volume). Although the practice of patrilineality usually ensures that the offspring of such unions are incorporated into the husband's kin group, sometimes Pygmy wives take their children back to their own community upon divorce. Yet the ability of that practice alone to account for the widespread gene flow under consideration is doubtful, and Verdu cautions that "an alternative explanation . . . should be considered" (this volume). For his part, Cavalli-Sforza considers the return of mixed offspring upon divorce to be an unlikely explanation and suggests that the flow of Bantu wives to Pygmy men under early conditions of equality is a much better one; the only other explanation he deems possible is the production of illegitimate offspring—fathered by Bantu men, but raised by Pygmy women (1986).
21. Often, the notion of food production conferring on its bearers a "competitive edge" in regional relations is not explicitly stated, but it is implicit in how evolutionary models are used by scholars to make sense of local-level relations. That is, just as food-producers achieved ascendancy over hunter-gatherers in the course of human history, the food-producing "advantage" is presumed to prevail in any micro-level interactions between hunter-gatherers and food-producers, when, in fact, the character of those relations is always an empirical question.
22. For her part, Klieman considers Pygmy inequality to have emerged in two distinct phases. First, the political and ritual power initially accorded to autochthons was usurped by Bantu territorial chiefs once chiefdoms began emerging in some areas in the period 500–1000 CE (Klieman 2003). Second, once political power came to be based largely on trading wealth during the Atlantic era (1500–1900 CE), autochthons were reduced to a position of inferiority, as were a great many equatorial peoples who became clients, slaves, and so on (Klieman 2003).
23. Vansina provides a rough estimate of c. 500 CE for the diffusion of banana cultivation in the region and considers a hierarchical social order to have taken shape, at least in many regions, over the course of the first half of the second millennium CE (1990).
24. Working from Vansina's (1990) date of 1500 BCE for Bantu arrival in the region, one can estimate the duration of the era of complementarity to be about two millennia (1500 BCE–500 CE), while working from Klieman's

(2003) date for Bantu arrival of 4000 BCE, it would be about four and one half millennia. Thus, according to Vansina's chronology, the era of complementarity would comprise approximately 57 percent of the period since Bantu arrival in the region, while, according to Klieman's, it would comprise approximately 75 percent of this period.

25. A traditional narrative about the origin of their low status is found among various Pygmy groups across the equatorial region: I collected it among Biaka (CAR) and Baka (Cameroon), as did Turnbull among Mbuti in the Ituri (1965a). Despite its diverse versions, its principal plot line is that the initial state of relations between Pygmies and Bantu was one of equality, in which both parties had full access to available technologies, including agriculture and iron production. However, once Bantu came to enjoy exclusive rights to these technologies—because they stole them, Pygmies abandoned them, etc.—Pygmies came to be at a disadvantage, resulting in their inferior social position and the disappearance of former relations of equality.
26. Once Pygmies gain a certain fluency in the cultural milieu of the wider society—by attaining a certain level of formal education, for example—they often abandon this deferent demeanor, as they come to feel comfortable outside their own milieu.
27. Such accounts of the "coveted" nature of Pygmies as alliance partners should be put in historical perspective. The accounts cited in the text draw on a particular historical moment—that of the mid- to late nineteenth century and the early years of the twentieth century. During this time, ivory was a key item of regional trade and, in many areas, Pygmies were its primary producers. Thus, alliances with Pygmies were undoubtedly sought out for the access they provided to this important resource. This is not to say that Pygmies were lacking in value in other times and places, but that their relative value in regional systems must have varied according to historical circumstances. Thus, whereas Pygmy males were generally considered unfit as spouses by Bantu women during the later phases of the precolonial era, Trilles notes that local Pygmy men are able to "easily marry Fang girls" in the area of northern Gabon and southern Cameroon around the turn of the twentieth century (1903, 431, translation mine). This exception to the general rule was likely a result of the position they occupied as producers of ivory for the regional economy during this time.
28. During the course of the year, a Pygmy residential group may live at a few different sites, but one of these is usually occupied more than the others and regarded more as a "home." It is such camps that I refer to as "base camps." Whereas other camps are often made with a particular purpose in mind—hunting, collecting honey, working for Bantu, etc.—base camps are places of ongoing residence that support a multiplicity of activities and act as nodes in regional social networks: as sites of visiting, dancing, ritual activity, etc. Among BaGyeli Pygmies in southwestern Cameroon, the base camp—kwaato—"is the place where one lives in the formal sense of the word" (Biesbrouck 1999), while, among Mbuti, it is the "place for social life" where people gather "on the basis of genuine social and kinship ties" (Ichikawa 1978).
29. In later phases of the precolonial era, the major items Pygmies produced for regional trade were meat and ivory (Grenfell in Johnston 1969; Harms

1981; Lloyd 1907; Stanley 1890; Trilles 1903; Ward [1890] in Schultz 1986), although camwood, used by many equatorial peoples for cosmetic and ritual purposes, was also important (Grenfell in Johnston 1969; Stanley 1885; Vansina 1990). In addition, for the eastern region in the late nineteenth century, Mounteney-Jephson (1891) mentions animal skins and feathers, while Battel notes that Pygmies were producing both ivory and elephant tails for trade with the Loango kingdom at the dawn of the seventeenth century ([1625] in Ravenstein 1901). Furthermore, Klieman cites linguistic evidence for other early trade items such as indigenous tobacco, animal skins, "cache-sexes" made of animal skins, natural dyes, and ankle shakers (1997, 2003), while, for the area of northern Republic of Congo and southern Cameroon, she considers autochthonous participation in regional trade to have begun as early as 500 BCE (1997).

30. In this regard, a statement by Grinker concerning the cultural difference attributed by Lese agriculturalists to their Efe Pygmy alliance partners in the Ituri is instructive: "The Lese hold that they are civilized and cultural because, among other things, they live in villages, cultivate food crops, and go to school and church, whereas the Efe are savages who live in the forest, hunt and gather, have only temporary settlements, and know nothing of God, mathematics, and the French language" (1994, 74).

31. There are two references in the early literature to Pygmy social formations that appear to have involved a higher-level coordination of local units. The first is in an account provided by a Zanzibari trader to Stanley (1890), in which he discusses his visit to a Pygmy trading "polity." Arising along the Lomami River in response to the establishment of a Zanzibari trading settlement at Nyangwe in 1869 (Vansina 1990), it was large-scale and produced prodigious amounts of ivory, while its territory was held exclusively by Pygmies and defended militarily. The trader explained that trading negotiations were all channeled through one leader, whom he refers to as its "king" (Stanley 1890). The second reference to a large-scale Pygmy social formation is by Mounteney-Jephson (1891), who participated in the Emin Pasha relief expedition led by Stanley. He describes a Pygmy settlement, south of the Mangbetu kingdom on the Nava River, that consisted of "two large and several small villages under a king called M'Galima" (1891, 371–372). My interpretation of the two "kings" who led these "polities" is that they were probably temporary leaders of temporary social formations, akin to the charismatic war leaders discussed for other equatorial societies during the late precolonial era. Such leaders were given temporary powers during a particular historical moment—e.g., a time of conflict—and, once it came to an end, the leader's authority would cease (Copet-Rougier 1997; Vansina 1990). The particular historical moment for both these social formations was the final decades of the nineteenth century, when trading networks in the eastern part of the forest were characterized by high levels of violence. It is within this context that Pygmy local groups in the area appear to have joined together to create exclusive, defensible territories based on the ivory trade (also see Kagabo and Mudandagizi 1974; Lewis 2000).

References

Abega, Severin C. n.d. "Princes et Chimpanzés: Pygmées BEDZAM." Unpublished paper.

Althabe, Gerard. 1965. "Changements Sociaux Chez le Pygmées Baka de l'est Cameroun." *Cahiers d'Etudes Africaines* 5:561–92.

Avelot, M. R. 1905. "Recherches sur l'histoire des migrations dans le bassin de l'Ogôoué et la région littorale adjacente." In *Bulletin de Géographie Historique et Descriptive.* Paris: Ministère de l'Instruction Publique et des Beaux-Arts.

Bahuchet, Serge. 1979b. "Utilisation de l'espace forestier par les Pygmées Aka, Chasseurs-cueilleurs d'Afrique Centrale." *Information sur les Sciences Sociales* 18:999–1019.

———. 1985. *Les Pygmees Aka et la Foret Centrafricaine.* Paris: SELAF.

———. 1986. "The Forest Ecology." In *On the Edge of the Forest: Cultural Adaptation and Cognitive Development in Central Africa*: 5–21. J. W. Berry, J. M. H. van de Koppel, C. Sénéchal, R. C. Annis, S. Bahuchet, L. L. Cavalli-Sforza, and H. A. Witkin. Berwyn, IL: Swets North America Inc.

———. 1991. "Les Pygmées d'Aujourd'hui en Afrique Centrale." *Journal des Africanistes* 61:5–35.

———. 1993a. "History of the Inhabitants of the Central African Rain Forest: Perspectives from Comparative Linguistics." In *Tropical Forests, People and Food: Biocultural Interactions and Applications to Development*, edited by C. M. Hladik, A. Hladik, O. F. Linares, H. Pagezy, A. Semple, and M. Hadley. Paris: UNESCO:37–54.

———. 1993c *La Rencontre des Agriculteurs: Les Pygmées parmi les Peuples d'Afrique Centrale.* Paris: Peeters-SELAF.

———. 2012. "Changing Language, Remaining Pygmy." *Human Biology* 81(1):9.

Bahuchet, Serge, and Henri Guillaume. 1982. "Aka-Farmer Relations in the Northwest Congo Basin." In *Politics and History in Band Societies*, edited by E. Leacock and R. B. Lee. Cambridge: Cambridge University Press, 189–211.

Balandier, Georges. 1968. *Daily Life in the Kingdom of the Kongo: From the Sixteenth to the Eighteenth Century.* New York: Pantheon Books.

Battell, Andrew [1625] In Ravenstein, E G 1901 *The Strange Adventures of Andrew Battell of Leigh, in Angola and the Adjoining Regions (reprinted from "Purchas His Pilgrimes").* London: The Hakluyt Society.

Bertaut, M 1943 Contribution à l'étude des Négrilles de la région du Haut-Nyong. *Bulletin de la Société d'Études Camerounaises* 4:73–97.

Biesbrouck, Karen 1999 Agriculture among Equatorial African Hunter-Gatherers and the Process of Sedentarization: The Case of the Bagyeli in Cameroon. In *Central African Hunter-Gatherers in Multidisciplinary Perspective: Challenging Elusiveness*, K Biesbrouck, S Elders, and G Rossel (eds.), pp. 189–206. CNWS, Leiden.

Birmingham, David 1977 Central Africa from Cameroun to the Zambezi. In *The Cambridge History of Africa* (3): 519–566.

Bruel, G 1911 *Notes Ethnographiques sur Quelques Tribus de l'Afrique Equatoriale Française. I: Les Populations de la Moyenne Sanga: Pomo, Boumali, Babinga.* Paris: Leroux.

Burrows, Guy 1898 *The Land of the Pigmies.* New York: Thomas V. Crowell & Company.

Capagory, M 1933 Apprivoisement des Babingas. Colonie du Moyen Congo, Circonscription de l'Alima.

Cavalli-Sforza, Luca L, ed. 1986 *African Pygmies.* Orlando: Academic Press.

Cavalli-Sforza, Luca L, L A Zonta, F Nuzzo, L Bernini, W W de Jong, P Meera Khan, A K Ray, L N Went, M Siniscalco, L E Nijenhuis, E van Loghem, and G Modiano 1969 Studies on African Pygmies I: A pilot investigation of Babinga Pygmies in the Central African Republic (with an analysis of genetic distances). *American Journal of Human Genetics* 21:252–274.

Clist, Bernard 1999 Traces de très anciennes occupations humaines de la forêt tropicale au Gabon. In: *Central African Hunter-Gatherers in Multidisciplinary Perspective: Challenging Elusiveness,* K Biesbrouck, S Elders, and G Rossel (eds.), pp. 21–40. CNWS, Leiden.

Copet-Rougier, Elisabeth 1987 "Du clan a la chefferie dans l'est du Cameroun." *Africa* 57 (3):345–363.

———. 1997 Histoire politico-économique de la Haute-Sangha. *Paper presented at Trinational Sangha River Region Conference,* New Haven.

De Heutsch, Luc 1962 Aspects de la sacralité du pouvoir en Afrique. *Annales du Centre d'Etude des Religions*:1. Brussels.

———. 1972 *Le roi ivre ou l'origine de l'Etat.* Paris: Editions Gallimard.

Delobeau, Jean-Michel 1977 Yamonzombo et Yandenga: les relations entre les villages monzombo et les campements Pygmées Aka dans la sous-préfecture de Mongoumba (Centrafrique). Thèse 3° cycles. Paris: EHESS.

Denbow, James R 1984 Prehistoric Herders and Foragers of the Kalahari: The Evidence for 1500 Years of Interaction. In Schrire, C (ed.) *Past and present in hunter gatherer studies.* Orlando, FL: Academic Press:175–194.

Dentan, Robert K 1988 Band-Level Eden: A Mystifying Chimera. *Cultural Anthropology* 4:276–284.

Deschamps, Hubert 1962 *Traditions Orales et Archives au Gabon: contribution à l'ethno-histoire.* Paris: Éditions Berger-Levault.

Diamond, Jared 1999 *Guns, Germs, and Steel: The Fates of Human Societies.* New York: WW Norton and Company, Inc.

Dodd, Robert 1984 Rituals and the Maintenance of Internal Cooperation Among the Baka Hunters and Gatherers. *Unpublished paper.*

———. 1986 The Politics of Neighbourliness. *Paper presented at the Fourth International Conference on Hunting and Gathering Societies, London.*

Douet, L 1914 Les Babingas. *Ethnographie: Bulletin de la Société d'Ethnographie de Paris* 2:15–32.

Eggert, M K H 2005 The Bantu problem and African archaeology. In: *African Archaeology: A critical introduction,* AB Stahl (ed.), pp. 301–326.

Ehret, Christopher 2001 Bantu expansions: re-envisioning a central problem of African history. *International Journal of African Historical Studies*: 34:5–40.

Fallers, Lloyd A 1973 *Inequality: social stratification reconsidered.* Chicago: University of Chicago Press.

Geschiere, Peter 1982 *Village communities and the state: changing relations among the Maka of south-eastern Cameroon since the colonial conquest.* London: Kegan Paul International Ltd.

Gordon, Robert J 1984 *The !Kung in the Kalahari Exchange: An Ethnohistorical Perspective.* In: Past and present in hunter gatherer studies, Carmel Schrire (ed.). Orlando: Academic Press.

Grinker, Richard R 1994 *Houses in the Rainforest: Ethnicity and Inequality among Farmers and Foragers in Central Africa.* Berkeley and Los Angeles: University of California Press.

Guillaume, Henri 2003 *Du miel au café, de l'ivoire à l'acajou: La colonisation de l'interfluve Sangha-Oubangui et l'évolution des rapports entre chasseurs-collecteurs pygmées Aka et agriculteurs (Centrafrique, Congo), 1880–1980.* Paris: Editions Peeters (SELAF).

Guille-Escuret, Georges 1998 La Révolution Agricole des Pygmées Aka: De la Structure dans l'Événement et Réciproquement. *Homme* 147:105–126.

Guthrie, Malcolm 1967–1971 *Comparative Bantu,* 4 vols. Farnham, England: Gregg Press.

Guyer, Jane I 1986 Indigenous Currencies and the History of Marriage Payments: a case study from Cameroon. *Cahiers d'Études africaines* 104:XXVI:4: 77–610.

Harms, Roy 1981 *River of Wealth, River of Sorrow: The Central Zaire Basin in the Era of the Slave and Ivory Trade, 1500–1891.* New Haven: Yale University Press.

Hewlett, Barry S 2009 Victims of Discrimination: An Anthropological Science Critique of Human Rights and Missionary Narratives of African Pygmy Marginalization. *Paper presented at workshop on hunter-gatherer marginalization at Kyoto University, August 25, 2009.*

Hewlett, Barry S, J M H Van de Koppel, and L L Cavalli-Sforza 1982 Exploration Ranges of Aka Pygmies of the Central African Republic. *Man* 17:418–430.

Hill, Kim R, R S Walker, M Bozicevic, J Eder, T Headland, B Hewlett, A M Hurtado, F Marlowe, P Wiessner, and B Wood 2011 Co-Residence Patterns in Hunter-Gatherer Societies Show Unique Human Social Structure. Science 331 (6022): 1286–1289.

Ichikawa, Mitsuo 1978 The Residential Groups of the Mbuti Pygmies. *Senri Ethnological Studies* 1:131–188.

Jackson, Dorothy 2003 *Twa Women, Twa Rights in the Great Lakes Region of Africa.* London: Minority Rights Group International.

Johnston, Harry H 1969 George Grenfell and the Congo; a history and description of the Congo Independent State and adjoining districts of Congoland, together with some account of the native peoples and their languages, the fauna and flora, and similar notes on the Cameroons and the Island of Fernando Pô. New York: Negro Universities Press.

Joiris, Daou V 1996 A Comparative Approach to Hunting Ritual among Baka Pygmies (Southeastern Cameroon). In *Cultural Diversity among Twentieth Century Foragers: An African Perspective,* S Kent (ed.), pp. 245–275. Cambridge: Cambridge University Press.

Kagabo, José & Mudandagizi, Vincent 1974 Complainte des gens de l'argile: les Twa de Rwanda. *Cahiers d'Etudes Africaines* 53:XIV:1:75–87.

Kalck, Pierre 1971 *The Central African Republic: a Failure in De-colonisation.* New York: Praeger.

Kisliuk, Michelle R 1998 *Seize the dance!: BaAka musical life and the ethnography of performance.* New York, Oxford: Oxford University Press.

Klieman, Kairn A 1997 Hunters and Farmers of the Western Equatorial Rainforest: Economy and Society, 3000 B.C. to A.D. 1880. *Unpublished Ph.D. dissertation.* University of California, Los Angeles.

———. 2003 *"The Pygmies Were Our Compass:" Bantu and Batwa in the History of West Central Africa, Early Times to c. 1900 C.E.* Portsmouth, New Hampshire: Heinemann.

Koch, Henri 1968 *Magie et Chasse dans la forêt Camerounaise.* Paris: Éditions Berger Levrault.

Kopytoff, Igor 1987 "The Internal African Frontier: The Making of African Political Culture." In: *The African Frontier: The Reproduction of Traditional African Societies,* Igor Kopytoff (ed.), 3–84. Bloomington: Indiana University Press.

Laburthe-Tolra, Philippe 1981 *Les seigneurs de la forêt: essai sur le passé historique, l'organisation sociale et les normes éthiques des anciens Bëti du Cameroun.* Paris: Publications de la Sorbonne.

Laman, Karl E 1953 *The Kongo.* Stockholm: Victor Pettersons Bokindustri Artiebolag.

Lavachery, P 1997–1998 De la pierre au metal: Archéologie des depots Holocènes de l'abri de Shum Laka (Cameroon). *Unpublished Ph.D. thesis,* Université Libre de Bruxelles.

Lee, Richard B & Devore, Irven (eds.) 1968 *Man the hunter.* Chicago: Aldine.

Leonhardt, Alec n.d. "History of Baka-Non Baka Relations." *Unpublished paper.*

Lewis, Jerome 2000 *The Batwa Pygmies of the Great Lake Region.* London: Minority Rights Group International.

Lewis, Jerome & Knight, Judy 1995 *The Twa of Rwanda: an assessment of the situation of the Twa and promotion of Twa rights in Post-War Rwanda.* Chadington, England: World Rainforest Movement.

Lloyd, A B 1907 [1899] In Dwarf Land and Cannibal Country. London: T. Fisher Unwin Adelphi Terrace.

Mauss, Marcel 1967 The gift: forms and functions of exchange in archaic societies. New York: Norton.

Moïse, Robert E 1992 "A Mo Kila!" ("I Refuse!"): Living Autonomously in a Biaka Community. *Unpublished MA thesis.* New York University.

———. 2003 Loved Ones and Strangers: society, history and identity in equatorial Africa. *Unpublished PhD dissertation.* New York University.

———. 2011 If Pygmies Could Talk: creating indigenous development in equatorial Africa. *Before Farming: the archaeology and anthropology of hunter-gatherers* 2011/4.

———. n.d. Naturvolk or Entrepreneurs: toward a history of political entrepreneurship among Pygmy peoples in equatorial Africa. *Unpublished paper.*

Montano, Valeria, G Ferri, V Marcari, C Batini, O Anyaele, G Destro-Bisol, and D Comas 2011 The Bantu expansion revisited: a new analysis of Y chromosome variation in Central Western Africa. *Molecular Ecology* 20:2693–2708.

Mosko, Mark S 1987 The Symbols of "Forest": A Structural Analysis of Mbuti Culture and Social Organization. *American Anthropologist* 89: 896–913.

Mounteney-Jephson, A J 1891 *Emin Pasha and the Rebellion at the Equator: a story of nine months experiences in the last of the Soudan provinces.* New York: C. Scribner's Sons.

Myers, Fred R 1986 *Pintupi Country, Pintupi Self: Sentiment, Place and Politics among Western Desert Aborigines.* Washington, D. C.: Smithsonian.

———. 1988c Critical Trends in the Study of Hunter-Gatherers. *Annual Review of Anthropology* 17:261–282. Palo Alto: Annual Reviews, Inc.

Oliver, Roland 1966 The Problem of the Bantu Expansion. *Journal of African History* 7(3):361–376.

Oslisly, R 1996 Archéologie et paléoenvironnement dans la Réserve de la Lopé: Rapport final. Groupement AGRECO/C.T.F.T. *Unpublished manuscript.*

Patin, E, G Laval, L B Barreiro, A Salas, and O Semino 2009 Inferring the demographic history of african farmers and pygmy hunter-gatherers using a multilocus resequencing data set. *PLoS Genet* 5:e1000448.

Pedersen, Jon and Waehle, Espen 1986 The Complexities of Residential Organization among the Efe (Mbuti) and the Bamgombi (Baka): a critical view of the notion of 'flux' in hunter-gatherer societies. *Paper presented at the Fourth International Conference on Hunting and Gathering Societies, London.*

Putnam, Anne E 1954 *Madami: My 8 Years of Adventure with the Congo Pigmies.* New York: Prentice Hall.

Putnam, Patrick 1948 The Pygmies of the Ituri Forest. In Coon, C (ed.) *A Reader in General Anthropology.* New York: H. Holt:322–342.

Regnault, M 1911 Les Babenga (Négrilles de la Sangha). *L'Anthropologie* 22: 261–288.

Rupp, Stephanie 2003 Interethnic Relations in Southeastern Cameroon: Challenging the "Hunter-Gatherer"–"Farmer" Dichotomy. *African Study Monographs, Supplement* 28:37–56.

Schadeberg, Thilo 1999 Batwa: The Bantu Name for the Invisible People. In: *Central African Hunter-Gatherers in Multidisciplinary Perspective: Challenging Elusiveness*, K Biesbrouck, S Elders, and G Rossel (eds.), pp. 21–40. CNWS, Leiden.

Schultz, Michael 1988 Forager-Farmer Relations: Some New Fieldwork on the Batúa and the Baotó, Equateur, Zaire. *Paper presented at the Fifth International Conference on Hunting and Gathering Societies, London.*

Schweinfurth, Georg 1874 *The Heart of Africa: three years' travels and adventures in the unexplored regions of Central Africa from 1868 to 1871.* New York: Harper and Brothers.

Seitz, Stefan 1993 *Pygmées d'Afrique Centrale.* Paris: SELAF.

Stanley, Henry M 1885 *The Congo and the Founding of its Free State: a Story of Work and Exploration.* New York: Harper and Brothers.

———.1890 *Through the Dark Continent or the sources of the Nile around the great lakes of Equatorial Africa and down the Livingstone river to the Atlantic ocean.* New York: Harper and Brothers.

Terashima, Hideaki 1985 Variation and Composition Principles of the Residence Group (Band) of the Mbuti Pygmies: Beyond a Typical/Atypical Dichotomy. *African Study Monographs, Supplemental* 4:103–120.

Tishkoff, S A, F A Reed, F R Friedlaender, C Ehret, and A Ranciaro 2009 The genetic structure and history of Africans and African Americans. *Science* 324:1035–44.

Trilles, R P 1903 Mille lieues dans l'inconnue. Les Ba-Yaga. *Les Missions Catholiques*: 35.

Turnbull, Colin M 1961 *The Forest People.* New York: Simon and Schuster.

———. 1965b *Wayward Servants: the Two Worlds of the African Pygmies.* New York: Natural History Press.

Vallois, H V & Marquer, P 1976 *Les Pygmées Baká du Cameroun: Anthropologie et Ethnographie avec une annexe Démographique.* Paris: Muséum National d'Histoire Naturelle.

van de Sandt, Joris 1996 Mutual perception and the struggle for control over natural resources in Bagyeli-Fang relations: five ways of coping with changing relations. Paper for the Colloquium: Hunter-Gatherers of Equatorial Africa, Leiden.
Vansina, Jan 1978 *The Children of Woot: A History of the Kuba Peoples.* Madison: University of Wisconsin Press.
———. 1984 Western Bantu Expansion. *Journal of African History* 25:129–145.
———. 1986 Do Pygmies Have a History? *Sprache und Geschichte in Afrika* 7(1): 431–445.
———. 1990. *Paths in the Rainforests: Toward a History of Political Tradition in Equatorial Africa.* Madison: University of Wisconsin Press.
———. 1994/95 A Slow Revolution: Farming in Subequatorial Africa. In: *The Growth of Farming communities in Africa from the Equator Southwards*, JEG Sutton (ed.), pp. 15–26. Azania special vol. 29–30. London/Nairobi: British Institute of eastern Africa.
———. 1995 New linguistic evidence and 'The Bantu Expansion'. *Journal of African History*: 36:173–195.
Verdu, Paul, F Austerlitz, A Estoup, R Vitalis, and M Georges 2009 Origins and genetic diversity of pygmy hunter-gatherers from Western Central Africa. *Current Biology* 19:312–318.
Waehle, Espen 1986 "Efe (Mbuti Pygmy) Relations to Lese-Dese Villagers in the Ituri Forest, Zaire: Historical Changes During the Last 150 Years." *Sprache und Geschicte in Afrika* 7(2): 375–411.
———. 1999 Introduction. In: *Central African Hunter-Gatherers in Multidisciplinary Perspective: Challenging Elusiveness*, K Biesbrouck, S Elders, and G Rossel (eds.), pp. 265–278. CNWS, Leiden.
Wijsman, Ellen M 1986 Estimation of Genetic Admixture in Pygmies. In: *African Pygmies*, L. L. Cavalli-Sforza (ed.), pp. 361–426. Orlando: Academic Press.
Wilmsen, Edwin 1983 The Ecology of Illusion: Anthropological foraging in the Kalahari. *Reviews in Anthropology* 10 (1): 9–20.
———. 1989b *Land Filled With Flies: a Political Economy of the Kalahari.* Chicago: University of Chicago Press.
Woodburn, James 1997 Indigenous discrimination: the ideological basis for local discrimination against hunter-gatherer minorities in sub-Saharan Africa. *Ethnic and Racial Studies:* 20(2):345–361.
Wrigley, Christopher 1992 The *Longue Durée* in the Heart of Darkness. *Journal of African History*: 33:129–134.

Film

BBC 1996 *A Caterpillar Moon.*

5

Human Biology and Health of African Rainforest Inhabitants

Alain Froment

Introduction

African hunter-gatherers are found primarily in two very different ecosystems. Some live in semidesertic lands, such as the Koesaan (or San) of Namibia and South Africa, or in savanna, such as the Hadza and Sandawe of Tanzania; others live in the central African rainforest. All these people have short stature, but genetic differences between them are important (Tishkoff et al. 2009). Only those inhabiting the forest are called Pygmies. "Pygmy" is the only ethnonym that refers to a morphologic character, a very short stature, in fact, the shortest in the world. The etymology comes from the Greek word *pugon* (cubit—about 30 cm) and refers to the legend of very small people fighting with cranes. Andrew Battel, an English traveler whose book was published in 1625, remembered this story when he met a group of small people, and the scientific tradition, beginning with Du Chaillu and Schweinfurth, kept this mythologic name, despite its inappropriate meaning. Many African legends also mention small peoples, even in areas where they are not expected to have ever lived, such as in the Niger Bend where Dogons describe *Tellem*—the cave dwellers they found when they arrived in the Bandiagara area. The existence of forest populations with a relatively short stature raises interesting problems about human variation and adaptation; the peculiarity of rainforest ecosystem also induces medical consequences. This chapter is based on updated earlier works reviewing these specific aspects of human biology (Froment 1993, 2001, 2003, 2008).

Historical Construction of a Category

Definition

Central African hunter-gatherer groups, episodically called *Négrilles*—"small negroes"—by French authors (Hamy 1872) and *Twiden* (from the local name of Twa) by Germans (Gusinde 1956) are generally named Pygmies.

However offensive this term may be, it is nonetheless used sometimes by those to whom it refers. Furthermore, the term *Pygmy* tends to be universal and will hence be used here despite the discriminatory connotations it may carry.

Until recently, Pygmies could be distinguished from their farmer neighbors by a seminomadic lifestyle. Aside from this cultural definition, which tends to be obsolete as these communities are now nearly sedentarized, is there now a biological definition to characterize them? Since the beginning, the problem was to assign to the newly discovered Pygmy group a taxonomic place within human variation and to link it to ancestral forms of mankind (Flower 1888; Virchow 1894). Cavalli-Sforza's definition (1986) is a biocultural one:

> It therefore seems reasonable to call Pygmies groups of Africans who live (or lived until a short while ago) as hunters-gatherers in the tropical forest, and have a stature smaller than the farming tribes with whom they are in contact and have a relation of servitude. . . . [I]n this book, we will keep the name "Pygmy" for all populations from Central Africa that have a small stature (in practice, well below 160 cm on average) and live, or presumably lived until a short while ago, exclusively as hunter-gatherers in or near the tropical forest. (1986)

Because several Bantu populations were morphologically close to the Twa, sometimes called "pygmoids" because they tend to be taller than the average Pygmy mean, Hiernaux et al. (1976) admitted that the pygmoid definition was based more on a social status and a way of subsistence than on morphological criteria.

Cultural anthropologists, linguists, and musicologists distinguish three groups of Pygmies: the Eastern, Western, and Southern. The Eastern group (Efe, Asua, and Mbuti) live in the east of the Democratic Republic of Congo. The Western group (Aka, Baka, BaKola or BaGyeli, Bedzan, and Bongo) inhabits the CAR, Cameroon, and Gabon. The Southern group, the Cwa or Twa, are mainly sedentarized in the DRC (Ntumba and Ekonda area), Rwanda, and Burundi. These Twa (or Cwa, Tua, Ka, Koa) are either hunters, sedentarized pot makers, or fishermen, as they are in Zambia. Though they sometimes settle on the highlands or in savannas and thus distinguish themselves from other pygmy groups (*Batèmbô ba màrungu* versus *Batèmbo ba manywémà*: Kazadi 1981), the Twa have no genetic differences and are not the result of Bantu admixture as Schebesta postulated (Hiernaux 1974).

How then can a Pygmy be physically distinguished from a non-Pygmy? In his description, Seligman enumerated the main stereotypical Pygmy features:

> Physically if they really are Pygmies, their stature varies from 130 to 145 cm, the average being 143 and 135, according to the sex. Their skin can be reddish, or reddish brown or very dark, the body being often

> covered with a slight down. The head tends towards brachycephaly, with a cephalic rating of 79; the nose is very wide with a slightly marked or absent bridge; the eyes are rather big and prominent; the face is short and tends to be wide; it's generally prognathic and often so much that Sir Henry Johnston proposed the word "prognathic-Pygmies" for a group including these little men and a certain inferior type of negro of the forest. (1930)

Vallois and Marquer (1976) in their work about the Baka of Cameroon, did not give any definition of Pygmies that can be differentiated from that given for "Blacks"—the Bantu farmers—neither did Lefrou (1943) or Lalouel (1950). This *Pygmy-Negro* duality is also present in classical texts (Lee and DeVore 1968). French authors insist on nonmetrical characters in search of morphological peculiarities: lighter skin and eye color, hairiness more abundant, eyes seeming deeper because of more salient brow ridges, "tuber-shaped" nose, thin and non-everted lips with a marked convexity of the upper lip. However, some Black groups living in the forest would also display "pygmoïd" features, which blends even more the physical differences. That is why a sensible colonial administrator wrote that

> somebody not aware would walk by "Negrilles" without taking any notice of them, however as soon as one has observed some of them, they are unmistakeable. But it would be difficult to say how they are distinguishable from other black people. . . . In short, nothing or very few things give the possibility of distinguishing Babingas from other Blacks. (Bertaut 1943)

Physical anthropologists then concluded: "we searched to know if there was a classic Negrille type to be opposed to the black type. We must at once say that we failed" (Olivier 1950), a view already proposed by Vallois (1940). Table 5.1 summarizes some anthropometrical data of different Pygmy groups compared with neighbor populations.

Mythology

The raciological and descriptive terminology could be just a matter of curiosity if it was not linked to a vivid discrimination. Some people sharing the same areas still consider Pygmies as half-apes ("speaking chimpanzees," in a myth collected among the Tikar of central Cameroon) or as a property, and owing to the belief in Pygmies' mastery over occult forces of the forest, they are also simultaneously despised and feared (Kazadi 1981; Vansina 1990). Odambo-Adone (1990) provides a summation of this sentiment in saying that, "Psychologically the word Pygmy has turned to a gangrene, a virus that one drags with oneself and that we can, as much as wanted, spread to one's offspring." The prejudices are further illustrated in a thesis written by a Cameroonian anthropologist:

Table 5.1. Height, weight, and BMI of eastern, western, and southern (Twa) Pygmy groups and their non-Pygmy neighbors.

Population	Country	n M	Height	Weight	BMI	n F	Height	Weight	BMI	Author	Time of Survey	Source
Eastern Pygmies												
Mbuti	DRC	115	141.7			26	135.7			Czekanowski	1938	Froment 1993
Mbuti	DRC	514	144.3	39.8	19.1					Hiernaux	1968	Froment 1993
Mbuti	DRC	71	144.4			38	136			Skolnick	1971	Cavalli-Sforza 1986
Mbuti	DRC	69	144.2	41.5	20	32	137.4	36.9	19.5	Turnbull	1957–1958	Cavalli-Sforza 1986
Efe	DRC	98	142.9							Schebesta	1933	Froment 1993
Efe	DRC	136	142			82	134			Schebesta	1938	Froment 1993
Efe	DRC	26	144.8	43	20.5	27	136.5	39.7	21.3	Dietz	1980	Dietz 1989
Efe	DRC	114	143			110	136			Bailey et al.	1980–1991	Bailey 1991
Efe	DRC	23	146.4	43.2	20.2	26	138.6	38.3	19.9	Bailey et al.	1980–1991	Shea 1996
Asua	DRC	50	144.6							Schebesta	1933	Froment 1993
Asua	DRC	170	144			108	136.8			Schebesta	1938	Froment 1993
Asua	DRC		144	37.7	18.2		138.4	37	19.3	Schebesta	1948	Vincent 1962
Asua	DRC					116	137.1			Vincent et al.	1956–1959	Vincent 1962
Western Pygmies												
Aka	CAR	75	156			12	143			Poutrin	1910	Gusinde 1955
Aka	CAR	33	155.7	48.4	20	19	144	41.6	20.1	Pales	1934	Pales 1938
Aka	DRC	565	155			475	144			Lalouel	1950	Gusinde 1955
Aka	CAR	87	152.3	46.4	20	56	144.9	42.6	20.3	Cresta	1961	Cresta 1965
Aka	CAR	427	152.9	47.9	20.5	392	144.3	42.2	20.3	Penetti et al.	1967–1968	Penetti 1986
Aka	CAR	48	152.7	48.3	20.7	43	145	42.7	20.3	Cavalli-Sforza	1969	Cavalli-Sforza 1986
Aka	CAR	125	155.1	47.9	19.9	104	145.9	41.7	19.6	Bahuchet and Pagezy	1977	unpubl.
Aka	CAR	127	153.6	47.3	20	105	143.8	41.4	20	Cavalli-Sforza	?	P. De Kneiff unpubl.
Aka	CAR	105	154.4	48.6	20.4	111	144.6	44.1	21.1	Cavalli-Sforza	?	P. De Kneiff unpubl.
Baka	Cam		145				130			Ternay	1948	Gusinde 1955

Baka	Cam	60	153.1							Vallois and Marquer	1976	Froment 1993
Baka	Cam	55	154.9							Vallois and Marque	1976	Froment 1993
Baka	Cam					206	146.1	46.1	21.6	Van Eijk	1978	Van Eijk 1986
Baka	Cam	75	154.4	49.6	20.8	73	146.6	44.4	20.7	Yamauchi	1996	Yamauchi 2000
Baka	Cam	182	155.6	50.3	20.8	184	146.5	43.4	20.2	Froment	2002–2007	Froment unpubl.
Baka	Cam	36	155.6	49.9	20.6	9	149.5	45.3	20.3	Froment	2002–2007	Froment unpubl.
Kola	Gabon	8	151.1			4	141.6			Castillo-Fiel	1949	Gusinde 1955
Bagyeli	Cam	28	158	49.9	20	38	148.5	43.3	19.6	Froment	1988	Froment 1993
Bagyeli	Cam	82	155.1			99	147.1			Pagezy	2004	Pagezy unpubl.
Koya	Gabon	20	156.9	49.8	20.2	11	145.8	46.5	21.9	Le Bomin	2006	Le Bomin unpubl.
Bedzan	Cam	37	158.3	53.9	21.5	19	152.1	47.3	20.4	Verdu	2011	Verdu unpubl.
Bongo	Gabon	25	159.7	51.6	20.2	19	151.2	46	20.1	Verdu	2006–2007	Verdu unpubl.
Bongo	Gabon	12	159.7	52.4	20.5	9	151.8	49.2	21.4	Verdu	2006–2007	Verdu unpubl.
Binga	Gabon	53	157.9	50.1	20.1	43	148.3	43.1	19.6	Fleuriot	1942	Froment 1993
Baka-Bagyeli	Cam	49	153.4	49	20.8	37	144.1	43.3	20.9	Ajoulat and Chabeuf	1950	Olivier 1950
Twa												
Twa	DRC?	64	157.9							Czekanowski	1938	Froment 1993
Twa	Rwanda	101	153			84	144.2			Gusinde	1949	Froment 1993
Twa	DRC	163	157.4							Hiernaux	1965	Froment 1993
Twa	DRC	100	160.8	46.1	17.8					Hiernaux	1968	Froment 1993
Twa	DRC	10	160.9	51	19.7					Austin	1976	Austin 1976
Twa	DRC					50	150.5	43.2	19.1	Pagezy	1978	Pagezy 1978
Twa	DRC	21	153.8	45.8	19.4					Ghesquiere and Karvonen	1981	Ghesquiere 1981
Twa	DRC	53	160.9	49.4	19.1					Pagezy	1988	Froment 1993

(continued)

Table 5.1. (continued)

Population	Country	n M	Height	Weight	BMI	n F	Height	Weight	BMI	Author	Time of Survey	Source
Neighbors (and the corresponding Pygmy group)												
Lese (Efe)	DRC	68	159.9	52.2	20.4	69	153.7	48.3	20.4	Dietz	1980	Dietz 1989
Lese (Efe)	DRC	76	162			96	152			Bailey et al.	1980–1991	Bailey 1991
Divers (Asua)	DRC					44	153.8				1962	Vincent 1962
M'Bimu (Aka)	RCA	139	164.2	56.2	20.8	124	156.8	51.6	21	Cresta	1961	Cresta 1965
Divers (Aka)	RCA	50	161.2	54.6	21	62	154.4	50.4	21.1	Cavalli-Sforza	1969	Cavalli-Sforza 1986
Bagandu (Baka)	Cam					122	156.9	55.6	22.6	Van Eijk	1978	Van Eijk 1986
Nzime (Baka)	Cam	42	168.5	62.7	22.1	58	155.8	51.9	21.4	Froment	2002–2007	Froment unpubl.
Bangando (Baka)	Cam	29	163.9	56.8	21.1	19	156	50.6	20.8	Froment	2002–2007	Froment unpubl.
Yassa (Bagyeli)	Cam	47	164.8	60.6	22.3	48	155.3	53	22	Froment	1996	Froment 1996
Mvae (Bagyeli)	Cam	32	165.1	60.1	22	33	155.8	54.1	22.3	Froment	1996	Froment 1996
Tikar (Bedzan)	Cam	22	167.2	64.3	23	16	156.7	58.4	23.8	Froment	2002–2007	Froment unpubl.
Ntomba (Twa)	DRC	10	168.5	58.6	20.6					Austin	1976	Austin 1976
Oto (Twa)	DRC					130	159	53.4	21.1	Pagezy	1978	Pagezy 197

Source: Compilation by Becker 2012.

> Expressions such as *Vungo mbpi ngieli* (as skillful as a Pygmy) and *Ke mpfunde mbpi* (to run like a Pygmy), used in Ngumba or in other close Bantu languages, say everything about the development of certain abilities characterizing hunters. Similarly, Pygmies' jaws which are very developed and solid in order to be able to tear meat; their upper limbs which are long and robust to enable Pygmies to climb trees, to confront big animals and to throw the spear as far and as skilfully as possible, also their abundant pilosity which allows them to resist the cold. (Ngima Mawoung 1993)

In reaction, Pygmy groups have started to organize themselves and to increase their international visibility, and they often introduce themselves as indigenous peoples, which in Africa is almost meaningless. The first anthropologists who studied them, however, never related them to any missing link between mankind and other primates: "In spite of their small size, their relatively long arms and their short legs, Akkas are definitely real men; and those who believed they were half apes must today be utterly disillusioned" (De Quatrefages 1887).

Legends or iconography about Pygmies in antiquity have been abundant since pharaonic times.[1] The assimilation of the Egyptian god Bes with a Pygmy is not established, but Bahuchet (1993) notes that the hieroglyph *dng* (*deneg*, or *danga*, only known by two inscriptions), identifies Pygmies and is different from *nmw*, which refers to a pathological, achondroplasic dwarf. Dapper, in his *Description of Africa*, published in 1668, says that groups of Bakkes-Bakkes are cited in Portuguese documents of the sixteenth century (De Quatrefages 1887), so travelers were in some way, and probably negatively, prepared to meet a special people. In Madagascar, a debate on a small-bodied and fair-skinned people called "Quimos" also persisted for decades during the Enlightment period. Edward Tyson's 1699 monograph, *Orang-outang sive Homo Sylvestris or the Anatomy of a Pygmie Compared with That of a Monkey, an Ape and a Man*, appears as the foundation of comparative anatomy (Tinland 1968) and describes the detailed autopsy of a chimpanzee, viewed as intermediary between monkey and man. At the time, delimitation between satyrs, bushmen (*Homo sylvestris*), pygmies, and apes was blurred. For Tyson, the Pygmy described by Homer is not human, while the "Hottentot" and Laps, though only in the lowest degree, belong to humankind. Buffon (1749) confirms this point but says nothing about the still legendary African Pygmies. New explorations in the "Heart of Africa," done in 1871 by Schweinfurth (1875), boosted the interest of anthropologists (Broca 1874; De Quatrefages 1874; Hamy 1879). Two Pygmy teenagers led to Italy and educated as young nobles by Count Miniscalchi-Erizzo became famous (Mantegazza and Zanetti 1874); the whole story is detailed in De Quatrefages (1887).

Myths live on. The geneticist Gérard Lucotte (1990), from a study of the Y chromosome, claims that Lobaye was the Garden of Eden and that Adam was a Pygmy. The exile from Paradise is an idea also found in Reichholf (1997), who identifies a perpetrator: sleeping sickness. Population genetics have since then been spectacularly developed and are detailed in Verdu (this volume). Not all Pygmy populations have been sampled, and drift effects due to the small size of groups do not facilitate the analysis, but a lot of progress is expected both in tracing the origins of Pygmy groups and in connecting their phenotype and their genotype. To date, the short stature has not been explained, though it has been determined to have nothing in common with genetic dwarfisms known in medicine (Becker et al. 2011). Geneticists agree that Pygmy groups display a high genetic heterogeneity, that they are completely African, even "ultra-African" (Cavalli-Sforza 1986) for some characters, that they share some traits with the San (erythrocyte acid phosphatase, phosphoglucomutase), and that some rare markers are unique to them, such as hemoglobin Babinga, which concerns only 0.8 percent of Babingas; hemoglobin Flatbush, peptidase C2, lactico-dehydrogenase Babinga, or 6-phosphogluconate dehydrogenase Mbuti. Regarding dermatoglyphic patterns, which have a strong genetic basis, there are clear differences between Pygmies and non-Pygmies (Dankmeijer 1947; Vecchi 1981), but also between Pygmy groups, such as Efe and Babinga (Glanville 1969).

Paleoanthropology

Before the start of haematogenetic surveys (Boyd 1963), some other populations, such as Negritos (Spanish diminutive of *Negro*, black), in Luçon Island, Philippinas, also called Aëtas (Genet-Varcin 1951), Andaman islanders (called Mincopies in the nineteenth century), Semang in the Malacca Peninsula, Veddas from Ceylon, and some groups in Papua New-Guinea and New-Britain, were thought to have a common stem with African Pygmies (Schmidt 1910, Leroi-Gourhan and Poirier 1953). The intriguing discovery of *Homo floresiensis* fueled more speculations about insular dwarfism (Falk et al. 2009). In zoology, this type of dwarfism supposes that prey size is reduced by predators, but for humans it is difficult to prove such a pressure.

Since the nineteenth century, two competing hypotheses have been imagined: either Pygmies are the first autochtonous people of Africa or they degenerated from earlier, taller populations. Kollmann (1894) thought that Pygmies were the earliest form of mankind, on the basis of both archaeological (skeletons found at Schweizerbil, Switzerland) and embryological considerations (Pygmies retain pygmoid features displayed by the human fetus). This curious theory was dismissed by Schwalbe (1906), and in 1912, Poutrin interpreted the small size as the consequence of isolation:

> In the same way as some animals did, they became smaller as the result of environmental conditions, such as the limited variety and availability of edible foods, and, because of their isolation (islands,

> impassable forests), they settled into a recognisable racial type, whose physical characteristics can no longer be easily influenced by a more or less abundant food supply. (p. 394)

Sir Arthur Keith (1948) viewed them as descending from "Dartians" (the name he used, in remembrance of paleontologist Raymond Dart, to call Australopithecines). In fact, the stature of early humans is not well-known, but if most Australopithecines were short, then *Homo ergaster*, like the 1.6-million-year-old Turkana boy of Kenya, was tall (Brown et al. 1985). At the moment it is impossible for geneticists to tell whether the common ancestral population of Pygmies and non-Pygmies was short (and then non-Pygmies' stature increased) or tall (in which case the Pygmy stock became shorter).

In regard to the more recent African population, paleontological records are very rare. At Iwo Eleru, Nigeria, from a skull and postcranial bones dated at 13,000 BP, stature has been estimated around 165 cm (Brothwell and Shaw, 1971). At Elmenteita (Gamble's Cave, Kenya), around 7,400 BP, stature is higher. In forest areas, prehistoric skeletons are very rare and not more ancient than 6,000 to 9,000 years. At Ishango (DRC), a complete humerus and tibia allowed paleontologists to estimate the stature at around 166 cm (Twiesselmann 1958). In the pre-Neolithic sites of northwest Cameroon, skeletons have middle to short stature (Ribot et al. 2001): at Shum Laka, two male burials of the early phase (7,000 BP) gave a relatively high stature (161 and 166 cm), while two female skeletons belonging to the late phase (3,200 BP) had statures of 145 and 154 cm; the Mbi Crater skeleton, belonging to the same late period, is even smaller (141 cm). These two sites are close, geographically and chronologically, to the cradle of the Bantu expansion. The diet of Shum Laka people of the late phase was rich in animal proteins, so they were probably hunter-gatherers. But multivariate analysis of their skull measurements did not show a specific Pygmy pattern, if any. A skeleton discovered in the Ituri Forest and dated at 810 BP (Mercader et al. 2001) has an estimated stature of 157 cm. The skull seems gracile and small (170 × 125 mm), but most of the face is missing, and the anthropological study was unable to trace its ancestry. Neither DNA nor isotope studies are available, but the presence of a piece of iron and pottery fragments in the grave casts some doubt on a Pygmy origin. Moreover, the presence of farmers is ancient in the area: botanical studies suggest that the central African rainforest has been cultivated for millenia (Vansina 1985); many archaeological sites in Gabon and Congo prove the existence of sedentary villages around 4,000 BP; and the apparition of Early Iron Age is attested around 800 BC (Essomba 1992; Oslisly 2006; Lavachery et al. 2010).

Many migrations have occurred since the Iron Age, which explains why some groups, such as the Baka, who migrated from the Great Lakes area, speak an Ubangian language while their neighbors are Bantu (Bahuchet 2012). In Cameroon, the Bakola-Bagyeli Pygmy group is probably not aboriginal; its linguistic proximity with the Ngumba and Mabi suggests that they belong to the migration of the *Kwasio* group that claims its origin in the Congo Basin and reached its present place less than two centuries ago. Linguistics is of no help because all the languages spoken by Pygmies have been borrowed from their neighbors or former neighbors (Bahuchet 1993, 2012), and some Bantu myths remember the role that Pygmies played in their migrations (Vansina 1990).

Not all Pygmies inhabit the deep rainforest, and to assume that they are strictly dependent to the forest would be a mistake. The most northern group, the Bedzan, live in a very open landscape in central Cameroon, while Twa groups in the DRC, Burundi, and Rwanda also live on the forest-savanna ecotone. Several other groups of hunter-gatherers living on savannas (Hadza and Sandawe in East Africa and San in Austral Africa) also bear remote genetic relationships to Pygmies (Lachance et al. 2012). A common origin between all these groups has been debated very often (Schweinfurth 1875; Dart 1937; Schebesta 1938; Ruggles Gates 1958; Toerien 1961; Vigilant et al. 1989; Campbell and Tishkoff 2010). Following Hamy, Montandon (1933) defined a pygmoid race that grouped "steatopyges" (Bushmen and Hottentots), Negrilles and Negritos. This position was still accepted three decades later by C. S. Coon (1963, 1965) but is now only of historical interest. Except the fact that all are or were nomadic and hunters, they have many genetic and cultural differences. Physically, San have a wide nose and short stature, features that they share with the Pygmies and reveal an adaptation to moist, not dry, areas. Hiernaux and Boedhi Hartono (1980) found some morphological resemblance between Hadza and Hukwe (extinct "black Bushmen"), but not at all with Pygmies. Thilmans (1968) calculated D^2 Mahalanobis distances between Philippine Negritos, who are brachycephalic and not very prognathic, and African Pygmies and rejected a real resemblance. However, the hypothesis of a common morphological pattern of adaptation to a moist tropical environment, either by convergence or shared remote ancestry, cannot be excluded.

Morphological Analysis

Head Shape

In their classical book *Crania Ethnica,* De Quatrefages and Hamy (1882) describe some skulls from central Africa, such as Bulu of Cameroon and Camma, Njavi, Apindji, and Adouma from Gabon, who are not Pygmies; neither is the famous series of ninety-three skulls collected in the Fernan-Vaz area of Gabon by Du Chaillu for the British Museum. The debate then focused on the existence of brachycephaly in Africa. Hamy's opinion (1872), based on a few skulls, was that Pygmies were round-headed, which related them to Negritos, but when Verneau (1896) described an adult woman measuring only 116 cm with a long head, this opinion was revised and two types of Pygmies were postulated, one round-headed with a moderate prognathism, the second long-headed and prognathic, with varying degrees of admixture between them (Poutrin 1910–1912). Marquer (1972) found that orbit depth is greater in Pygmies than for other central African populations, but that other skull measurements, in particular prognathism and nose width, are the same.

On average, Pygmies' heads are characterized by a very low face and a wide nose, one of the widest in the human species, displaying an equilateral shape (width and height equal). According to a personal craniometric database derived from literature, 28 percent of 1,084 European skulls and 81 percent of 338 sub-Saharan Africans (mean 27.3 ± 2.0 mm) have a nasal aperture equal or greater than 26 mm. With a skeletal mean of 27.1 ± 2.2 mm for 43 individuals, Pygmies do not have a wider nasal aperture than Bantus, but nasal height is lower: respectively 45.2 ± 2.8 mm for Pygmies and 48.2 ± 3.4 for non-Pygmies. Only Australian aborigines have a wider nose, but they also have a bigger skull. On a worldwide scale (but strangely not for Khoisan and Australians), nose width is correlated with a moist and hot climate (Weiner 1954; Glanville 1968; Hiernaux and Froment 1976; Noback et al. 2011). Figure 5.1 compares the Pygmy skulls with skulls of other African populations.

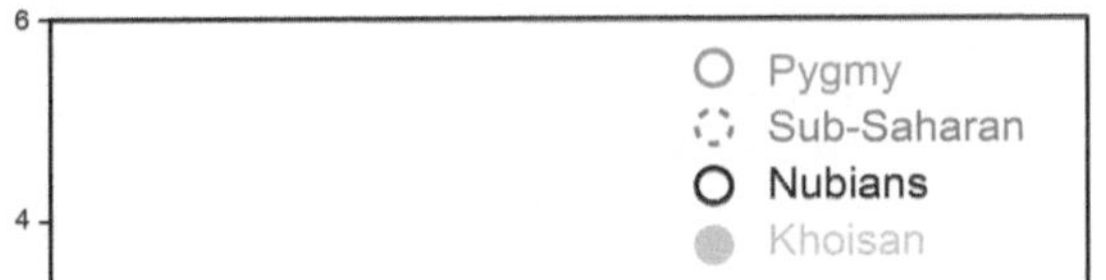

Figure 5.1. Multivariate analysis of skull dimensions comparing Pygmies, Koeesan, and sub-Saharan Africans, and Nubians.
Note: There is some overlap between Pygmies and other sub-Saharan groups. Koeesan are not close to Pygmies.

A Pygmy facial pattern is easier to detect on the living than on the skeleton (Froment 1993). Nose width is visibly larger among Pygmies because nostrils are swollen (Heymer 1992). Lips are thinner than in other African populations. Out of seventy-one populations, seventeen have a lip thickness of less than 23 mm; the ten Pygmy populations sampled all belong to this group (mean 17.8 ± 3 mm, versus 25.1 ± 2 in the rest of Africa). The European mean is 13.4 ± 1. Ear dimensions are also smaller. Though stature is reduced, head size seems proportionally less reduced, a character already visible at birth (Vincent et al. 1962). In several areas, Pygmies are named "big heads"—Poutrin (1910) reports a legend from Angola saying that "the Amboellas of Cubango say that there is downstream . . . a population of dwarfs named Ba-Quaneituata ('children of the forest'), having such a big head that they raise it with difficulty." The existence of men with big heads is also reported by Father Des Avanchers in 1866 and by Du Chaillu in 1867 (De Quatrefages 1887).

Morphological analysis of global cephalic variables is displayed in figure 5.2. It shows that the Pygmies hold a rather peripheral place in terms of the shape of the head, mainly because of their wide nose, which resembles that of Australian Aborigines. Nevertheless, they stand within the African

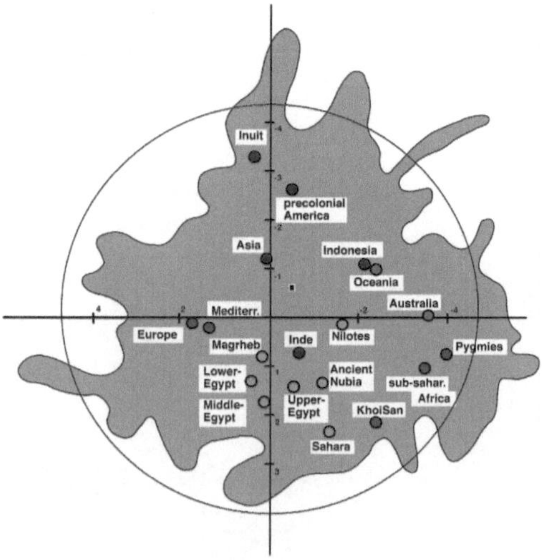

Figure 5.2. Principal component analysis of 920 masculine world populations, using Mahalanobis distances D^2 (SPSS-PC).
Note: The horizontal axis, which represents 65 percent of variance, is strongly correlated with nose breadth. African Pygmies are, within mankind variation, one of the groups with the widest nose.
Source: See Froment 1988 and 1992 for technical details.

Table 5.2. Pygmy skull measurements.

Measure mean + sd(mm)	Skull Length	Skull Breadth	Skull Height	Bizyg	Face Height	Nose Height	Nose Breadth	Frontal Diameter	Inter Orbit	Nasion Basion	Basion Prosth.
M Pygmies	1759	1346	1313	1286	640	462	274	923	210	981	1020
	58	61	57	57	41	24	20	60	29	46	57
M Non-Pygmy	1834	1341	1333	1302	700	487	275	974	249	1016	1021
	39	30	28	25	74	15	7	23	17	25	13
F Pygmies	1709	1298	1266	1209	617	440	266	906	206	955	995
	60	50	47	44	43	30	25	39	27	45	26
F Non-Pygmy	1743	1299	1277	1200	632	461	266	933	231	956	962
	41	29	21	18	18	11	9	23	8	21	29

variation, and our analysis, confirming Twiesselmann's (1942) and Marquer's (1972) observations, concludes that at the *individual* level, it was impossible to distinguish morphologically with certitude a Pygmy skull from another African skull. Table 5.2 reports skull measurements taken from only forty-one individuals (compiled from Thilmans 1968; Twiesselmann 1942; and Marquer 1972), compared to the means of thirty-three masculine and fourteen feminine sub-Saharan African non-Pygmy populations.

Stature and Body Proportions

The short stature of Pygmies has a genetic origin, but the genes related to this phenotype are still unknown (Becker et al. 2011; Jarvis et al. 2012). A conventional definition proposed by Schmidt (1905) and repeated by Schreider (1968) says that Pygmies are people measuring less than 150 cm for men and 140 cm for women; the male threshold is 148 cm in Seligman (1930) citing Haddon. Other groups, taller than 150 cm, are sometimes called pygmoids or even pygmoforms, and would be mixed-blood people (Schebesta 1940). Olivier (1950) notes that the oriental Mbuti group (east Zaïre) is shorter than the occidental group (CAR, Cameroon). This threshold is completely arbitrary, and as there is no gap between body morphology of Pygmies and other African populations, rather a progressive transition (Hiernaux 1974; Froment 1993); stature cannot be the only biological character defining Pygmies. Moreover, some farming groups in the rainforest, but also on the savanna (such as the Suku of Kasai), are shorter than pygmoids. However, because of genetic differences between "true" Pygmies (i.e., Mbuti) and pygmoids, Hiernaux maintains the distinction under the hypothesis that the reduction of stature would be proportional to the time of adaptation since the entrance in the forest (Hiernaux 1966, 1977): 174 cm among Sara of Chadian savanna, 166 among Oto, 157 among Twa, 153 among Baka, and 143 among Efe. He also states that, according to their oral traditions, the ancestral branches of Mbutis, Bingas, and Twas differed from each other more than their current descendants before two evolutionary processes made them closer: the adaptive convergence following their occupation of the same particularly constraining environment and the absorption of genes coming from dominant, slightly differentiated populations. Not only do these conclusions make superfluous the hypothesis of a unique ancestral migration of all the Pygmies and pygmoids of whom Mbutis would be the less mixed-race population, but they also dispute this hypothesis. Instead, they indicate that there was independent migration in the forest for at least three branches, respectively ancestral to Mbutis, Bingas, and Twas. There is absolutely no reason to believe these three branches migrated at the same epoch. (Hiernaux 1975)

On a series of eighty Baka men, Olivier found that only 20 percent of individuals measured less than 150 cm. Individual variation is important. In the same Bakola village, I found that one elder woman measured 130.7 cm, and a young man of the same family 170.9 cm; 11 percent of Pygmies were taller than the mean stature of their Bantu neighbors, and conversely 12 percent of villagers had a stature lower than the Pygmy average. Out of a list of 1872 world male populations (Froment, personal compilation of anthropological literature, with African data mainly from Hiernaux 1968), 298 have a stature below 160 cm in Central America (Huasteca, Otomi, Totonaque), South America (Aymara, Yanomami, Guarani), Asia (aborigenes of Taiwan, Yunan, Moi), India (Santal, Khasi), Indonesia (Sarawak, Flores, Timor, Java, Sepik), New Guinea, Melanesia, Africa (Fulero, Bira, Humu, Tembo, Mbonge, Konjo, Komo, Chokwe, Bandaka, etc.), or Labrador (Inuit), who are not perceived as related to Pygmies. All African Pygmy samples were under the 160 cm threshold, except the Kuba Cwa of the DRC measuring 160.8 cm. Compared to the rest of mankind, Pygmies display a global diminution of size affecting stature, biacromial, and bicristal diameters, with a proportionally higher bust because of shorter legs and shorter forearms. Among Konda, body proportions are harmoniously reduced, while among Binga and Mbuti, this reduction is less proportional (Hiernaux 1977). On the skeleton, though pelvis and limb bone studies are scarce, some anatomical details have been noticed (clavicle torsion, tibial retroversion, scapula shape: Matiegka and Maly 1938). Considering all body dimensions, it becomes feasible to identify the morphology of Ituri Pygmies as being at the extreme of human variation (smallest stature, widest nose), with the other subgroups (in order) Binga, Kola, and Twa linking them to other Africans and beyond, to the rest of mankind.

According to Gomila (1980), multivariate morphological studies done by Desmarais (1978) in Rwanda led to

> a picture which singularly looks like what the classic anthropologists drew a priori but which resulted from a long analysis of not only biological but socio-historical data. The starting point would be an initial collection of units of agriculturalists strongly grouped around the average. Then one or several of ancient groups of small sized hunter-gatherers, ostracized by the other groups, moved away, forming a crescent from the Sans (bushmen) in the South and after a gap, Twas and Mbutis and finally Bingas. (194)

But what can have shaped Pygmy morphology? Three main selective forces generated by the environment can be considered: food availability, transmissible diseases, and climate.

Diet and Nutritional Status

Nutritional status is the result of a complex interaction of factors: availability of food resources, energy expenditure, quality and amount of food, disease load, and social rules about food sharing.

Diet and Energy Balance

A quantitative food consumption survey was carried out by our team in 1984 among Bakola Pygmies from Cameroon and their Bantu neighbors. The main results are given in table 5.3.

Seasonal variations of weight are less marked among forest populations than among savanna populations where food shortage can be very severe because of lack of rains. However, seasonal variation in food availability exists in the rainforest, and when game is less abundant, people suffer from meat hunger (Garine 1990), which has a greater psychological impact than metabolic. Pagezy and Hauspie (1989) showed a correlation between seasonal variation in the nutritional status of children and infectious epidemics such as measles, whooping cough, and diarrhea. Regarding energy balance, Mvae farmers of southern Cameroon devote twenty minutes per day to harvest, one hour to hunt, and two hours to cultivation; women in that group devote thirty minutes to fishing activities and three hours to cultivation, but we were not able to conduct such observations among Bakola (Froment et al. 1996). Yamauchi et al. (2000) evaluated nutritional status, activity pattern, and dietary intake among the Baka of east Cameroon. He estimated a daily energy intake ranging between 5.3 MJ/d (1,260 kcal/day) and 8.2 MJ/day (1,952 kcal/day); protein intake varied from 47 to 73 g/day and fat only 6 to 10 g/day.

Table 5.3. Food consumption survey in southern Cameroon.

Units = grams/day	Iyasa Fishermen	Mvae Farmers	Bakola Pygmies
Meat	18	165	201
Fish	176	37	18
Starchy foods	723	611	744
Vegetables and fruits	33	100	54
Nuts and leguminous plants	2	26	17
Fats	22	43	10
Calories	1923	1785	1635
Ratio cal/ body weight	***34.0***	***31.3***	***35.1***

Note: A large part of the diet comes from wild forest products: bushmeat; fish; unconventional proteins (insects, caterpillars); edible leaves, fruits, and nuts (palm tree Elaeis); and tubers (wild yams). Stable isotope proportions of carbon and nitrogen, measured on organic material, such as hair or bone, discriminate very well the nature of diet according to environment and food strategies (figure 5.3). The fact that these isotopes are stable over time allows measurements on fossil bones, which is one of the bases of paleonutrition.
Source: Koppert et al. 1993.

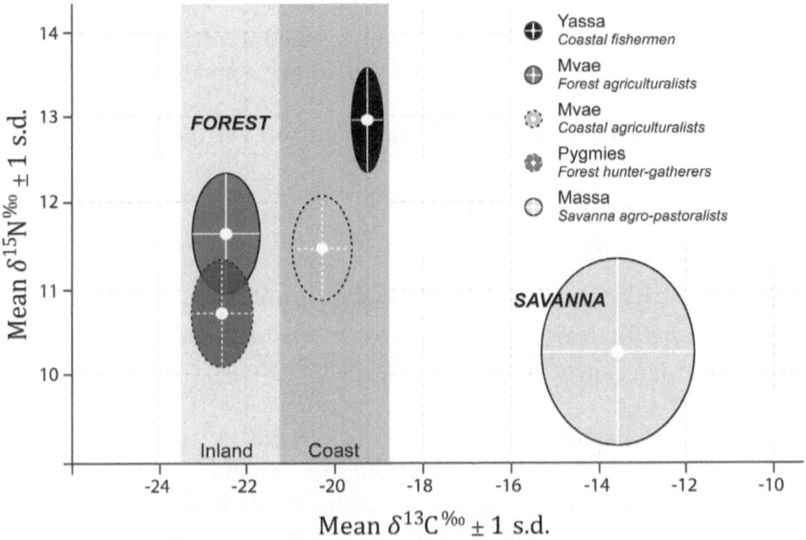

Figure 5.3. Stable isotopes of carbon and nitrogen analysis of hair in various Cameroonian populations.
Note: Carbon 13 (horizontal axis) is mainly of vegetal origin and separates clearly savanna and forest inhabitants and, to a lower extent, between open (coast) and dense forest. Nitrogen 15, which depends on animal proteins, discriminates the different diets of forest people.
Source: Froment and Ambrose 1995.

In the earliest conjectures, including Stanley's (1890), the small size of Pygmies was attributed to diet deficiencies, especially in starchy foods and proteins, but Vallois (1954) easily showed that this was not true. In our own food surveys of the Bakola group, we found a daily average consumption of meat of 200 g, reaching 285 g in adult men (Koppert et al. 1993). The debate about the dependence of Pygmies on starchy foods, namely, cassava imported from America and banana imported from Asia, is a political issue discussed by Bahuchet et al. (1991); if Pygmies rely on these cultivated crops to survive, it means that their survival independent of farmers is possible (Sato, this volume). Bahuchet (1979) noted that with 20 percent of their time devoted to cultivating villagers' fields, Akas were yielding 50 percent of their food, in fresh weight, and that 40 percent of the time was devoted to foraging. Among Bakola, who cultivate cassava themselves, our survey shows that the caloric input brought from outside is only 20 percent, but it is 65 percent among Efe (Bailey and Peacock 1988). Cassava is an inhibitor of iodine absorption. The frequency of endemic goiters among Efe in the Ituri is much lower than in villagers, 9 percent versus 43 percent, even among Efe who live in villages, which suggests a genetic mechanism; a small sample of fourteen mixed-blood subjects displayed an intermediary frequency of 29 percent (Dormitzer et al. 1989). Goiter frequencies and spleen and liver enlargement (often related to malaria and to some other liver diseases) observed among Aka are given in table 5.4.

Table 5.4. Clinical features among Pygmies and non-Pygmies in CAR.

	AKA		NGANDU	
	Children	**Adults**	**Children**	**Adults**
Goiter	24/109 22.0%	45/116 38.8%	29/54 53.7%	34/52 65.4%
Liver	28/109 25.7%	38/116 32.8%	3/54 5.6%	10/52 19.2%
Spleen	18/109 16.5%	29/116 25.0%	6/54 11.1%	2/52 3.8%

Source: Hewlett unpublished health survey 1987.

Taste sensitivity tests showed that Bakola did not perceive sweetness strongly (Hladik et al. 1986), which could be explained by a relaxation of selection because of the abundance of edible fruits in the forest compared to savanna. Regarding the adaptation to a diet rich in protein, Paolucci et al. (1973) observed a serum amino acid profile among Aka living in farmers' villages close to what is seen in phenylketonuria, but that was quickly corrected with a diet enriched in animal proteins. The gene coding for alanine glyoxylate aminotransferase (AGT), an enzyme involved in detoxication of proteins, is related to the food history of mankind (Caldwell et al. 2004). Lactase persistance has been selected with the domestication of cattle during the Neolithic—and Pygmies are lactose intolerant.

The "thrifty genotype" hypothesis proposed by Neel in 1962 (Neel 1999) aims to explain why natural selection retained diabetes genes. If mankind experienced an alternance of food shortages and abundance, a better use of glucose would reinforce insulin resistance, which is advantageous in the absence of a large amount of sugar in food, but which is at risk for diabetes if this amount increases. The "carnivore trail" hypothesis (Colagiuri and Miller 2002) is based on the low amount of carbohydrates in the diet of hunter-gatherers and herders, where proteins and lipids are dominant; in this context, insulin resistance would be an advantage because glucose is better used. But there is no clear evidence that hunters-gatherers experienced food shortages in the past, and they do not store fat during affluent periods. Man is a meat eater (Kaplan et al. 2000), and several evolutionary hypotheses have been imagined to explain this taste: *Expensive Tissue Hypothesis* (Aiello and Wheeler 1995), *Meat-adapted Genes Hypothesis* (Finch and Stanford 2004), *Reduced Muscle Mass Hypothesis* (Leonard et al. 2003), *Cooking Hypothesis* (Wrangham 2009), and *Omnivore Hypothesis* (Ungar et al. 2006). The "paleolithic diet," rich in animal proteins and fibers and poor in gluten, with no sugar, salt, and milk, is the diet that mankind ate during 99 percent of its history and is the best adapted to our physiology (Lindeberg 2012).

Nutritional Status

Data from table 5.1 allow estimation of nutritional status. It can be noted that Pygmies' adult stature represents 93 percent of non-Pygmies' height, but weight is proportionally lower, only 79 percent. The body mass index (BMI) of male Bakola Pygmies is 93.6 percent of the international standard and 91 percent of their Bantu neighbors. For women, the BMI is even lower: 19.6, that is, 88 percent of the Bantus. Bakola women, who measure 148 cm and weigh 44 kg, have a BMI of 20, a relatively low value because their fat mass is only 19 percent compared to 25 percent in Bantu women; their height is 7 cm lower than Bantu women, but their weight is 20 percent lower and their skinfolds 35 percent lower. As workload and food consumption do not seem to be very different between the two groups, it can then be supposed that diseases could explain the nutritional deficit observed in Bakola women. But overall, Pygmies have a poorer nutritional status than farmers.

Growth

The existence of small-bodied populations raises the question of a universal growth pattern in mankind. To explain adult short stature, three mechanisms can be postulated: (1) growth stops at an earlier age, (2) growth rate is reduced, or (3) growth stops or slows down during a limited period. Debates about Pygmy growth have been hampered by the remoteness of places where Pygmies live, which has two consequences: the difficulty of obtaining measurement of babies at birth and the absence of age records. For these reasons studies are rare and contradictory. Only two studies concerning growth pattern have been conducted among Pygmy children from birth to five years, a longitudinal one among Efe (Bailey 1991) and a transversal one among Baka (van Eijk 1986). A longitudinal study follows the same children for up to five years, while a transversal study observes many children, but only once. Efe babies are smaller at birth than farmers' babies, while Baka children are not different. Among the Twa, Vincent et al. (1962) also showed a small size at birth (length 44.7 ± 3.2 cm, weight 2.60 ± 0.6 kg compared to 47.9 ± 2.0 cm and 2.92 ± 0.5 kg among villagers, a highly significant difference due to in utero development. The pelvis size of mothers, estimated from the bicristal diameter measured on the living, cannot explain a small baby size, as there is only 1 cm of difference between four Pygmy populations (237 mm) and six farmer populations (248 mm).

Growth impairment is not only detected during puberty (Van de Koppel and Hewlett in Cavalli-Sforza 1986) but also early in infancy (Bailey 1991) at a much earlier period than in high-stature groups such as Tutsi (Hiernaux and Vanderborght 1956). This reduction of body size works in a negative allometry way that does not affect skull and teeth (Shea and Gomez 1988) through an endocrine mediation. There are two forms of malnutrition, wasting and stunting (Gorstein et al. 1994), and in Africa they have a different prevalence in dry and wet areas (Froment and Koppert 1999). The former is observed in savanna areas and is characterized by a deficit in weight compared to height. The latter, seen in forest areas, is a growth retardation: height and weight seem normal, but in fact the child is older than his or her aspect would suggest. It is usually difficult to score malnutrition among Pygmy children, first because their age is usually unknown and nutritional indicators of stunting rely on age, and second because a part of this stunting is not because of malnutrition but to their genetically small size. The weight-to-height ratio, which has the advantage of being independent of age, varies in Bakola children between 96 and 103 percent of the WHO standard between infancy and puberty. Later on, the nutritional status gets worse and reaches 85 percent in adults. Arm circumference and tricipital skinfold are both 10 percent lower than other Cameroonian populations.

Hiernaux's surveys among Twa, and ours among Bedzan of Tikar area in Cameroon, have noted cases of infant malnutrition in sedentarized groups; it is then possible that changes in way of life make Pygmies more vulnerable to the risks of an unbalanced diet. Because of the heavy load of parasitic diseases (Mann et al. 1962), they develop more anemias, in spite of their important meat consumption, than vegetarian populations of the Cameroonian savanna (Froment et al. 1993). Their fat mass, estimated by skinfolds, and their lean arm diameter are significantly lower in both sexes than among Bantu farmers (personal observation and Dietz et al. 1989). However, their muscular development and their physical performances are good (Pagezy 1978; Ghesquière and Karvonen 1981).

Diseases

Diseases among Pygmies are the consequence of a highly pathogenic environment, but also of great poverty (Ohenjo et al. 2006). There is a strong correlation between biodiversity and climate, and this correlation is verified for transmissible diseases, either viral, bacterial, parasititic, or mycositic (Guernier et al. 2004). Dunn (1977) noted that Australian aborigenes of the central desert, or San from Kalahari, are infected by no more than one to three parasites; this figure goes up to twenty among African Pygmies and up to twenty-two among Malayan Semang. Because of its biodiversity, the rainforest then appears dually abundant—in both food and disease (Colfer et al. 2006). Moist and hot, it acts as an incubator for transmissible diseases. In Cameroon, a country stretched between Saharan and equatorial zones, we were able to show a spectacular negative correlation between latitude and intestinal worm infections such as whipworm and roundworm: their prevalence is about 2 percent in the driest part of the country (12° north), and reaches 98 percent near the equator (see the maps in Froment and Koppert 1999). In the Ituri Forest, Mann et al. (1962) found that 36 percent of subjects were infected with a pathogenic amoeba (*Entamoeba histolytica*), while 13 percent of children were suffering another intestinal protozoon, *Giardia lamblia*, a cause of severe diarrhea.

A strong body odor is mentioned by Régnault (1911), and strangely interpreted by Coon (1965) as being selected as insect repellent, but these observations are just anecdotic. What is probable is that nomadic life inhibits crowd diseases: smallpox, measles, mumps, cholera, rubella, diphtheria, influenza, or tuberculosis are more efficiently transmitted within dense concentrations of humans, as are airborne and foodborne parasites and fecal pollution. An excessive amount of parasites (fleas, chigoes, louses, ticks) or the death of someone in the camp was a strong motivation to move to another place, which may explain that parasitic load was lighter in the past and could have increased with sedentarization. Diarrheal diseases, mainly caused by poor drinking water quality, are the main cause of children's mortality and failure to thrive. Though Pygmies' short stature is genetic, one can expect that growing in a more protected environment would lead to a gain in stature of several centimeters. Observations have shown that forest people migrating to urban environments, such as Mayan children who have emigrated from Yucatan to the United States, can experience dramatic changes in both stature (nearly 12 cm) and body proportions (Bogin et al. 2002). Regarding Pygmies, no secular trend is clear (Travaglino et al. 2011), but everywhere they live is far from urban areas where hygiene conditions could be improved; if such improvements were to be made, it is possible that the resulting reduction in parasitic load would lead to a general gain in stature.

Current hunter-gatherer societies' transition to sedentarism has been marked with a host of social and medical difficulties that are almost invariably interlinked. Their diet is altered, partly because of the scarcity of game, their limited access to arable land, and deficient experience in agriculture. Women are burdened with extra work in the fields, an activity not widely practiced in the past. Women still have low muscle mass and very limited fat mass (estimated by skinfold measurements), which constrains their energy storage capacity. Such storage is necessary for agricultural work, gathering forest products, and breast-feeding. Precarious housing and unsanitary conditions, pneumonia caused by cold and damp nights, exposure to new risks (alcoholism, smoking, venereal disease), and limited access to health care (especially by reason of poverty and segregation) are all evils that threaten the long-term existence of Bakola, as a recent outbreak of tuberculosis has recently and cruelly reiterated.

A widely accepted idea is that traditional medicine is no longer effective because healers are now unable to cope with diseases they are unfamiliar with (Voeks and Sercombe 2000). But one can wonder whether there are such new diseases, with the exception of AIDS. It is, in fact, impossible to make a complete list of diseases affecting Pygmy populations because the usually crude statistics are provided by local dispensaries, which Pygmies do not attend very often. Hewlett et al. (in Cavalli-Sforza 1986) made the interesting attempt to list 669 fatality cases reported by the Akas, and though traditional diagnoses could not fit exactly with clinical categories, some trends did emerge: the main killers are clearly respiratory (116, including 80 cases of measles) and digestive infections (140); accidents accounted for 30 deaths, hernias for 8, murders for 4, and suicides for 2. Strangely enough, fatal malaria was cited in only 11 cases.

A traditional diet—rich and varied—as well as relative isolation and mobility are all favorable features of the hunter-gatherer lifestyle that have protected Pygmies from many diseases transmissible either by direct human contact or insect vector bite. Conversely, the poverty and bad hygiene brought about in the course of sedentarization cause high mortality rates. The epidemiological and economic transition Pygmies are currently undergoing renders them vulnerable to the disadvantages of both lifestyles. Their progressive integration as citizens increased access to health care and schools, and the benefits accrued through various development projects should gradually improve the presently precarious state of health. Regular demographic monitoring would verify this hypothesis by measuring the growth of the population and the reduction of infant mortality. Objective evaluations of traditional medicines with careful case-control designs, are impossible to carry out. Regardless, if Pygmies are renowned witch doctors, their knowledge addresses only those diseases caused by nonorganic means such as witchcraft while the victims may be suffering from other ailments that are organic and curable with Western medicines. Recent social and environmental changes have undoubtedly modified the pattern of these diseases, but it is yet unclear whether the loss of rainforest habitat will ultimately increase the costs, or the benefits, to the health of Pygmy populations (Dounias and Froment 2006, 2011).

Malaria

Malaria, as well as some other diseases, may cause liver or spleen enlargement, but, as reported in table 5.5, this condition was more frequent among Akas than among their Bantu neighbors.

Because of its high mortality rate, malaria caused by *Plasmodium falciparum* has resulted in the selection of many hemoglobin variants. The most studied is a mutation called "S," the gene for normal hemoglobin synthesis being called "A." Sickle-cell disease (sickle-cell anemia) is a disease displayed by SS homozygotes, who receive the S mutation from both parents, and usually die in infancy or childhood. *Plasmodium*, the disease agent transmitted by a mosquito bite, lives in red blood cells and has less affinity for hemoglobin S than for Hb A. This explains why a lethal gene in the homozygous state is very common in some African populations. We can then predict that S gene frequency is much higher in a population exposed to the disease. The mosquito breeding sites are favored by forest clearing—so it is logical to expect a sickle cell frequency higher among farmers than among hunters. Wiesenfeld (1967) showed a relationship between farming, malaria, and sickle-cell anemia: the lower S gene frequency would indicate a lower exposition to malaria during

Table 5.5. Malaria indicators (percentage of individuals by age).

	Splenomegaly					Malaria (blood smear)				
Population	**Aka**	**Bayaka**	**Bagyeli**	**Mbuti**	**Farmers**	**Aka**	**Bayaka**	**Bagyeli**	**Mbuti**	**Farmers**
Number of individuals	*351*	*183*	*149*	*505*	*321*	*351*	*183*	*149*	*505*	*321*
0–2 years	51	40	57	15	64	44	20	57	24	55
3–12 years	79	44	84	54	69	59	50	52	44	74
Adults	66	30	73	55	42	34	20	26	24	48
Total	**67.8**	**35.5**	**75.8**	**51.9**	**55.8**	**41.8**	**30.6**	**36.2**	**29.3**	**60.7**

Source: Pampiglione and Ricciardi, in Cavalli-Sforza 1986.

the past millenia. If this gene was less frequent among Pygmies than among farmers, it would mean an increase of malaria related to deforestation since the Neolithic in cultivated areas (Coursey and Alexander 1968); however, gene S is very frequent among Mbuti. In personal surveys conducted between 1984 and 2009 among 404 Bakola (193 men and 211 women, with a mean age of twenty-nine years), 10 AS heterozygotes were found, which correspond to a low prevalence of 2.5 percent (1.7 percent in children less than sixteen years and 2.9 percent in adults). Among 629 Bantu farmers (mainly Ewondo, Banen, Bassa, Ngumba, Fang from central, coastal and southern areas of Cameroon), the carriers of the S gene were 23.5 percent. Conversely, 15 out of 100 Baka were carriers of the gene, compared to 18.3 percent of 82 Bantu of east Cameroon. There is then no clear pattern to explain such a difference between Baka and Bakola.

Other Parasites

Intestinal parasites are common in hot and humid tropical climates. In the case of roundworm or whipworm, the egg or larval stage of their life cycle often takes place in the soil, and they are then absorbed through the ingestion of food contaminated by the parasites' eggs. Hookworms and threadworms are contracted transdermally. These parasites cause or exacerbate child malnutrition and are a source of fatigue and sometimes of severe anemia in adults. The problem of fecal pollution is more acute in sedentary villages, where the areas surrounding houses are highly contaminated, mainly because latrines are seldom used, if ever. Among Bakola, although defecation takes place in nature, and unfortunately sometimes in rivers, the mobility of their small groups spares them from excessive contamination. However, we can predict that their settlement will cause a spread of these parasitic diseases, increased by the fact that deworming drug treatments are less accessible for them than for villagers. Table 5.6 shows that among Bakola compared with Mvae villagers living in the same forest, parasitism rates were actually slightly lower. It is the same for malaria and for filariae, which are

Table 5.6. Compared haematologic and parasitologic results between Bagyéli (or Bakola) and a neighboring Bantu forest population.

		n	A. Bagyeli Pygmies	n	B. Mvae Bantu
Haematology	Haemoglobin g%	76	11.8 ± 1.8	173	11.4 ± 1.6
	G-immunoglobulins g%		33.7 ± 9.2		24.2 ± 11.0
Bloodparasites	Malaria	84	2%	193	8%
	Filariae		8%		13%
Gutparasites	Roundworm	68	51%	213	69%
	Whipworm		85%		89%

also parasites transmitted by insects. The result is that anemia rates are not more severe than among villagers (although Pygmies eat more meat); however, the immunoglobulin levels, which are serum antibodies, show that the Pygmies are more exposed to bacterial and viral infections.

Table 5.7 gives the prevalence of intestinal parasites among Aka and their neighbors. Neither for protozoal infection—here pathogen amoeba *Entamoeba histolityca*—nor for roundworm and whipworm are differences between Pygmies and sedentary villagers significant, which confirms that the protection against fecal contamination offered by a nomadic lifestyle has vanished.

Regarding sleeping sickness (trypanosomiasis), a possible tolerance among Pygmies has been postulated. They are free of the disease for Bertaut (1943), yet Vallois and Marquer (1976) indicated a fatal case near Abong-Mbang in Cameroon. Cases among Bakola were also observed a decade ago in the Bipindi area of Cameroon, where a resurgence of the disease occurred. The dispersion of people explains a low infection rate in the past, but because flies circulate along the roads, risks are now increasing in perturbated areas where the disease is not controlled. It is easier for the tsetse fly (*Glossina palpalis*) to infiltrate the villages than small scattered populations. In addition, flies are attracted by the color blue, a color that does not exist in traditional Pygmy clothing, but it is frequent in secondhand clothes commonly worn now.

Among Aka, 14 percent of people were affected by cutaneous mycosis (Pennetti et al., in Cavalli-Sforza 1986); scabies is widespread among children; and chiggers is common (1986), though the agent, *Tunga penetrans*, or sand flea, is not indigenous but imported from the Americas. Serological tests showed that two parasites of animal origin, toxoplasmosis and toxocariasis, affected nearly half of the population (1986). The former is transmitted by cats, but as there are no cats in Aka villages, it may be that it comes from wild feline meat insufficiently cooked. The latter comes from dogs, the only animal domesticated by Pygmies. Among Mbutis, pathogenic filariae *Loa loa* and *Onchocerca volvulus* affected 46 percent and 79 percent of people respectively (1986), and 16 percent of Mbuti children had pinworm (*Enterobius vermicularis*).

Yaws and Other Bacterial Diseases

Venereal syphilis, like AIDS, is uncommon among pygmies because of their sexual isolation. Conversely, nonvenereal endemic treponematosis, called yaws or framboesia, a disease that could be ancestral to venereal syphilis (Froment 1994) transmitted by direct contact or flies, is highly contagious during childhood. In Bakola, in a sample of 415 subjects (198 men and 217 women) with a mean age of twenty-nine years, 13 had a weakly positive serology, and 21 had a strongly positive serology, which correspond to an infection prevalence of 8.2 percent (table 5.8). Among Bantu villagers—mainly Bassa, Ngumba, and Fang—out of 70 persons (29 men and 41 women) with a mean age forty-five, only 2 (2.9 percent) were positive at a low rate.

Table 5.7. Prevalence (percentage of individuals by age) of intestinal parasites among Aka and a neighboring farmer population.

	Amoebiasis *(E. histolytica)*					Worms ascaris / trichuris				
Popul.	**Aka**	**Bayaka**	**Bagyeli**	**Mbuti**	**Farmers**	**Aka**	**Bayaka**	**Bagyeli**	**Mbuti**	**Farmers**
Number of individuals	*354*	*183*	*149*	*505*	*321*	*299*	*155*	*111*	*586*	*400*
0–2 years	32	0	0	30	14	26 74	40 60	100 100	15 69	42 28
3–12 years	32	7	17	10	22	18 82	53 88	71 93	42 78	62 73
Adults	37	7	33	18	35	16 77	52 90	61 91	37 75	39 66
Total	**35.8**	**6.4**	**27.0**	**16.4**	**27**	**17 78**	**52 88**	**65 92**	**38 76**	**53 70**

Source: Pampiglione and Ricciardi, in Cavalli-Sforza 1986.

Table 5.8. Treponema serology, by age and sex, among Bakola.

Age Category	M	F	Total	High Rate
< 16 y	2/61	1/59	3/120 = 2.5%	3/3
16–30 y	9/61	4/66	13/127 = 10.3%	11/13
31–45 y	5/42	1/51	6/93 = 6.5%	2/6
45–60 y	4/24	6/28	10/52 = 19.2%	5/10
> 60 y	1/10	1/13	2/23 = 8.7%	0/2
Total	21/198 = 10.6%	13/217 = 6.1%	34/415 = **8.2%**	

Treponemal infection is then 2.8 times more frequent among Bakola with higher antibody levels, mainly in people younger than thirty. This difference can be explained both by poorer hygiene conditions surrounding Pygmies (the disease is transmitted by direct infected skin lesions or by flies) and by greater difficulty in accessing medical care. In both colonial and postcolonial times, large campaigns of yaws eradication have been undertaken, but only sedentary populations were easily reached. A single injection of antibiotics is sufficient to cure the infection. In our survey, however, only children older than five were sampled, which introduces a slight bias in prevalence estimates; also, the proportion of children under fifteen was 120/415 (29 percent) among Bakola and only 1/70 (1.4 percent) among Bantus. In the Likouala area of Congo, Salomone and Taglioni (2000) took this disease as an example of social discrimination; worse, in that area, gorillas affected by yaws were treated while Pygmy guides were not.

Bertaut (1943) did not see any cases of leprosy, though Trilles (1932) thought this a result of leprosy victims being abandoned; however, the latter author is known to be very unreliable. Among settled Bedzan Pygmies, I identified several cases between 1987 and 1993, and Hewlett (in Cavalli-Sforza 1986), in his census of mortality, notes four cases among the Aka.

Viral Infections

The equatorial rainforest is considered a "viral forest" (Zerner 2003). There are hundreds of pathogenic viruses, and no epidemiological survey has ever tried to study them exhaustively. Some of the most widespread and dangerous infections, such as measles and poliomyelitis, are easy to diagnose and are reliably reported by dispensaries—but there again, Pygmies are underdeclared. Retrospective serological surveys are biased by the fact that they concern survivors only. What has been observed is that measles epidemics follow the roads and induce a very high mortality, mainly, but not only, in very young children. For its survival, this virus needs to spread in large populations because a survivor becomes immunized and is no longer a transmitter. Pygmies living in small and remote camps have long been relatively safe from measles and other infections, but their migration along the roads sharply raises this risk. Studies on malaria and yellow fever, both transmitted by biting insects, also argue for geographic isolation (Jadin 1938; Chippaux and Chippaux-Hyppolite 1965). Before its control, smallpox played an important role in population survival; this disease disappeared in 1977, but it killed eleven Aka before this date. A related disease, monkey pox, is much milder, but is uncontrolled because of its wild reservoir; between January and August 2012, 727 cases, including fifteen deaths, have been declared in Democratic Republic of Congo. Preliminary results regarding hepatitis C, which is mainly transmitted by transfusions and injections, showed a 8 percent prevalence (among 86 Bakola with a mean age of thirty-eight years), identical to that of fishermen of the coastal region. Mvae farmers had an infection rate three times higher, and this could be because of a prophylactic campaigns (vaccination or anti-yaws injections) done with insufficiently sterilized syringes (Njouom et al. 2003).

The direct contacts between Pygmies (Baka and Bakola) and some Bantu with wild animals, especially primates, are frequent. These contacts can be wounds and bites incurred during hunting or by other injuries sustained during the animals' butchery, especially when women cut the meat in the kitchen (Aghokeng et al. 2010). AIDS appeared some decades ago from several viral strains present in apes, but among Pygmies it is still rare as a cause of death. A prevalence of 4 percent has been signaled both in Baka and Bakola groups, while other forest populations may have twice this level (Jackson 2007). The foamy viruses present in the saliva of great apes are transmitted by bite and infect approximately 5 percent of Pygmies (Calattini et al. 2007). A preliminary plasma survey also established that about 10 percent of Bakola had antibodies against Ebola virus, which has never been seen in the region. It is therefore an unapparent contact, possibly with weakly pathogenic strains that are potentially protective against more severe forms. Infection with HTLV-2 appears to be specific to Bakola, with a prevalence of 5 percent in the elderly (because the risk is cumulative with age); a similar virus is also endemic in some native Amazonian communities, but the reason for this coincidence is unknown. In 2005, a new human retrovirus (the fifth known), named HTLV-3, not related to HIV, was discovered in an elder Bakola living far in the forest, and two years later a second case was discovered in an elder Baka, both apparently healthy. The long term clinical consequences of such an infection are still unknown; although no epidemic threat is expected, a screening test has been developed (Calattini et al. 2011).

Chronic Diseases

Arterial hypertension is often related to stress and supposedly not frequent in foraging societies. However, in a personal survey conducted in Bakola of south Cameroon, its frequency, 15 percent of adults, was not negligible.

Salt is a traditional gift offered by visitors, and a high salt intake combined with alcohol and heavy tobacco consumption is responsible for this presumably recent condition. Malaria can impair renal glomerular function and cause blood pressure disorders. Stress itself is not absent in hunter-gatherers' life, despite the idyllic view popularized by such anthropologists as Sahlins (1972) or Turnbull (1965, 1971). Mann et al. (1962) found that only 5 percent of Mbuti adults had excessively high blood pressure, while Kesteloot et al. (1996) found no blood pressure difference between Bantus and Baka Pygmies and observed a positive correlation between high BP and increasing age. Pennetti et al. (in Cavalli-Sforza 1986) did not observe this trend among 168 Aka, but did diagnose 10 percent of individuals with high blood pressure. Hunters are said to have remarkable powers of endurance in following game, though few cardiac tests have been performed to check this claim and heart disease epidemiology is not well-known. Measures of energy expenditure by calorimetry during effort (Ghesquière and Nkiama 1993; Ferretti et al. 1993) showed great endurance in children as well as adults.

Regarding dental health, Walker and Hewlett (1990) observed that social differentiation may exist inside foraging societies and that some chronic diseases may reflect, if not an emerging hierarchy, at least some differences originating in hunting success. Most hunter-gatherers show an important dental attrition, abrasion, calculus deposit, and gingivitis; cavities are usually rare in Paleolithic foragers, but nowadays tooth health is decaying (Kent 1991). Pennetti et al. (in Cavalli-Sforza 1986) note that caries frequency is 23 percent (27 percent in men and 18 percent in women) and that 53 percent of men and 69 percent of women have no missing teeth, but that most were suffering from atrophic gingivitis.

Many other chronic diseases are not well assessed. For example, no systematic study of ophtalmological diseases is available. For hunters, good sight is vital, and color blindness is rare. Mental disorders are not absent in nonacculturated foraging societies but are very difficult to study (Doob 1965).

Combining various indicators of global health on 575 Aka of all ages, Pennetti et al. (in Cavalli-Sforza 1986) concluded that only 23 percent of males and 14 percent of females could be considered as healthy. On the basis of various health surveys confirming these poor results (Froment 2001), Bogin (2011) questioned the notion of affluence among hunter-gatherer societies—though focused on the San, his reflection can be extended to Pygmies. However, the very degraded environment and social conditions we see today did not exist in the past.

Consequences of Sedentarization

The nature of Pygmies' transition to settlement along roads has ranged from being very enthusiastic to almost forced by external pressures, though most often it is rather voluntary. Cases of forced settlements ordered by governments have existed in the past, but have failed. In some cases, it is not the Pygmies who come to the road, but the road or the logging track that comes to the Pygmies. Densification of the road network serves as a stimulus to migration, but there is no equation between road development and sedentarization because there are true "villages," not camps, settled far from roads. However, the road is a significant draw, especially as it facilitates selling game meat as well as access to health or education structures. What sedentarization means for health is

- A redistribution of activities and energy balance (less hunting = less walking, more fieldwork = postural pathologies: sciatics and hernias are frequent).
- A change in diet: less meat and fish, less fiber and vitamins, more fat and carbohydrates, less food diversity.
- An increased food security brought by agriculture; Pygmies are not very involved in cash crops such as cocoa or coffee production.
- An easy access to bars and shops selling alcohol and tobacco, leading to direct pathologies (intoxication) and indirect (weakening of the body facilitating tuberculosis and other diseases).
- Prostitution and the spread of classic sexually transmitted infections and AIDS in communities previously protected by their social isolation.
- The appearance of epidemics such as measles, transmitted by direct contact but requiring a critical mass of people around (these direct contact diseases follow the roads).
- A greater "visibility" of men for vectors of diseases (mosquitos transmitting malaria and arboviroses, flies transmitting filariae and trypanosomiasis).
- A better education, but at the expense of traditional knowledge.
- An increase in the size of the "camps," which become villages. The result is an increased promiscuity, with easier transmission of diseases, accumulation of garbage, increased intestinal parasitism (anemia, growth retardation, mental feebleness), and bacterial and viral intestinal infections and diarrheas.
- A greater exposure to malaria while living in clearings, as mosquito breeding is greatly favored by deforestation. Also, the villagers are culturally and financially more likely to sleep under mosquito nets than Pygmies
- An increase in the Pygmy population on the demographic level because of the decreasing mortality rates and a greater access to health care (clinics closer, social barriers weaker), but then accelerated population growth will also unavoidably accelerate the pressure on wild resources.

In sum, epidemiological studies have shown that sedentarization has brought some benefits to the Pygmies, such as the marketing of forest products or access to schools and population growth, but it has also exposed them to new health hazards. The advantage of hunters-gatherers' societies is that their low demographic density is insufficient for transmission of acute infections such as smallpox, influenza, polio, and measles; conversely, intimate contact with wild animals exposes them to potentially dangerous new viruses. It also protects them by "dilution" in the environment from arthropod vectors of parasites (malaria, onchocerciasis, trypanosomiasis). Nomadism in the rainforest avoids the exposition to epidemic waves that spread along the roads, and it prevents the accumulation of waste and fecal pollution around the habitat. However, the fitness of adults is balanced by a high rate of infant mortality. Overall, their life expectancy is significantly lower than in the neighboring village communities.

In Yaoundé, the capital of Cameroon, as in many urban centers in Africa, one-third of men and one-half of women in all social classes are overweight, and true obesity is seen in 5 percent and 20 percent, respectively (Pasquet et al. 2003). A few cases of overweight Baka have already been observed, and though it still concerns less than 0.5 percent of the community, even this proportion is unprecedented and is expected to grow in the near future.

Demographic Evolution

Pygmies represent a very tiny proportion of present-day African population; in Cameroon, where figures are more precise than elsewhere, they are probably less than 40,000 out of 17 million inhabitants. Nevertheless, Pygmies are the largest group of hunter-gatherers in the world. In equatorial rainforest, vegetal biomass is enormous and allows a carrying capacity of one ton of mammals by square kilometer. However, the population density is low, around one inhabitant/km^2 for hunter-gatherers, 7–10/km^2 for farmers. The average size of a Bantu village is 230–330 people for a territory of 170–300 km^2, while the Pygmy camps rarely exceed 40 people (Bahuchet 2000). Because of the high disease burden, life expectancy is short and infant mortality high. Dhellemmes (1986), a missionary who spent thirty years with Baka and built an updated census, estimated that their life expectancy at birth was around twenty-four years, less than half that of other Cameroonian populations. From an adaptive perspective, this high infant mortality is easily given to Malthusian interpretation—as a means of regulating access to finite food resources, which are collected but not reproduced. To evaluate biological adaptability where technological development is limited, WHO recommended in 1964 an urgent genetic study of hunter-gatherer groups because they were vanishing quickly. However, this statement contained an ambiguity, as it was not clear whether it was the groups who were disappearing, or their way of life. Clearly, there is no threat of extinction regarding Pygmies. On the contrary, their increasing numbers because of better access to health care has mechanically provoked a passage to agriculture (Kitanishi 2003; Leclerc 2012). Today there is no Pygmy society that relies exclusively on hunting and gathering; they are all settled in permanent villages, even if some spend the majority of their time in the forest. Prolonged breast-feeding lowers fertility by increasing birth interval, which is about four years among Akas (Cavalli-Sforza 1986). Pagezy (1978) was not able to demonstrate any fitness difference between taller Bantu women and shorter Twa pygmoid women. Among San, the fertility of malnourished women was higher (Kirchengast and Winkler 1994). Observations of many hunting-gathering societies concluded that juvenile survival increases with decreasing population density (Walker et al. 2006; Walker and Hamilton 2008).

The results of a census of Bakola people, which was based on age estimation, are described in figure 5.4. About 80 percent of the estimates were made by comparing the apparent age of individuals with others of known age in the same hamlet. Though this type of census contains an unavoidable degree of inaccuracy, a general pattern does emerge. The sex ratio is exactly 1.00: 1,439 men to 1,439 women. The Pygmy population is relatively young—about 40 percent are less than fifteen years old—which is essentially identical to the surrounding population (40–41 percent, according to the national Cameroonian census of 1987, the latest available). In fact, when one looks into the details, the Bakola population tends to die at a young age. Among the Akas, Cavalli-Sforza (1986) notes that 47 percent of the population dies before age fourteen (17 percent in the first year and 20 percent between one and six), and that people older than fifty account for only 7 percent. The number of elderly among the Bakola people is also low: only 2.7 percent are older than sixty-five, with twice as many women as men above this age. According to

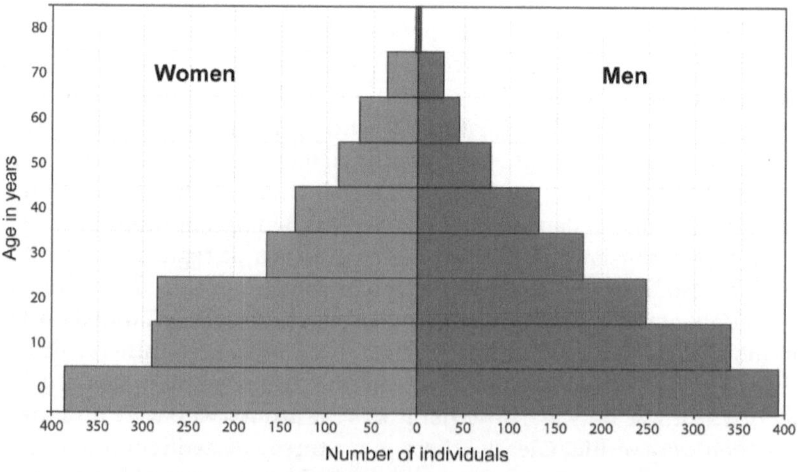

Figure 5.4. Age pyramid for the majority of the Bakola group.
Note: The importance of children is visible, though their number is limited by a heavy mortality rate before five years of life.

the 1987 census, in neighboring Bantu populations the percentage of elderly people over sixty-five years is 10.1 percent, which is almost four times more than in Pygmies.

Adaptation to Rainforest

Mechanism of Body Size Reduction

We do not know much about endocrinological mechanisms responsible for morphology, though the thyroid and pituitary gland are suspected (Crockford 2003). It has been thought that the short stature was the result of a lack of responsiveness to growth hormone because the serum levels are normal (Merimee et al. 1981; Baumann et al. 1989; Cortez et al. 1996; Bozzola et al. 2009). Only a few Babinga and Mbuti were tested for growth hormone levels. The growth factor IGF-1 (insulin-like growth factor 1), which is released under the influence of growth hormone (GH), is lower in the serum of African Pygmies but normal in a population of low stature (height 151.0 ± 3.3 cm male, weight 50.3 ± 3.4 kg) from New Guinea (Merimee et al. 1987). The circulating levels of the growth hormone binding protein (GHBP) is decreased in all Pygmy populations (Baumann et al. 1991). Some authors interpreted the short adult stature as primarily due to the absence of an adolescent growth spurt (Merimee et al. 1990), but measures have often been collected on subjects whose ages were not precisely known and on which only cross-sectional analyses were made, making a serious analysis of growth pattern impossible. A spurt is visible in transversal curves published by Van de Koppel and Hewlett (in Cavalli-Sforza 1986) among Aka and Hagino et al. (2012) among Baka. Longitudinal data that we are currently collecting in both populations thus far confirms this point (Ramirez-Rozzi and Froment, unpublished).

Experiments on cell lines taken on Efe Pygmies (Geffner et al. 1995; Hattori et al. 1996) suggested that the basis for short stature was the IGF-1 resistance and found a molecular variation at IGF-1 receptors level that they associate with this reduced responsiveness to IGF-1 (Hwa et al. 1999). Dulloo et al. (1996) found no differences in serum IGF-1 between individual Baka Pygmies and control populations and suggested that low levels of IGF-1 in Pygmies reside in differential exposure or responsiveness to environmental challenge rather than in an inherited defect in the systemic GH-IGF-1 axis. Furthermore, Bowcock and Satorelli (1990) failed to observe abnormalities of IGF-1 allelic distribution or of only 330 base pairs of the IGF-1 promoter in Pygmies. We recently found several polymorphisms associated with stature in GH receptor (GHR) and insulin-like growth factor 1 (IGF-1) genes, which have experienced divergent natural selection pressures between Pygmies and non-Pygmies. Several mutations in the GHR intron 1 show genetic differentiation between the Baka Pygmies and the Nzime Bantus: Baka have a higher proportion of the ancestral allele compared to non-Pygmies, and its derived alleles are associated with a taller stature (Becker et al. 2012). But the physiological pathways, including epigenetic processes, influencing final stature are still elusive. Lachance et al. (2012) also found some loose associations between Pygmy stature and some genes involved in pituitary function, such as TRHR (thyrotropin-releasing hormone receptor), among others.

Climate and Body Size

Reduction of body mass is observed among forest-dwelling cattle (*Bos brachyceros, Ovis jubata*), buffalo (*Syncerus caffer nanus*), antelopes (*Neotragus pygmaeus, N. batesi, Cephalophus monticola*), hippopotamus (*Hip. liberiensis*), bonobo chimpanzee (*Pan paniscus*), elephant (*Loxondota cyclotis*), crocodiles, and so on. In humans, size reduction is observed in all African hunter-gatherer populations (Pygmies, San, Hadza) and in some other tropical groups such as Andaman islanders or Negritos. If nutritional deficiencies are excluded, several explanatory hypotheses (reviewed in Perry and Dominy 2009) have been advanced, all of which are disputed. An adaptation to the moist and hot rainforest climate implies thermoregulatory rules (Hiernaux et al. 1975; Ruff 1993). The application of Bergmann's rule, stated in 1847, predicts that a smaller body is expected in moist and warm areas like tropical rainforests, where perspiration is not efficient because the air is saturated with humidity and there is no wind to help evaporation of sweat. Body mass reduction is a way to optimize the weight-to-surface, ratio, as mass varies as a cube and surface as a square. However, there are exceptions; Papua highlanders living in a fairly cold environment (Diamond 1991) also display this morphology, while San hunters are very short in a dry hot climate, where, according to Allen's rule, they should be elongated. Austin and Ghesquiere (1976) showed that larger-bodied Bantu populations had a better heat tolerance when exposed to a walk under the sun or to the immersion of an arm and a leg in warm water than Twa Pygmies. The thermoregulatory theory, which works when comparing Inuit and Nilotes, can also account for the advantage of a smaller body in equatorial forest, though various exceptions still await explanation.

Other Hypotheses

Mobility: A smaller body makes for easier mobility in dense forest, and this can explain what is observed in animals (Hiernaux 1974). A study of nineteen hunter-gatherer groups, however, failed to demonstrate any relation between relatively lower limb length and space mobility (Holliday and Falsetti 1995). As for thermoregulation, the small size of San in a very open landscape is puzzling (Diamond 1991), except if they first adapted to a forest ecosystem and moved recently in savanna, which is not demonstrated by archaeology.

Energetic economy (Hiernaux 1975; Shea and Bailey 1996): Rainforest has limited wild starchy resources, and a smaller body is more economic in energy requirements: an average Pygmy weighing 37 kg and measuring 142 cm has a basal metabolism of 5.0 MJoules (1200 kcal), while a man weighing 65 kg and measuring 172 cm requires 6.9 MJoules (1647 kcal) (Mann et al. 1962). This reduction of needs allows more people to live on the same resources and the resulting increase in population size allows for greater genetic diversity.

Vitamin deficiency (O'Dea 1993): Because of the dense canopy, there is a deficit of ultraviolet light in the rainforest. Growth impairment may result from this deficiency, as UV rays are necessary for vitamin-D synthesis. This hypothesis is untenable for several reasons: first, any lack of vitamin D would manifest itself as rickets or osteoporosis, neither of which are present among Pygmies; second, there are vitamin-D stocks in the livers of hunted animals. Finally, a deficit in UV would drive selection toward very light skin, when in fact Pygmy skin color, measured by reflectometry, has no detectable differences. Cavalli-Sforza (1986) provides a simple explanation in which less intense tanning occurs under the canopy; an explanation supported by the fact that forest Twas are also lighter than savanna Twas (1986). However, Hiernaux (1967) showed that Oto farmers were significantly darker than Twas on the inside of the arm, a place minimally exposed to sun, though Oto and Twa share the same living conditions. In Cameroon, we did not find significant differences between Baka and Nzimé Bantus.

Behavior: Alice Brues (1959) postulated that there was an affinity between body shape and hunting weapons like the spear or bow; Lee (1979) showed that small !Kung men had better hunting results than tall ones, with a strong difference after the age of thirty-five. In the same group, Winkler and Kirchengast (1994) found that slender men had a lower social position, less children with a higher offspring mortality, and a higher ratio of daughters than robust men. If shorter men can feed more children, dissemination of the short-stature genes can be expected.

Life history theory: Migliano et al. (2007), studying Negritos, proposed that an adaptation to high mortality because of environmental pressure could drive a quicker maturation through an acceleration of life history processes, with an earlier menarche allowing an increase infertility rate compromised by a reduced life span. This hypothesis triggered a debate (Becker et al. 2010; Migliano et al. 2010) that clearly requires new data to supply more indicators of maturation. Among African Pygmies data on children who have a precise age are scarce, and we are currently collecting more information on longitudinal growth among Baka and Aka.

Conclusion

Some cultural anthropologists have challenged the idea that Pygmies form a separate ethnic category. For instance, in east Cameroon, Rupp (this volume) says that nothing, either physically or in their way of life, can distinguish Baka and Bantus, except when Baka are smiling and display their chiseled teeth (Rupp 2011). This opinion is not, however, shared by all ethnologists (Bahuchet 2012). Biological anthropologists have shown that physical and genetic characteristics are not the same among Pygmies and non-Pygmies, but that they vary in a progressive, clinal way. One of the reasons for this is that many genetic exchanges exist between the two communities, even if some are hidden. Pygmies' particularism predates scientific classification, colonial or otherwise, and is clear both in the minds of the Bantu villagers and among Pygmies themselves: "Boguiel are conscious to belong to a special breed" (Loung 1959). In the Democratic Republic of the Congo, however, Oto villagers despise Twa but also refer to a common ancestry myth justifying a severe marriage interdiction (Pagezy 1988). So we see that specific nouns hide rather vague group definitions. In a provocative way, Bahuchet (1993) stated: "verily, the Pygmies do not exist" as a whole, because it is difficult to determine what is really shared, culturally and biologically, by the different groups named Baka, Babongo, Bakola, BaAka, Basua, Efe, Asua, Batwa, and so on. Ethnic identification is usually a cultural process (Amselle 1987), but for Pygmies it also involves a genetic determinism that locks them into a devaluating social system.

Today cultural changes go fast (Loung 1959; Althabe 1965). The former hunter-gatherer is now a farmer and will tomorrow be a worker, a clerk, or any other professional choice, including those of physician, architect, or lawyer. However, if sexual segregation is enduring, their physical appearance will not change and discrimination will persist. The future of Pygmy societies is a compromise between conflicting intrinsic and extrinsic factors. Intrinsic factors lie on a divergence between conservative people, who want to keep the way of their ancestors, or at least know nothing else than that way of life, and those who desire more thorough integration into the state. In the latter group, difficulties arise from interpersonal conflicts over leadership in a society that has no leaders. Extrinsic factors are numerous: the environmental changes (deforestation, rarefaction of game), evolving relationships with farmers, paternalism of churches, and the increasing role of NGOs, which often offer stereotypical solutions to "development." Pygmies can play the card of indigenous people, but that is a choice wrought with ambiguity, binding them to a stereotype of primitive or prehistoric people. Difficulty in adjusting to rapid change has led to the quick spread of alcohol and drug use. As Guillaume (1989) forecast, "Pygmies and forests are bound to be socialized and submitted to the cultural order . . . to inexorably suffer 'socialization' and 'domestication.'" Even long ago, Pales (1938) observed that "the relationships with medical doctors are the only ones which seem favorable for the future of the Babingas." Of course, closer proximity to schools and health centers means encountering and accepting changes to culture and perhaps losing oneself. The paradox lies in the fact that Pygmies are obviously isolated, genetically as well as economically (predation of resources, absence of metallurgy, pottery, and weaving). At the same time, however, they are so linked to local farmers that they speak their language, give them their allegiance, and are also to some extent reliant upon them for food. The issues surrounding Pygmies would certainly be less pronounced if, even keeping their morphology, they did not live on the fringe of the economy. Still mute, and poor despite living on rich lands, it will not be long before they raise their voices either in acceptance or refusal of the state institutions' proposals for them, proposals akin to the "taming" of the colonial age, sedentarization and integration under the current administration, and all at the cost of their freedom.

Note

1. Letter of Pharaoh Pepi II of Sixth Dynasty to his general Herkhouf (2360 av.JC); drawing on the wall of a tomb of Fifth Dynasty with the name "Aka"; Homer *Iliade* III, 6; Strabo *Geogr.* I, VII, XV, 57; Nosonnus *Dyonysos* I, II; Ovide, *Metamorphoses* I, VI, 90, Fast. I, VI; Juvenal *Sat.* XIII, 173; Stat. I, I *Styl.*6; Claudian. *Carm.*40; Aristotes *De Animalibus Historiae* VII, 12; Photius *Narrat.*40; Pliny *Hist. Nat.* VI, 35; Pomponius Mela I.III, 4 et VII, 8; Aulu Gelle *Noct. Att.* IX, 4; Herodote II, 32 et IV, 43; Philostrates *Vit. Apoll.* III, 47. Aristotle says, "It is in this district (the Nile Springs) that Pygmies live, whose existence is not a fable. There really is a species of small sized men and their horses are small too. They spend their lives in caves."

References

Aghokeng, A. F., A. Ayouba, E. Mpoudi-Ngolé, S. Loul, F. Liégeois, E. Delaporte, and M. Peeters. 2010. "Extensive Survey on the Prevalence and Genetic Diversity of SIVs in Primate Bushmeat Provide Insights into Risks for Potential New Cross-species transmissions." *Infect. Genet. Evol.* 10:386–96.

Aiello, L., and P. Wheeler. 1995. "The Expensive-tissue Hypothesis: The Brain and the Digestive System in Human and Primate Evolution." *Current Anthropology* 36:199–221.

Althabe, G. 1965. "Changements sociaux chez les Pygmées Baka de l'Est-Cameroun." *Cahiers d'Etudes Africaines* 20:561–92.

Amselle, J. L. 1987. "L'ethnicité comme volonté et comme représentation: à propos des Peul du Wasolon." *Annales ESC* 42:465–89.

Austin, D. M., and J. Ghesquiere.976. "Heat Tolerance of Bantu and Pygmoid Groups of the Zaire." *Human Biology* 48:439–53.

Bahuchet, S. 1979. "Utilisation de l'espace forestier par les Pygmées Aka." *Social Science Information* 18:999–1019.

Bahuchet, S. 1993. "L'invention des Pygmées." *Cahiers d'Etudes Africaines* 129:153–81.

Bahuchet, S. ed. 2000. *Les Peuples des Forêts Tropicales Aujourd'hui.* Rapport de Synthèse, Programme Avenir des Peuples des Forêts Tropicales, European Union, Brussells, Folon s.a., volume II.

Bahuchet, S. 2012. "Changing Language, Remaining Pygmy. *Human Biology* 84:11–43.

Bahuchet, S., D. Mac Key, and I. De Garine. 1991. "Wild Yams Revisited: Is Independence from Agriculture Possible for Rain-forest Hhunters-gatherers?" *Human Ecology* 19:213–43.

Bailey, R. C. 1991. "The Comparative Growth of Efe Ppygmies and African Farmers from Birth to Age 5 Years." *Annals of Human Biology* 18:113–20.

Bailey, R. C., and N. R. Peacock. 1988. "Efe Pygmies of Northeast Zaïre: Subsistence Strategies in the Ituri Fforest." In *Coping with Uncertainty in Fod Supply*, edited; by I. de Garine and G. A. Harrison. Oxford: Clarendon Press, 88–117.

Baumann, G., M. A. Shaw, and T. J. Merimee. 1989. Low levels of high-affinity growth hormone-binding protein in African Pygmies. *New England Journal of Medicine* 320:1705–9.

Baumann, G., M. A. Shaw, R. C. Brumbaugh, and J. Schwartz. 1991. Short stature and decreased serum Growth Hormone-Binding Protein in the mountain Ok people of Papua New Guinea. *Journal of Clinical Endocrinology and Metabolism* 72:1346–9.

Becker, N. S. A. 2012. *La faible stature des populations pygmées d'Afrique centrale: une approche évolutive.* PhD thesis anthropology, University Paris, 6.

Becker, N. S. A., P. Verdu, B. Hewlett, and S. Pavard. 2010. Can life history trade-offs explain the evolution of short stature in human pygmies? A response to Migliano et al. (2007). *Human Biology* 82:17–27.

Becker, N., P. Verdu, A. Froment, S. Le Bomin, H. Pagezy, S. Bahuchet, and E. Heyer. 2011. Indirect evidence for the genetic determination of short stature in African Pygmies. *American Journal of Physical Anthropology* 145:390–401.

Becker, N., P. Verdu, M. Georges, P. Duquesnoy, A. Froment, S. Amselem, Y. Le Bouc, and E. Heyer. 2012. The role of GHR and IGF1 genes in the genetic determination of African pygmies' short stature. *European Journal of Human Genetics* (under the press).

Bertaut, M. 1943. Contribution à l'étude des Négrilles de la région du Haut-Nyong. *Bulletin de la Société d'Etudes Camerounaises* 4:73–95.

Bogin, B. 2011. !Kung nutritional status and the original "affluent society"—a new analysis. *Anthrop. Anz.* 68:349–66

Bogin, B., P. Smith, A. B. Orden , M. I. Varela Silva, and J. Loucky. 2002. Rapid change in height and body proportions of Maya American children. *American Journal of Human Biology* 14:753–61.

Bowcock A., and V. Sartorelli. 1990. Polymorphism and mapping of the IGF1 gene, and absence of association with stature among African Pygmies. *Hum. Genet.* 85:349–54.

Boyd, W. C. 1963. Four achievements of the genetical method in physical anthropology. *American Anthropologist* 65:243–52.

Bozzola, M., P. Travaglino, N. Marzilian, C. Meazza, S. Pagani, M. Grasso, M. Tauber, M. Diegoli, A. Pilotto, E. Disabella, P. Tarantino, A. Brega, and E. Arbustini. 2009. The shortness of Pygmies is associated with severe under-expression of the growth hormone receptor. *Molecular Genetics and Metabolism* 98:310–13.

Broca, P. 1874. Les Akka, race Pygmée de l'Afrique Centrale. *Revue d'Anthropologie* III: 279–87 and 462–70.

Brothwell, D., and T. Shaw. 1971. A late Upper-Pleistocene proto-west African negro from Nigeria. *Man* 6:221–7.

Brown F., J. Harris, R. Leakey, and A. Walker A. 1985. Early *Homo erectus* skeleton from West Lake Turkana, Kenya. *Nature* 316:788–92.
Brues, A. 1959. The spearman and the archer: an essay on selection on body build. *American Anthropologist* 61: 457–9.
Buffon, G. L. L. 1749. *Histoire Naturelle, générale et particulière: De l'Homme. Bibiothèque d'Anthropologie*, reedited by F. Maspéro, Paris, 1971.
Calattini, S., A. Betsem, E. Betsem, A. Froment, Ph. Mauclère, P. Tortevoye, C. Schmitt, R. Njouom, A. Saib, and A. Gessain. 2007. Simian foamy virus transmission from apes to humans, rural Cameroon. *Emerging Infectious Diseases* 13:1314–20.
Calattini, S., E. Betsem, S. Bassot, S. A. Chevalier, P. Tortevoye, R. Njouom, R. Mahieux, A. Froment, and A. Gessain. 2011. Multiple retroviral infection by HTLV type 1, 2, 3 and simian foamy virus in a family of Pygmies from Cameroon. *Virology* 410:48–55.
Caldwell, E. F., L. R. Mayor, M. G. Thomas, and C. J. Danpure. 2004. Diet and the frequency of the alanine: glyoxylate aminotransferase Pro11Leu polymorphism in different human populations. *Human Genetics* 115:504–9.
Campbell, M. C., S. A. and Tishkoff. 2010. The evolution of human genetic and phenotypic variation in Africa. *Current Biology* 20:R166–R173.
Cavalli-Sforza, L. L. (ed.). 1986. *African Pygmies*. New York: Academic Press.
Chippaux, A., and Cl. Chippaux-Hyppolite. 1965. Immunologie des arbovirus chez des Pygmées Babinga de Centrafrique. *Bulletin de la Société de Pathologie Exotique* 58:820–33.
Colagiuri, S., J. C. and Miller. 2002. The "carnivore connection"—evolutionary aspects of insulin resistance. *European Journal of Clinical Nutrition* 56:30–5.
Colfer, C. J. P., D. Sheil, and M. Kishi. 2006. *Forests and Human Health. Assessing the Evidence*. CIFOR Occasional Paper; No. 45, Bogor, Indonesia: Center for International Forestry Research.
Coon, C. S. 1963. *The Origin of Races*. New York: A. A. Knopf.
Coon, C. S. 1965. *The Living Races of Man*. New York: A. A. Knopf.
Cortez, A. B., C. Van Dop, R. C. Bailey, N. Bersch, M. Scott, D. W. Golde, and M. E. Geffner. 1996. IGF-I resistance in virus-transformed B-lymphocytes from African Efe Pygmies. *Biochem. Mol. Med.* 58:31–6.
Coursey, D. G., and J. Alexander. 1968. African agricultural patterns and the sickle cell. *Science* 160:1474–5.
Cresta, M. 1965. Contributo alla conoscenza antropologica dei Babinga. *Quaderni de la Ricerca Scientifica* 28:81–101.
Crockford, S. J. 2003. Thyroid rhythm phenotypes and hominid evolution: a new paradigm implicates pulsatile hormone secretion in speciation and adaptation changes. *Comparative Biochemistry and Physiology—Part A: Molecular and Integrative Physiology* 135:105–29.
Dankmeijer, J. 1947. Finger prints of African Pygmies and Negroes. *American Journal of Physical Anthropology* 5:453–84.
Dart, R. A. 1937. The physical characters of the Auni= Khomani Bushmen. *Bantu Studies* 11:176–246.
De Quatrefages, A. 1874. Observations sur les races naines africaines, à propos des Akkas. *Bulletins de la Société d'Anthropologie de Paris* 2, IX:500–6.
De Quatrefages, A. 1887. *Les Pygmées*. Paris, Baillère, 238–74.

De Quatrefages, A., and E. T. Hamy. 1882. *Crania Ethnica. Les crânes des races humaines*. Paris: Baillère.
Desmarais, J. C. 1978. Le Rwanda des anthropologues: archéologie de l'idéologie raciale. *Anthropologie et Sociétés* 2:71–93.
Dhellemmes, R. P. 1985. *Le père des Pygmées*. Paris, Flammarion.
Diamond, J. M. 1991. Why are pygmies small? *Nature* 354:111–2.
Dietz, W. H., B. Marino, N. R. Peacock, and R. C. Bailey. 1989. Nutritional status of Efe Pygmies and Lese horticulturists. *American Journal of Physical Anthropology* 78:509–18.
Doob, L. W. 1965. Psychology, *in*: R. A. Lystad (ed.), *The African World*. New York: Praeger.
Dormitzer, P. R., P. T. Ellison, and H. H. Bode. 1989. Anomalously low endemic goiter prevalence among Efe Pygmies. *American Journal of Physical Anthropology*, 78:527–31.
Dounias E., and A. Froment. 2006. When forest-based hunter-gatherers become sedentary: consequences for diet and health. *Unasylva* 57:26–33.
Dounias, E., A. and Froment. 2011. From foraging to farming among present-day forest hunter-gatherers: consequences on diet and health. *International Forestry Review* 13:1–11.
Dulloo, A. G., Y. Shahkhalili, N. Mensi , J. Jacquet, and J. Girardier. 1996. Dissociation of systemic GH-IGF-I axis from a genetic basis for short stature in African Pygmies. *European Journal of Clinical Nutrition* 50:371–80.
Dunn, F. L. 1977. Health and disease in hunter-gatherers: epidemiological factors. *In*: D. Landy (ed.), *Culture, Disease, and Healing*, New York: Macmillan, 99–113.
Essomba, J. M. (ed.). 1992. *L'Archéologie au Cameroun*. Paris: Karthala.
Falk, D., C. Hildebolt, K. Smith, W. Jungers, S. Larson, M. Morwood, T. Sutikna, Jatmiko, E. Wahyu Saptomo, and F. Prior. 2009. The type specimen (LB1) of *Homo floresiensis* did not have Laron Syndrome. *American Journal of Physical Anthropology* 140:52–63.
Ferretti, G., G. Atchou, B. Grassi, C. Marconi, and P. Cerretelli. 1993. Energetics of locomotion in African pygmoids. *European Journal of Applied Physiology* 62:7–10.
Finch, C. E., and C. B. Stanford. 2004. Meat-adaptive genes and the evolution of slower aging in humans. *Quarterly Review of Biology* 79:3–50.
Fleuriot, A. 1942. Les Babinga de Mekambo. *Bulletins et Mémoires de la Société d'Anthropologie de Paris* 9, 3:101–16.
Flower, W. H. 1888. The Pygmy races of men. *Nature* 38:44–6 and 66–9.
Froment, A. 1988. Body morphology and the savanna-forest transition: a west African example. *International Journal of Anthropology* 4:61–74.
Froment, A. 1992. La différenciation morphologique de l'Homme moderne: congruence entre forme du crâne et répartition géographique du peuplement. *Comptes-Rendus de l'Académie des Sciences*, t.315, série III:323–9.
Froment, A. 1993. Adaptation biologique et variation dans l'espèce humaine: le cas des Pygmées d'Afrique. *Bulletins et Mémoires de la Société d'Anthropologie de Paris* 5:417–48.
Froment, A. 1994. Les tréponématoses: une perspective historique. *In:* O. Dutour, G. Pàlfi, J. Berato and J-P Brun (dirs.), *L'origine de la Syphilis en Europe*, Paris: Editions Errance, 260–8.

Froment, A. 2001. Evolutionary biology and health of hunter-gatherer populations. *In:* C. Panter-Brick, R. Layton and P. Rowley-Conwy (eds), *Hunters-gatherers: an interdisciplinary perspective*, Cambridge University Press, 239–66.
Froment, A. 2003. Survivre en forêt équatoriale. *In:* A. Froment and J. Guffroy (eds), *Peuplements anciens et actuels des forêts tropicales*, Editions IRD, Paris, 21–39.
Froment, A. 2008. Biodiversity, environment and health among rainforest-dwellers: an evolutionary perspective. *In:* Carol J. P. Colfer (ed.), *Human Health and Forests: A Global Overview of Issues, Practice and Policy*. Earthscan, London, 263–78.
Froment, A., and S. Ambrose. 1995. Analyses tissulaires isotopiques et reconstruction du régime alimentaire en milieu tropical. Implications pour l'archéologie. *Bulletins et Mémoires de la Société d'Anthropologie de Paris* 7:77–98.
Froment, A., and G. Koppert. 1999. Malnutrition chronique et gradient climatique en milieu tropical. *in*: S. Bahuchet, D. Bley, H. Pagezy and N. Vernazza-Licht (eds), *L'homme et la forêt tropicale*. Société d'Ecologie Humaine-APFT, Marseille, Editions de Bergier, 639–59.
Froment, A., G. Koppert, and J. F. Loung. 1993. "Eat well, live well": nutritional status and health of forest populations in southern Cameroon. in: Hladik C. M., Hladik A., Linares O., Pagezy H., Semple A. and Hadley M. Editors, *Tropical Forests: People and Food*, Man and the Biosphere Series vol.13, Parthenon-UNESCO, Paris, London, 357–64.
Froment A., I. de Garine, Ch. Binam Bikoï, and J-F Loung (eds). 1996. *Bien Manger et Bien Vivre : Anthropologie Alimentaire et Développement en Afrique intertropicale: du Biologique au Social*, Paris: L'Harmattan.
Garine, I. de. 1990. Adaptation biologique et bien-être psycho-culturel. *Bulletins et Mémoires de la Société d'Anthropologie de Paris* 2: 151–74.
Geffner, M. E., N. Bersch, R. C. Bailey, and D. W. Golde. 1995. Insulin-like growth factor I resistance in immortalized T cell lines from African Efe Pygmies. *J. Clin. Endocrin. & Metab.* 80:3732–8.
Genet-Varcin, E. 1951. *Les Négritos de l'île de Luçon (Philippines)*. Paris, Masson.
Ghesquiere, J. L., and M. J. Karvonen. 1981. Some anthropometric and functional dimensions of the pygmy (Kivu Twa). *Annals of Human Biology* 8:119–34.
Ghesquiere, J. L., and E. Nkiama. 1993. Implications of small body size to strength and endurance of Pygmy and village children in the Ituri Forest. *4th World Conference on Human Ecology*, Merida, Mexico, July 1993.
Glanville, E. V. 1968. Nasal shape, prognathism and adaptation in man. *American Journal of Physical Anthropology* 30:29–38.
Glanville, E. V. 1969. Digital ridge-counts of Efe Pygmies. *American Journal of Physical Anthropology* 31:427–8.
Gomila, J. 1980. L'Afrique subsaharienne. *In*: Hiernaux J. (ed.), *La Diversité Biologique Humaine*, Masson, Paris, 107–95.
Gorstein, J., K. Sullivan, R. Yip, M. de Onis, F. Trowbridge, P. Fajans, and G. Clugston. 1994. Issues in the assessment of nutritional status using anthropometry. *Bulletin of the World Health Organization* 72:273–83.
Guernier, V., M. E. Hochberg, and J. F. Guégan. 2004. Ecology drives the worldwide distribution of human diseases. *PLoS Biology*, 2:740–6.
Guillaume, H. 1989. "L'Etat sauvage . . .": Pygmées et forêts d'Afrique centrale. *Politique Africaine* 34:74–82.
Gusinde, M. 1956. *Die Twiden*. Vienne-Stuttgart: Wilhelm Braumüller.

Hagino, I., K. Hayashi, K. Kawamura, H. Sato, and T. Yamauchi. 2012. Adolescent growth spurt and growth pattern factors related to the short stature of Pygmy hunter-gatherers of southeast Cameroon. *Annals of Human Biology, Early Online*: 1–6.

Hamy, E. 1872. Note sur l'existence de Nègres brachycéphales sur la côte occidentale d'Afrique. *Bulletins de la Société d'Anthropologie de Paris*, 2, VII: 208–10.

Hamy, E. T. 1879. Essai de coordination des matériaux récemment recueillis sur l'ethnologie des Négrilles ou Pygmées de l'Afrique équatoriale. *Bulletins de la Société d'Anthropologie de Paris* 3, II:79–100.

Hattori Y., J. C. Vera, C. I. Rivas, N. Bersch, R. C. Bailey, M. E. Geffner, D. W. Golde. 1996. Decreased insulin-like growth factor I receptor expression and function in immortalized African Pygmy T cells. *J. Clin. Endocrinol. Metab.* 81:2257–63.

Heymer, A. 1992. Die physische Erscheinungsform der afrikanischen Pygmäen und Gedanken zur Evolution. *Mitteilungen der Anthropologischen Gesellschaft in Wien* 122: S 155–90.

Hiernaux, J. 1966. Les Bushong et les Cwa du royaume Kuba (Congo-Kinshasa): pygmées, pygmoïdes et pygméisation; anthropologie, linguistique et expansion bantoue. *Bulletins et Mémoires de la Société d'Anthropologie de Paris*, 9, XI: 299–336.

Hiernaux, J. 1967. Skin color and climate in central Africa: a comparison of three populations. *Journal of Human Ecology* 4:69–73.

Hiernaux, J. 1968. *La diversité humaine en Afrique subsaharienne. Recherches biologiques*. Ed.de l'Institut de Sociologie, Université Libre de Bruxelles.

Hiernaux, J. 1974. *The People of Africa*. New York, Scribner's Sons.

Hiernaux, J. 1975. L'adaptation biologique de l'homme à la forêt équatoriale africaine. *14th Yugoslav Congress of Anthropologists*, Zagreb.

Hiernaux, J. 1977. Long-term biological effects of human migration from the African savanna to the equatorial forest: a case study of human adaptation to a hot and wet climate. *In*: G. A. Harrison Ed., *Population Structure and Human Variation*, IBP Monograph N°11, Cambridge University Press, 187–217.

Hiernaux, J., and D. Boedhi Hartono. 1980. La position anthropologique des Hadza de Tanzanie. *L'Anthropologie*, 84:440–7.

Hiernaux, J., and A. Froment. 1976. The correlation between anthropobiological and climate variables in sub-Saharan Africa: revised estimates. *Human Biology* 48:757–67.

Hiernaux, J., P. Rudan, and A. Brambati. 1975. Climate and the weight/height relationship in sub-Saharan Africa. *Annals of Human Biology* 2:3–12.

Hiernaux, J., and H. Vanderborght. 1956. Croissance pondérale du nourrisson pendant la première année à Astrida (Ruanda). *Bull. Soc. Royale Belge d'Anthropologie et de Préhistoire* LXVII:133–9.

Hiernaux, J., E. Vincke, and D. Commelin. 1976. Les Oto et les Twa des Konda (zone de l'Equateur, Zaïre). *L'Anthropologie* 80:449–64.

Hladik, C. M., B. Robbe, and H. Pagezy. 1986. Sensibilité gustative différentielle des populations pygmées et non pygmées de forêt dense, de Soudaniens et d'Eskimos en rapport avec l'environnement biochimique. Paris, *Comptes-rendus de l'Académie des Sciences* 303:453–8.

Holliday, T. W., and A. B. Falsetti. 1995. Lower limb length of European early modern humans in relation to mobility and climate. *Journal of Human Evolution* 29:141–53.

Hwa, V., Y. Oh, and R. G. Rosenfeld. 1999. The insulin-like growth factor-binding protein (IGFBP) superfamily. *Endocrinological Review* 20:761–87.

Jackson, D. 2007. La situation sanitaire des femmes et des enfants chez les Pygmées d'Afrique centrale. *In*: Gitpa-iwgia (ed.), *Peuples autochtones d' Afrique et les objectifs du millénaire*. Paris: L'Harmattan, 71–88.

Jadin, J. 1938. Aperçu sur l'état sanitaire des Pygmées de l'Ituri. *Anthropologie* (Prague) 16:69–83.

Jarvis, J. P., L. B. Scheinfeldt, S. Soi, C. Lambert, L. Omberg, B. Ferwerda, A. Froment, J. M. Bodo, W. Beggs, G. Hoffman, J. Mezey, and S. A. Tishkoff. 2012. Patterns of ancestry, signatures of natural selection, and genetic association with stature in western African Pygmies. *PLoS Genetics* 8(4): e1002641.

Kaplan, H., K. Hill, J. Lancaster, and A. M. Hurtado. 2000. A theory of human life history evolution: diet, intelligence, and longevity. *Evolutionary Anthropology* 9:156–85.

Kazadi, N. 1981. Méprisés et admirés: l'ambivalence des relations entre les Bacwa (Pygmées) et les Bahemba (Bantu). *Africa* 51:836–48.

Keith, A. 1948. *A New Theory of Human Evolution*. London: Watts and C°.

Kent, S. 1991. Cause and effect of dental health, diet, and status among foragers. *American Anthropologist* 93:942–3.

Kesteloot, H., N. Ndam, S. Sasaki, M. Kowo, and V. Seghers. 1996. A survey of blood pressure distribution in Pygmy and Bantu populations in Cameroon. *Hypertension* 27:108–13.

Kirchengast, S., and E. M. Winkler. 1994. Nutritional status as indicator for reproductive success in !Kung San and Kavango females from Namibia. *Anthropologische Anzeitung* 54:267–76.

Kitanishi, K. 2003. Cultivation by the Baka hunter-gatherers in the tropical rain forest of central Africa. *African Study Monographs* Suppl.28:143–57.

Kollmann, J. 1894. Das Schweizerbild bei Schaffausen und Pygmäen in Europa. *Zeitschr. für Ethnol.* 26:189–251.

Koppert, G., E. Dounias, A. Froment, and P. Pasquet. 1993. Food consumption in three forest populations of the southern coastal Cameroon. In: Hladik C. M., Hladik A., Linares O., Pagezy H., Semple A., and Hadley M. Editors, *Tropical Forests: People and Food,* Man and the Biosphere Series vol.13, Parthenon-UNESCO, Paris, London, 295–311.

Lachance, J., B. Vernot, C. C. Elbers, B. Ferwerda, A. Froment, J. M. Bodo, G. Lema, W. Fu, T. B. Nyambo, T. R. Rebbeck, K. Zhang, J. M. Akey, and S. A. Tishkoff. 2012. Evolutionary history and adaptation inferred from whole-genome sequences of diverse African hunter-gatherers. *Cell* 150:457–69.

Lalouel, J. 1950. Les Babingas du Bas-Oubangui. *Bulletins et Mémoires de la Société d'Anthropologie de Paris* I, X:60–98.

Lavachery P., S. MacEachern, T. Bouimon, and C. Mbida. 2010. Kome-Kribi: rescue archaeology along the Chad-Cameroon oil pipeline, 1999–2004. *Journal of African Archaeology Monograph Series*, vol. 5.

Leclerc, C. 2012. *L'adoption de l'agriculture chez les Pygmées Baka du Cameroun. Dynamique sociale et continuité structurale.* Paris: Editions de la Maison des Sciences de l'Homme.

Lee, R. B. 1979. *The !Kung San; men, women and work in a foraging society.* Cambridge: Cambridge University Press.

Lee, R. B., and I. Devore (Eds). 1968. *Man the Hunter*. New York, Aldine, 16.

Lefrou, G. 1943. *Le Noir d'Afrique. Anthropobiologie et Raciologie.* Paris: Payot.
Leonard, W. R., M. L. Robertson, J. J. Snodgrass, and C. W. Kuzawa. 2003. Metabolic correlates of hominid brain evolution. *Comparative Biochemistry and Physiology Part A: Molecular & Integrative Physiology* 136:5–15.
Leroi-Gourhan, A., and J. Poirier. 1953. *Ethnologie de l'Union Française*, vol. 1, Paris, P.U.F, 404–21.
Lindeberg, S. 2012. Paleolithic diets as a model for prevention and treatment of Western disease. *American Journal of Human Biology* 24:110–5.
Loung, J. F. 1959. Les Pygmées de la forêt de Mill. Un groupe de Pygmées camerounais en voie de sédentarisation. *Cahiers d'Outre-Mer* XII:362–79.
Lucotte, G. 1990. *Introduction à l'Anthropologie Moléculaire. Eve était noire.* Paris: Lavoisier.
Mann, G. V., A. Roels, D. L. Price, and J. M. Merrill. 1962. Cardiovascular disease in African Pygmies: a survey of the health status, serum lipids, and diet of Pygmies in Congo. *Journal of Chronic Diseases* 15:341–71.
Mantegazza, P., and A. Zanetti. 1874. I due Akka del Miani. *Archive per l'Antropologia e l'Etnologia* IV:137–57.
Marquer, P. 1972. *Nouvelle contribution à l'étude du squelette des Pygmées occidentaux du Centre africain comparé à celui des Pygmées orientaux.* Mémoires du Muséum National d'Histoire Naturelle, série A, vol. LXXII.
Matiegka, J. and J. Maly. 1938. Etude de quatre squelettes de Pygmées centre-africains du Bassin de l'Ituri. *L'Anthropologie* 48:237–48 and 521–638.
Mercader, J., M. D. Garralda, O. M. Pearson, and R. C. Bailey. 2001. Eight hundred-year-old human remains from the Ituri tropical forest: Democratic Republic of Congo: the rock shelter site of Matangai Turu Northwest. *American Journal of Physical Anthropology* 115:24–37.
Merimee, T., G. Baumann, and W. Daughaday. 1990. Growth hormone–binding protein: II. Studies in pygmies and normal statured subjects. *J. Clin. Endocrinol. Metab.* 71:1183–8.
Merimee, T. J., J. Zapf, and E. R. Froesch. 1981. Dwarfism in the Pygmy: an isolated deficiency of insulin-like growth factor I. *New England Journal of Medicine* 305:965–8.
Merimee, T. J., J. Zapf, B. Hewlett, and L. L. Cavalli-Sforza. 1987. Insulin-like growth factor in Pygmies: the role of puberty in determining final stature. *New England Journal of Medicine* 316:906–11.
Migliano, A. B., L. Vinicius, and M. M. Lahr. 2007. Life history trade-offs explain the evolution of human pygmies. *Proc. Natl. Acad. Sci. USA* 104:20216–9.
Migliano, A. B., L. Vinicius, and M. M. Lahr. 2010. Why are Pygmies so short? A defense of Migliano's hypothesis. *Human Biology* 82:109–13.
Montandon, G. 1933. *La Race, les Races.* Paris: Payot.
Neel, J. V. 1999. The "thrifty genotype" in 1998. *Nutrition Reviews* 57:S2–9.
Ngima Mawoung, G. 1993. *Le système alimentaire des groupes pygmées Bakola de la* Région de Campo (Sud-Ouest du Cameroun), Université de Paris-Sorbonne, PhD thesis in Anthropology.
Njouom, R., C. Pasquier, A. Ayouba, A. Gessain, A. Froment, J. Mfoupouendoun, R. Pouillot, M. Dubois, K. Sandres-Saune, J. Thonnon, J. Izopet, and E. Nerrienet. 2003. High rate of hepatitis C virus infection and predominance of genotype 4 among elderly inhabitants of a remote village of the rain forest of south Cameroon. *Journal of Medical Virology* 71:219–25.

Noback, M. L., K. Harvati, and F. Spoor. 2011. Climate-related variation of the human nasal cavity. *American Journal of Physical Anthropology* 145:599–614.

O'Dea, J. D. 1993. Possible contribution of low ultraviolet light under the rainforest canopy to the small stature of Pygmies and Negritos. *Homo* (Stuttgart) 44:284–7.

Odambo-Adone, L. F. 1990. Quand un Pygmée proclame sa pygmitude et réclame ses droits. *Africa International* 229:77.

Ohenjo, N., R. Willis, D. Jackson, C. Nettleton, K. Good, and B. Mugarura. 2006. Health of indigenous people in Africa. *The Lancet* 367:1937–46.

Olivier, G. 1950. Anthropologie physique des Négrilles du Cameroun. *Bulletin de la Société d'Etudes Camerounaises* III:113–8.

Oslisly, R. 2006. Les traditions culturelles de l'Holocène sur le littoral du Cameroun entre Kribi et Campo. *In*: H-P. Wotzka (ed.), Grundlegungen: Beiträge zur europäischen und afrikanischen Archäologie für Manfred K. H. Eggert. Tübingen, Francke, 303–17.

Pagezy, H. 1978. Morphological, physical and ethoecological adaptation of Oto and Twa women living in the equatorial forest. *Journal of Human Evolution* 7:683–92.

Pagezy, H. 1988. *Contraintes nutritionnelles en milieu forestier équatorial liées à la saisonnalité et la reproduction: réponses biologiques et stratégies de subsistance chez les Ba-Oto et les Ba-Twa du village de Nzalekenga (Lac Tumba, Zaïre).* PhD thesis, Aix-Marseille University, France.

Pagezy, H., and R. Hauspie. 1989. Longitudinal study of growth in weight of African babies: an analysis of seasonal variations in the average growth rate and the effects of infectious diseases on individual and average growth patterns. *Acta Paediatrica Scandinavia* suppl. 350: 37–43.

Pales, L. 1938. Contribution à l'étude anthropologique des Babinga de l'AEF. *L'Anthropologie* 48:503–20.

Paolucci, A. M., M. A. Spadoni, and V. Pennetti. 1973. Modification of serum-free amino acid patterns of Babinga adult Pygmies after short-term feeding on a balanced diet. *American Journal of Clinical Nutrition* 26:429–34.

Pasquet, P., L. Temgoua, F. Melaman-Sego, A. Froment, and H. Rikong-Adié. 2003. Prevalence of overweight and obesity for urban adults in Cameroon. *Annals of Human Biology* 30: 551–62.

Perry, G., and N. J. Dominy. 2009. Evolution of the human Pygmy phenotype. *Trends in Ecology and Evolution* 24: 218–25.

Poutrin, L. 1910, 1911, 1912. Contribution à l'étude des Pygmées d'Afrique: les Négrilles du Centre africain. *L'Anthropologie* 21:435–504; 22: 421–549; 23: 349–415.

Régnault, M. 1911. Les Négrilles de la Sangha. *L'Anthropologie* 22:261–88.

Reichholf, J. 1997. *Das Rätsel der Menschwerdung: Die Entstehung des Menschen im Wechselspiel mit der Natur*. Verlag: Dtv.

Ribot, I., R. Orban, and P. De Maret. 2001. The prehistoric burials of Shum Laka rockshelter (north-west Cameroon). *Annales du Musée Royal des Sciences naturelles de Belgique*, vol. 164.

Ruff, C. B. 1993. Climatic adaptation and hominid evolution: the thermoregulatory imperative. *Evolutionary Anthropology* 2:53–60.

Ruggles Gates, R. 1958. The African Pygmies. *Acta Genetica Medica et Gemellologica*, 7:159–218.

Rupp, S. 2011. *Forests of Belonging: Identities, Ethnicities, and Stereotypes in the Congo River Basin.* University of Washington Press.
Sahlins, M. 1972. *Stone Age Economics.* New York: Transaction Publishers.
Salomone, G., and F. Taglioni. 2000. La marginalisation sanitaire des îlots Pygmées de la Likouala (Congo). http://fig-st-die.education.fr/actes_2000/salomone/article.htm.
Schebesta, P. 1938. *Die Bambuti Pygmäen vom Ituri.* Mémoire de l'Institut Royal Colonial Belge, vol.1, Falk, Bruxelles.
Schebesta, P. 1940. *Les Pygmées.* Paris, Gallimard.
Schmidt, E. 1905. Die Grösse der Zwerge und die sogenannten Zwergvölker. *Globus* 87:121–5.
Schmidt, W. 1910. Stellung der Pygmäenvölker. *Studien und Forsch. z. Menschen,* vol. VI and VII.
Schreider, E. 1968. Growth Hormone in African Pygmies. *The Lancet,* January 13, 1968 (letter to the editor).
Schwalbe, G. 1906. Zur Frage der Abstammung des Menschen. *Zeitsch. für Morph. und Anthrop.*, 9–80.
Schweinfurth, G. 1875. *The Heart of Africa.* Leipzig and London.
Seligman, C. G. 1930. *Races of Africa.* University Library of Modern Knowledge. Thornton Butterworth, London (NB: the quotation has been translated from the French edition of 1935).
Shea, B. T., and A. M. Gomez. 1988. Tooth scaling and evolutionary dwarfism: an investigation of allometry in human Pygmies. *American Journal of Physical Anthropology* 77:117–32.
Shea, B. T., and R. C. Bailey. 1996. Allometry and adaptation of body proportions and stature in African Pygmies. *American Journal of Physical Anthropology* 100:311–40.
Stanley, H. M. 1890. In Darkest Africa, or the Quest, Rescue, and Retreat of Emin Governor of Equatoria. London: Sampson Low, Marston, Searle and Rivington.
Thilmans, G. 1968. Recherches craniométriques sur l'origine des pygmées d'Afrique. *Bulletin de l'IFAN,* ser. B, XXX:401–28.
Tinland, F. 1968. *L'Homme Sauvage. Homo ferus et Homo sylvestris, de l'animal à l'Homme.* Paris: Payot.
Tishkoff, S. A., F. A. Reed, F. R. Friedlaender, C. Ehret, A. Ranciaro, and A. Froment, et al. 2009. The genetic structure and history of Africans and African Americans. *Science* 324:1035–44.
Toerien, M. J. 1961. Bush-Bantu hybrids and central African Pygmies. *South African Journal of Science* 57: 215–7.
Travaglino, P., C. Meazza, S. Pagani, G. Biddeci, and M. Bozzola. 2011. Secular trends in growth of African Pygmies and Bantu. *Hormones* 10:144–8.
Trilles, R. P. 1932. *Les Pygmées de l'Afrique Equatoriale.* Paris: Bloud et Gay.
Turnbull, C. M. 1965. *Wayward servants, the two worlds of the African Pygmies.* New York: The Natural History Press
Turnbull, C. M. 1971. *The Forest People.* New York: Simon and Schuster.
Twiesselmann, F. 1942. *Contribution à l'étude anthropologique des Pygmées de l'Afrique Occidentale.* Mémoires du Musée Royal d'Histoire Naturelle de Belgique, 2è série, fasc.27.

Twiesselmann, F. 1958. *Les ossements humains du gîte mésolithique d'Ishango.* Mémoires de l'Exploration du Parc National Albert, fasc.5, Bruxelles, 125 p. (summary in *Zeitung für Morphologie und Anthropologie* 1959, 50, 402).

Ungar, P. S., F. E. Grine, M. F. Teaford, and S. El Zaatari. 2006. Dental microwear and diets of African early *Homo. Journal of Human Evolution* 50:78–95.

Vallois, H. V. 1940. New research on the western Negrilles. *American Journal of Physical Anthropology* 26:449–71.

Vallois, H. V. 1954. Carence en viande et "pygméisation." *L'Anthropologie* 58:571–3.

Vallois, H. V., and P. Marquer. 1976. *Les pygmées Baka du Cameroun: anthropologie et ethnographie, avec une annexe démographique.* Mémoires du Muséum National d'Histoire Naturelle Paris, vol. 100.

Van Eijk, R. 1987. *Naître et Croître au Cameroun.* Doctoral dissertation in Medicine, University of Utrecht.

Vansina, J., 1985. L'homme, les forêts et le passé en Afrique. *Annales ESC* 6:1307–34.

Vansina, J. 1990. *Paths in the Rainforests: Towards a History of Political Tradition in Equatorial Africa.* London: J. Currey.

Vecchi, F. 1981. Geographical variation of digital dermatoglyphics in Africa. *American Journal of Physical Anthropology* 54:565–80.

Verneau, R. 1896. De la pluralité des types ethniques chez les Négrilles. *L'Anthropologie* 7: 153–67.

Vigilant, L., R. Pennington, H. Harpending, T. D. Kocher, and A. C. Wilson. 1989. Mitochondrial DNA sequences in single hairs from a southern African population. *Proceedings of the National Academy of Sciences of the USA* 86:9350–4.

Vincent, M., C. Jans, and J. Ghesquiere. 1962. The new-born Pygmy and his mother. *American Journal of Physical Anthropology* 20:237–47.

Virchow, R. 1894. Über Zwergrassen. *Mitteilungen der Anthropologischen Gesellschaft in Wien*, XXIV, 134–8.

Voeks, R. A., and P. Sercombe. 2000. The scope of hunter-gatherer ethnomedecine. *Social Science & Medicine* 51:679–90.

Walker, P. L., and B. S. Hewlett. 1990. Dental health, diet and social status among central African foragers and farmers. *American Anthropologist* 92: 383–98.

Walker, R. S., and M. J. Hamilton. 2008. Life-history consequences of density dependence and the evolution of human body size. *Current Anthropology* 49:115–22.

Walker, R., M. Gurven, K. Hill, A. Migliano, N. Chagnon, R. De Souza, G. Djurovic, R. Hames, A. M. Hurtado, H. Kaplan, K. Kramer, W. J. Oliver, C. Valeggia, and T. Yamauchi. 2006. Growth rates and life histories in twenty-two small-scale societies. *American Journal of Human Biology* 18:295–311.

Weiner, J. S. 1954. Nose shape and climate. *American Journal of Physical Anthropology* 12: 615–8.

Wiesenfeld, S. L. 1967. Sickle cell trait in human biological and cultural evolution. *Science* 157:1134–40.

Winkler, E. M., and S. Kirchengast. 1994. Body dimensions and differential fertility in !Kung San males from Namibia. *American Journal of Human Biology* 6:203–13.

Wrangham, R. 2009. *Catching Fire. How Cooking Made Us Human.* New York: Perseus Books.

Yamauchi, T., H. Sato, and K. Kawamura. 2000. Nutritional status, activity pattern, and dietary intake among the Baka hunter-gatherers in the village camps in Cameroon. *African Study Monographs* 21:67–82.

Zerner, C. 2003. The viral forest in motion. *In*: Candace Wade, ed., *In Search of the Rainforest.* Durham: Duke University Press, 246–83.

6

Foraging Lifestyle in the African Tropical Rainforest

Hiroaki Sato

Introduction

In the late 1980s, researchers of hunter-gatherer societies engaged in a lively discussion about the hypothesis of the "Wild Yam Question," that it was impossible for human beings to live without agricultural products in a tropical rainforest (Headland and Bailey 1991). Based on ecological surveys of the Ituri Forest, Hart and Hart (1986) pointed out that the interiors of rainforests were poorer in both animal and botanical food resources than either their fringe or the forest-savanna zone. Headland (1987) made the assertion that human beings could not live independently of agricultural products in tropical rainforests, where starchy foods like wild yam tubers were scarce. Bailey et al. (1989) argued that there were neither pure hunter-gatherer groups in tropical rainforests in the present day, nor evidence that they had lived there in the past. In addition, Headland (1997) mentioned that modern-day hunter-gatherer societies have long histories of trade and interaction with the dominant societies around them. In contrast, Bahuchet et al. (1991), Brosius (1991), Dwyer et al. (1991), Endicott et al. (1991), and Stearman (1991) countered the hypothesis on the bases of their respective fieldwork in a special issue of *Human Ecology* concerned with the Wild Yam Question. With regard to Africa, Bahuchet et al. (1991) made the point that the tropical rainforests in the west of the Republic of Central Africa were composed of diverse vegetation and that in some areas the productivity of wild yam tubers was sufficient to support a local group of hunter-gatherers, offering the Aka hunter-gatherers as an example. Because of limited data on a pure foraging lifestyle in tropical rainforests, however, the argument did not reach any conclusion. In the late 1990s, the research team of Mercader excavated one site after another in the African tropical rainforests (Mercader et al. 2000; Mercader et al. 2001a; Mercader et al. 2001b; Mercader 2002; Mercader et al. 2003a; Mercader 2003b). They found evidence that the sites in the African equatorial areas, which have been dense forest environments since 10,000 years BP, and open forest before that—even during the

Leopoldvillian cold-climate period—were used continuously. The works of Mercader's team raised the possibility of a purely foraging lifestyle, at least in African tropical rainforests. However, the archeological and ecological data associated with the foraging life in tropical rainforests are still too few to establish its possibility. We have no clear idea of how hunter-gatherers could live independently of farmed agricultural products in tropical rainforests. Further archeological records and ecological and behavioral data of foraging life in tropical rainforests are needed, as noted by Bailey and Headland (1991).

Since 1995, in order to examine the potentiality of tropical rainforests as a human habitat, we have surveyed the distribution and reserves of yams and yam-like plants in the tropical rainforest of southeastern Cameroon. Our findings support Bahuchet's view that tubers in the area are sufficient to support at least a small hunter-gatherer group (Sato 2001; Sato 2006). Even if, however, yam and yam-like tubers are plentiful, we do not have practical and concrete data on how hunter-gatherers act and live in a tropical rainforest. Indeed, there are currently no hunter-gatherer groups who live a pure foraging lifestyle in the African tropical rainforest, nor any historical record of it, as Bailey et al. (1989) observed. Pygmy people, who have been considered the most likely candidates among the various indigenous hunter-gatherer groups in this forest, presently engage in diverse subsistence activities including ones involving agricultural products to maintain their livelihood. Even when going on a hunting trip in the forests, they usually take field crops. Therefore we designed observational surveys of controlled foraging trips, during which no agricultural or commercial food (except salt and pepper) could be used, to verify the wild yam question and to reveal details of a "pure" hunting and gathering life without agricultural products. In 2001, after first gaining the cooperation of the Baka people, a Pygmy group inhabiting southern Cameroon, we conducted a preliminary ten-day survey to gain the knowledge necessary for carrying out controlled foraging trips. We then made two twenty-day trips, one in 2003 and another in 2005, and one fourteen-day trip in 2010 in the forests of southeastern Cameroon.

In this paper examining which food resources the Baka cooperators acquired as well as the costs of acquisition, we test the wild yam question hypothesis. In the case of negative findings, we will discuss what features of the foraging lifestyle have been revealed and their implications for reconstructing the Stone Age foraging lifestyle in the African tropical rainforest.

Controlled Foraging Trips

All the Baka cooperators lived in a small sedentary settlement near Ndongo, a farming village in the Moloundou Subprefecture inhabited mainly by the Bakwele. The Baka inhabitants, about 300 people, are the largest ethnic group around Ndongo. At present they settle along a main road, keep banana and cacao fields, and get some money, local wine, and food by helping local farmers

(Kitanishi 2003). They also customarily enter the forest to catch animals, chiefly using snares, and collect seed nuts such as wild mango seeds for one to two months a year. Field crops such as plantains, bananas, or cassava tubers are also taken along during these trips.

We chose the cooperators from among the research assistants who had helped with our wild yam distribution surveys in 1995 and 1996, and others were chosen based on nominations from this initial group. All cooperators had been accustomed to camping in forests and were familiar with each other intimately. These conditions were of critical importance as camping in the forests without field crops, which most cooperators had never experienced, would last for two or three weeks during our controlled foraging trips.

The cooperators in the first trip consisted of six married couples and four children between three and seven years old; in the second were eight married couples and seven children three to ten years old; and in the third, eight couples, a boy and a girl both over twelve years, and two other children under ten years old (table 6.1). Couples with very small children were excluded in consideration of the potential risks related to the group's inexperience in extended camping trips without agricultural products as a back-up food source.

Before each trip, we informed the cooperators on the details of our research and that they would be able to withdraw from the trip in the event of some problem occurring. We asked for each cooperator's voluntarily participation in this study and gained their consent.

The first trip was conducted in August 2003 (Sato et al. 2006), the second in October 2005 (Sato et al. 2012), and the third in April 2010. On the first two trips, the Baka cooperators set up their campsite on the bank of the Leke River, which flows through the foot of Mt. Bek, two days on foot from their settlement (figure 6.1). The campsite of the third trip, however, was established on the bank of the stream Mesimebem, only two hours on foot from their settlement, as the previous one could not be used because of new zoning laws in Cameroon. During our first two trips, people rarely visited the forest around Mt. Bek, whereas the forest around the campsite of the third trip was used daily by the Baka people for snaring or short hunting trips. The forests around both Mt. Bek and the third campsite were semi-deciduous, receiving annual rainfall of less than 1,600 mm (Letouzey 1985). As shown in figure 6.2, this survey area has four seasons: a minor rainy season (April to May), a minor dry season (June to August), a major rainy season (September to November), and a major dry season (December to March). The mean monthly temperature is 25°C year-round.

The survey periods were as follows: the first trip lasted August 16–September 5, 2003, (minor dry season); the second was October 2–23, 2005 (major rainy season); and the final trip was April 14–28, 2010 (minor rainy season). The third trip was limited to fourteen days out of concern that the forest was under high pressure from hunting and gathering and because the

Table 6.1. Baka cooperators in three controlled foraging trips.

Couple no.	**Individual**	**Sex**	**Age**	**Child.**	**Couple no.**	**Individual**	**Sex**	**Age**	**Child.**	**Couple no.**	**Individual**	**Sex**	**Age**	**Child.**
Aug. '03	**no.**		**estimated**		**Oct. '05**	**no.**		**estimated**		**Apr. '10**	**no.**		**estimated**	
DF1	DF11	M	45–50	1 boy	RF1	RF11	M	47–52	1 boy	SRF1	SRF11	M	52–57	
	DF12	F	45–50			RF12	F	47–52			SRF12	F	52–57	
DF2	DF21	M	25–30	1 boy[1], 1 girl	RF2	RF21	M	27–32	1 boy[1], 1 girl	SRF2	SRF21	M	32–37	1 boy, 1 girl
	DF22	F	25–30			RF22	F	27–32			SRF22	F	32–37	
DF3	DF31	M	35–40		RF3	RF31	M	37–42		SRF3	SRF31	M	42–47	
	DF32	F	35–40			RF32	F	37–42			SRF32	F	42–47	
DF4	DF41	M	30–35	1 girl	RF4	RF41	M	32–37		SRF4	SRF41	M	52–57	
	DF42	F	25–30			RF42	F	27–32			SRF42	F	52–57	
DF5	DF51	M	25–30		RF5	RF51	M	45–50		SRF5	SRF51	M	50–55	
	DF52	F	20–25			RF52	F	45–50			SRF52	F	50–55	
DF6	DF61	M	35–40		RF6	RF61	M	25–30		SRF6	SRF61	M	45–50	
	DF62	F	35–40			RF62	F	35–40			SRF62	F	40–45	
					RF7	RF71	M	25–30	2 girls	SRF7	SRF71	M	20–25	
						RF72	F	25–30			SRF72	F	15–20	
					RF8	RF81	M	30–35	2 boys[2]	SRF8	SRF81	M	42–47	1 boy, 1 girl
						RF82	F	25–30			SRF82	F	42–47	

DF1, RF1, and SRF1, DF2, RF2 and SRF2, DF3, RF3 and SRF3, DF4 and RF4, DF6 and SRF8, and RF5 and SRF5 are the same families respectively.
[1]an infant; [2]including an infant.

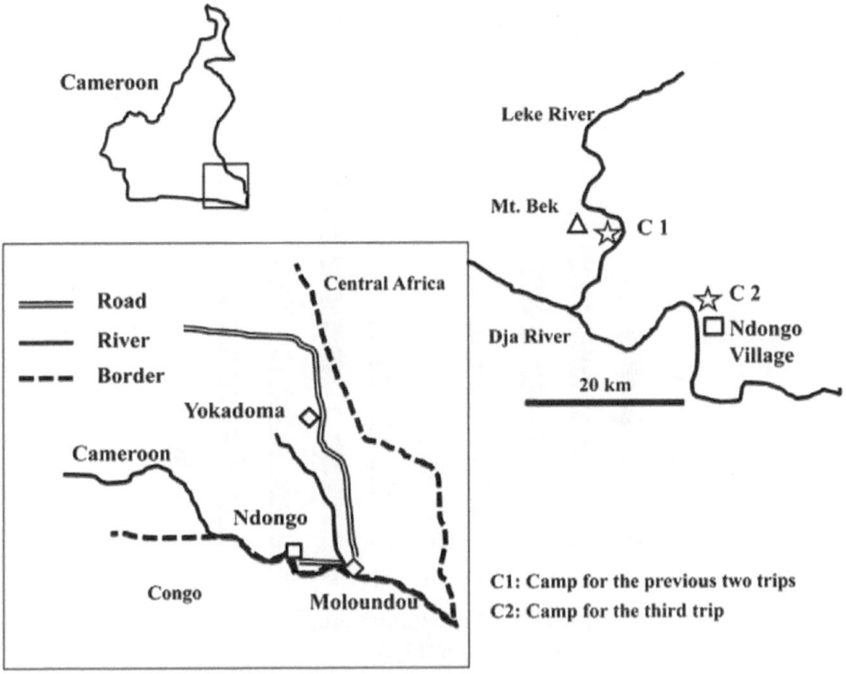

Figure 6.1. Survey areas.

use of wire snares was prohibited. During the survey periods, the research team members lived off agricultural and commercial foods exclusively.

On each of the trips, the cooperators used their usual tools for foraging, including machetes, iron spears, wooden digging sticks (instantly fitted with iron attachments when needed for digging up yam tubers), wire snares, iron axes, hook and line, fishing nets, carrying baskets, and so on. On the third trip, however, the cooperators did not use wire snares because they had become forbidden by a domestic law.

During the survey period, the cooperators collected wild yam and yam-like tubers, wild honey, termites, edible fungi, nuts, and the like. In addition, they set snares (on the first two trips) and fished. The work done exclusively by men included honey collecting, snaring, net fishing, and hook-and-line fishing, while only women performed dam-and-bail fishing. Both men and women collected yam tubers, termites, nuts, and edible fungi.

Usually adult men had meals together at a meeting place called *mbanjo* in the center of the campsite, both in the morning and evening. Women cooked the men's meals but ate separately with their children in front of their own huts. As husbands shared the food at the *mbanjo* and wives often exchanged plates with each other, it was unlikely that any individual suffered a lack of food.

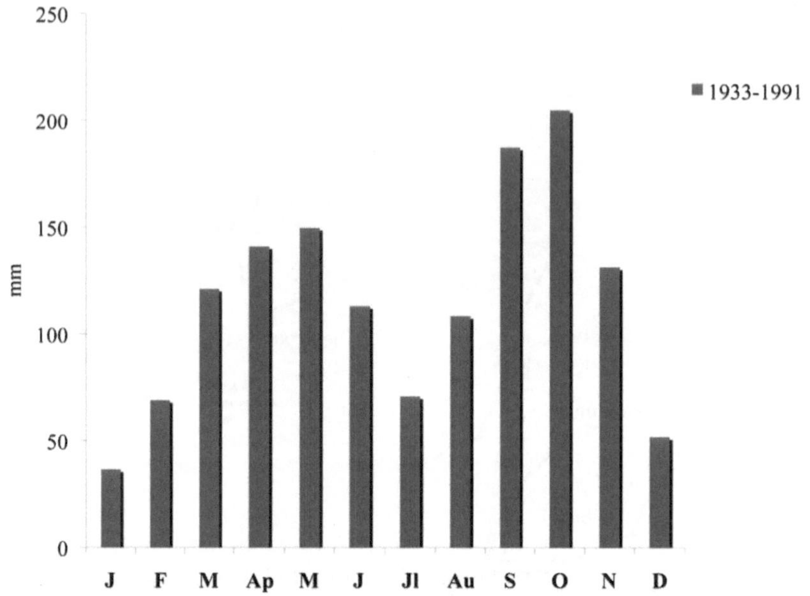

Figure 6.2. Mean monthly rainfall in Moloundou.
Source: Adapted from Sigaha-Nkamdiou 1993.

We checked the cooperators' body weight every morning before breakfast. All food brought back to the camp was identified and weighed. All rotten and discarded food was weighed in addition to that which remained at the end of the survey. Estimating the total food intake of all participants in the camp, we calculated both the total and per capita energy intake. We also timed each cooperator's trips away from camp from exit to return, and measured the number of steps each walked every day with pedometers.

Possibility of Foraging Lifestyle in the African Tropical Rainforest

Food brought back to the camp during the three survey periods consisted of various food types: yams and yam-like tubers (six species), mammals (twelve species), freshwater fish (nineteen vernacular names), termites (one vernacular name), honey (one species and two vernacular names), nuts (four species), edible fungi (eighteen vernacular names), and so on. Among these, yam and yam-like tubers ranked highest in weight in all three seasons. Mammals ranked second in both the minor dry and the major rainy seasons, followed by fish in the former and nuts in the latter. Honey ranked second, and fish ranked third in the minor dry season, when very few mammals were caught (table 6.2).

We derived the total food intake of cooperators and their children from the weight of all food acquired, less any leftovers—the bulk of which was

Table 6.2. The fresh weight of food brought into the camp.

Food type	8/16–9/5/'03 (16 persons)		10/2–23/'05 (23 persons)		4/14–28/'10 (20 persons)	
	Weight kg	Composition %	Weight kg	Composition %	Weight kg	Composition %
Yam tubers	660.7	59.4	1092.9	74.7	576.4	87.2
Game	343.1	30.9	290.6	19.9	2.7	0.4
Nuts	23.0	2.1	44.3	3.0	9.7	1.5
Fungi	3.0	0.3	8.7	0.6	1.3	0.2
Reptiles	3.6	0.3	0.0	0.0	7.7	1.2
Fish	49.5	4.5	12.9	0.9	17.9	2.7
Crutaceae	0.3	0.0	0.1	0.0	4.5	0.7
Termites	5.9	0.5	9.3	0.6	0.0	0.0
Snails	0.8	0.1	1.7	0.1	11.7	1.8
Honey	8.0	0.7	3.1	0.2	26.4	4.0
Others	13.9	1.3	0.2	0.0	2.9	0.4
Total	1111.8	100.0	1463.8	100.0	661.0	100.0

preserved dry meat taken back to their settlement—and the amount of food discarded during each survey period. We estimated the total energy intake at 2,528–2,864 kcal per consumption day in the minor dry season, at 2,479–2,777 kcal in the major rainy season, and at 2,000–2,184 kcal in the minor rainy season (table 6.3). The total energy intake per consumption day in the minor rainy season was much smaller than those in the minor dry and major rainy seasons because very little game meat and a fairly small amount of nuts were brought into the camp. Yam and yam-like tubers supplied dietary energy of over 1,500 kcal and more than 60 percent of the total energy intake in all three seasons. There was no significant difference in food energy between the amount of edible food brought into camp during the first half of each survey compared to that brought in during the second; however, we did observe a tendency for food energy to decrease during the latter half of the minor rainy season survey, contrastingly to the surveys in the minor dry and major rainy seasons (table 6.4).

A comparison of the cooperators' body weights between the first and second halves of the survey periods showed that on the first two trips most either kept their weight consistent or experienced a modest gain during the second half. The third trip differed in that six (two men and four women) of the sixteen cooperators on the third trip experienced a significant loss of

Table 6.3. Dietary energy intake.

Food type	8/16–9/5/'03 (16 persons)			10/2–23/'05 (23 persons)[3]			4/14–28/'10 (20 persons)				
	Total food intake kg[1]	Daily energy intake kcal[2]	%	Total food intake kg	Daily energy intake kcal	%	Total food intake kg	Daily energy intake kcal	%	Energy per 100 g kcal[4]	Edible ratio[5]
Yam tubers	427.8	1618.9–1792.3	62.9	650.1	1712.8–1896.3	68.4	379.6	1565.4–1733.1	84.2	112–124	0.7
Nuts	16.4	275.6	10.2	26.9	315.2	11.9	6.8	124.9	4.5	498	0.9
Fungi	2.7	2.9	0.1	6.9	5.2	0.2	1.1	1.3	0.0	32	0.9
Game meat	119.4	427.2–560.4	18.8	136.4	339.7–445.6	15.3	1.9	7.6–10.0	0.3	109–143	0.68
Reptiles	2.5	8.0	0.3	0.0	0.0	0.0	5.4	18.7	0.7	94	0.7
Fish	20.6	59.3–89.3	2.8	7.7	15.5–23.3	0.7	8.9	27.9–41.9	1.2	85–128	0.6
Crutaceae	0.2	0.4	0.0	0.1	0.1	0.0	3.1	7.2	0.3	63	0.3
Snails	0.3	1.2	0.0	0.7	1.7	0.1	4.7	18.4	0.7	107	0.4
Termites	5.9	71.0	2.6	8.6	72.0	2.7	0.0	0.0	0.0	356	1.0
Honey	6.0	63.0	2.3	2.3	17.0	0.6	19.8	226.3	8.1	311	0.75
Leaves	?	?	–	0.1	0.1	0.0	2.7	2.5	0.1	25	0.9
Total	601.9	2527.5–2864.1	100.0	839.8	2479.4–2776.6	100.0	433.9	2000.1–2184.2	100.0		

[1] Estimated total food intake: It was got, if discarded, and remaining food was subtracted from food brought into the camp.

[2] Estimated energy intake per consumption-day.
Consumption-day: Adjusted number of participants × research days; 296 in Aug '03, 425.1 in Oct. '05, and 271.6 in Apr. '10.
A child less than 10 years old was converted to 70 percent of an adult based on estimated BMR (basal metabolic rate) from body weight by sex and age grade (FAO, 2001).

[3] The survey period was from October 2 to 23, 2005, for six families and from October 3 to 23 for two families because of their tardy arrivals.

[4] The energy values for yam tubers, nuts, leaves, game meat, snakes, fish, snails, termites, and honey are derived from Leung (1968), and those for fungi, crutaceae, and fruit are from Standard Tables of Food Composition in Japan (2005).

[5] The edible ratio for yam tubers, nuts, leaves, game meat, snails, and termites were measured by the author; that for snakes, fish, and honey are derived from Kitanishi (1995); and that for fungi, crutaceae, and fruit are from Standard Tables of Food Composition in Japan (ibid.).

Table 6.4. Comparison of mean daily food energy[1] between the first half and the last half of survey period by family.

Aug. '03	First	Last	Oct. '05	First[2]	Last	Apr. '10	First	Last
Family no.	Kcal/ day	Kcal/ day	Family no.	Kcal/ day	Kcal/ day	Family no.	Kcal/ day	Kcal/ day
DF1	6896	8544	RF1	7428	7122	SRF1	5966	3691
DF2	6223	6769	RF2	9046	12521	SRF2	8518	8752
DF3	6980	5574	RF3	4854	7484	SRF3	4435	3436
DF4	8030	11029	RF4	9243	7668	SRF4	3943	5229
DF5	3078	10511	RF5	7347	13893	SRF5	6363	2964
DF6	8151	8363	RF6	4947	4537	SRF6	6027	4513
			RF7	10281	9802	SRF7	6550	3051
			RF8	9979	5204	SRF8	8191	4822
Median	6938	8454		8237	7576		6195	4102
Wilcoxon's test	ns			ns			0.0547	

First: The first half of survey period.
Last: The last half of survey period.
[1]Food energy was estimated from the edible portion of food brought into the camp but not from actually ingested food.
[2]The first half of survey period includes 11 days except for RF6 and RF7.

between 0.5–0.9 kg over the second half. Despite these surveys' limiting conditions, such as small population sample, short duration, and narrow forest area, the above findings indicate no negative evidence for the possibility of a hunter-gatherer lifestyle in a tropical rainforest during the minor dry and major rainy seasons. In the minor rainy season survey period of two weeks, foraging possibilities appeared to be maxed out. Considering, though, that the accessible forest seemed to have been under a high degree of hunting and gathering pressure and that the use of wire snares was restricted, we did not view the results of that survey as necessarily negative evidence. Rather, it was amazing that the amount of yam and yam-like tubers collected in the minor rainy season was no less than that in the minor dry and the major rainy seasons despite the harsh conditions such as the possible gathering pressure and the seasonality of yam productivity, which we will discuss later.

Recently, a long-term foraging trip (*molongo* in Baka language) of a different Baka group in a forest close to this survey area was reported on (Yasuoka 2006). According to Yasuoka, a molongo lasts for two to three months or longer and includes the participation of women and children as well as hunters, and the participants depend considerably more on plants gathered during these trips than on hunted animals for food. Yasuoka recorded all forest food that the molongo participants—about one hundred persons—gathered and hunted from March 5–April 23, 2002, (the major dry season)and estimated the total energy intake at 2,322 kcal per consumption day, 68 percent of which were from the tubers of wild yams and yam-like plants (2006). This is closely compatible with our results. Although under limited circumstances, we have four cases of a foraging life independent of farming products in all seasons in a southeastern Cameroon forest. In addition, the Late Pleistocene sites excavated by Mercader et al. (2003a) were near the region of these surveys. These findings suggest a high likelihood that a purely hunting and gathering lifestyle is possible, at least in the forest at the northwest margin of the Congo Basin.

Carrying Capacity

In the short term, the forest resources in southeastern Cameroon are able to support the foraging life independent of farming products in all four seasons. The question remains as to whether they can do so in the long term. The most important food for a foraging life in the African tropical rainforest is certainly yams and yam-like plants, which supplied over 60 percent of energy intake in three controlled foraging trips and the molongo observed by Yasuoka. How productive are tubers in the African tropical rainforest, and can they meet the requirements of foraging life in the long term? Because the available data on the productivity of tuber-producing plants are very few, we examine this question based on research we conducted in the survey area using a line-transect method to determine the productivity and density of yams and yam-like plants (Sato 2001). The tuber biomass of yams and yam-like plants in six sites was estimated at 5.3 to 17.0 kg per hectare, 15.3 to 17.0 kg per hectare at three sites (total length of line-transects: 4 m × 6 km) in the secondary forest, and 5.3 to 8.7 kg per hectare at three other sites (total length of line-transects: 4 m × 7 km). The sites were characterized by semi-deciduous vegetation and had few anthropogenic disturbances, similar to the campsites in the controlled foraging trips around Mt. Bek. Referring to the third value listed above and that recorded by Hladik et al. (1993) in southeastern Cameroon of 3.0 kg/ha, we consider 5.0 kg/ha to be a representative value for tuber biomass in our survey area. Is this value sufficient to maintain a long-term foraging life? Based on the dietary energy intake in three foraging trips we estimated the fresh weight of consumed edible tubers per consumption day at 2.06 kg in August 2003, 2.18 kg in October 2005, and 2.00 kg in April 2010. Adopting the value for October 2005, we calculated the annual requirements of edible tubers at about 800 kg (2.18 × 365) per person. Setting the utilization ratio of tuber biomass as one-fifth (based on Bahuchet's 1991 data), we estimated annual tuber requirements per person at 4,000 kg, which would require a forest area of 8 km^2 to produce. Thus, carrying capacity in the survey area was calculated at 0.125 person/km^2, similar to the 0.14 persons/km^2 in the area of Yasuoka's molongo (Yasuoka 2006) and the 0.10 persons/km^2 in the Lobaye Region (Bahuchet 1991). This value is reasonable and indicates that the tuber biomass of 5.0 kg/ha is sufficient to maintain a long-term foraging life in the southeastern Cameroon forest.

There are still some questions to be answered as for the productivity of tuberous plants, particularly the *Dioscorea praehensilis*, which accounted for more than 80 percent of collected tubers in the four foraging trips and is definitely a key food involved in the Wild Yam Question. Initially, there is a question concerning the seasonality of *D. praehensilis*, which renews annually both stem and tuber (Hladik et al. 1993). According to Dounias (2001), *D. praehensilis* uses its tuber for the sprouting and growth of new stems from April to July, stores tuber reserve from August to November, and then keeps its tuber reserve at maximum for the next sprouting after March. Therefore, November to March has been regarded as the best period for harvesting tubers and April to May the poorest (Dounias 2001; Yasuoka 2006). However, the findings on our three foraging trips differed from such accepted knowledge. As shown in table 6.5, *D. praehensilis* tubers collected per consumption day in October weighed the most, and those collected in April weighed the least, in accordance with the earlier findings; however, the tubers collected during the molongo from February to April—which should have been among the most massive—were instead the least. Although we were unable to directly compare these values, as the circumstances were different, we should account for the fact that the most unfavorable season easily surpassed the best season in the amount of yield.

As shown in figure 6.2, the graph of rainfall in this region has bimodal peaks: two rainy seasons, April–May and September–November, and two dry seasons, June–August and December–March. This seasonality raises the possibility that the beginning of the major rainy season in September is a second starting point for the biological cycle of *D. praehensilis*, in addition to the starting point considered by Dounias (2001) to begin in the minor rainy season, April–May. If this is correct, the best period for harvesting tubers of this type could be April–August, and it is likely that the sympatric presence of two types of *D. praehensilis*, which have different biological cycles, account for the relatively large amount of tubers collected in August and April.

The next question is the distribution of *D. praehensilis*. Perennial yam plants such as *D. minutiflora*, *D. smilacifolia*, and *D. burkiliana* are distributed evenly over any forest (Sato 2001), whereas annual *D. praehensilis* grows in clumps, which are scattered about the forest (Yasuoka 2006). Although a large clump of *D. praehensilis* on the top of Mt. Bek—around which almost all of the yam tubers collected in both the August 2003 and October 2005 foraging trips—had an estimated biomass of 118 kg/ha (Sato, 2006), the density and tuber biomass of this plant in other forests were very slight (Sato 2001). It is difficult to clearly determine the density or productivity of plants distributed

Table 6.5. Comparison of harvesting tubers of wild yams and yam-like plants.

	Our studies									**Yasuoka's study (2006)**		
Study season	**Aug.–Sep. '03 (CD = 280)**			**October '05 (CD = 388)**			**April '10 (CD = 266)**			**Feb.–April '02 (CD = 3325)**		
Tubers of yams and yam-like plants	**Weight** **kg**	**%**	**W/CD**	**Weight** **kg**	**%**	**W/CD**	**Weight** **kg**	**%**	**W/CD**	**Weight** **kg**	**%**	**W/CD**
Dioscorea praehensilis	636.6	96.4	2.27	1030.6	94.3	2.66	525.1	91.1	1.97	4519.8	83.0	1.36
D. semperflorens	10.9	1.6	0.04	27.0	2.5	0.07	2.3	0.4	0.01	654.6	12.0	0.20
D. mangenotiana	7.7	1.2	0.03	22.4	2.0	0.06	31.0	5.4	0.12	61.8	1.1	0.02
D. burkilliana	4.8	0.7	0.02	12.3	1.1	0.03	16.5	2.9	0.06	184.7	3.4	0.06
D. minutiflora	0.7	0.1	0.00	0.6	0.1	0.00	0.8	0.1	0.00	22.7	0.4	0.01
Dioscoreophyllum cumminsii							0.7	0.1	0.00	3.1	0.1	0.00
Total	660.7	100.0	2.36	1092.9	100.0	2.82	576.4	100.0	2.16	5446.7	100.0	1.65

Weight: The fresh weight of tubers of wild yams and yam-like plants brought into the camp.
CD: consumption-day
W/CD: Weight per consumption-day
The conversion factors of consumption-day in Yasuoka's study were adopted to compare our studies with Yasuoka's study.
Conversion factors in Yasuoka's study: boys or girls 12 or over and adults = 1, boys or girls between 2 and 12 = 0.5, and infants under 2 = 0.

unevenly with ordinal methods such as a line-transect. Indeed, we have very few available data on *D. praehensilis*. However, such data is absolutely necessary to solve the Wild Yam Question. Here we will present some points for future research.

The cooperators named three clumps other than Mt. Bek, all of which were the names of hills. Much more sunlight can reach the surface of the top and slopes of hills than the surface of flat forest. Such hilly areas are suited to light-demanding plants of the genus *Dioscorea*, especially annual species. Extensive geomorphological surveys focusing on hilly areas may result in new findings concerning tuber biomass and the distribution of yam plants.

Such clumps of annual *Dioscorea* species are possibly formed by human impacts (Yasuoka 2009). Recently, Yasuoka (2013) returned to yam-gathering campsites he visited in 2002 and 2005 and measured the density of wild yams near and far away from the abandoned huts of the previous field seasons. He found a high density of annual yams in the living areas. According to the Baka, these stems germinated from not planted tubers but discarded ones. This study clearly demonstrates the human impact on wild yam dispersal; the patches of annual yams in camp sites are likely to increase the yam productivity and to help the Baka to live sustainably in the forest.

Although the annual species of *Dioscoreaceae* are commonly distributed in dry forest and woody savanna, *D. praehensilis* and *D. semperflorens* are particularly found in tropical rainforests (Hamon et al. 1995). However, these two species are not mentioned in the two botanical reports of importance in the Ituri Forest (Hart and Hart 1986; Tanno 1981). This may result from that forest's tendency toward evergreen vegetation; less productivity of *D. praehensilis* would therefore be expected (Hladik et al. 1984). According to White (1983), however, drier peripheral semi-evergreen rainforest, otherwise known as semi-deciduous forest, occurs in two bands running transversely across Africa, to the north and south of the more moist forests. As in this survey area and the area of Yasuoka's molongo, large tuber productivity of *D. praehensilis* and *D. semperflorens* could be expected in semi-deciduous forests. Distributional studies of *D. praehensilis* and *D. semperflorens* in this vegetation type will be helpful in this matter and need to be conducted.

Game, Nuts, Fish, and Perennial Wild Yams as Key Foods

Second in contribution only to edible tubers, game meat contributed over 15 percent of the total energy intake on the first two controlled foraging trips; on the third trip, however, when very little game was caught because of the restriction on wire snaring, total energy intake was far less (table 6.3). Game meat, which was also the primary source of animal protein, was certainly essential to a foraging lifestyle in the African tropical forest. Almost all game was caught with wire snares. On the first two trips, ten to twenty of these traps were set per male cooperator, which resulted in a daily catch of over 1.8 kg per person. Snaring was efficient as well as effective, requiring the men only a couple of days to set and then no more than half an hour to check on each day afterward. All male cooperators were thus able to engage in yam tuber and nut collection together with their wives. If there were Paleolithic hunters in the African tropical forest, they did not have wire snares, but only stone spears as their main hunting gear. Although such tools can also bring considerable results (Yasuoka 2006), Paleolithic stone spear hunters must have been devoted to hunting and would have had little if any time for yam and nut collecting.

Almost all nuts collected in this survey were of the variety *Panda oleosa*. Because this nut has a hard shell, it takes eight months to three years to germinate after falling (Vivien and Faure 1996). Furthermore, because this tree is common in the survey area, it is not difficult to collect seeds anytime and anywhere. Beside *Panda oleosa*, several species in the genus *Irvingia* also are common in the survey area. Supplying a fair amount of dietary energy and enhancing the flavor of food as a source of dietary oil, Panda and Irvingia nuts, among others, were of substantial importance for foragers in the survey area. A particular problem is that it takes a lot of time and effort to collect and then shell the nuts. In this survey, couples usually collected seeds and shelled nuts in pairs. If men had not participated, the work, although light, would have placed a considerable burden on the women.

A small, but indispensable, amount of fish was procured in both the minor dry and minor rainy seasons. Although the Baka cooperators used three fishing methods—net fishing, hook-and-line fishing, and dam-and-bail fishing—only the last method should be considered in relation with the Wild Yam Question. Dam-and-bail fishing is steady and effective; engaging in it for one hour in a neighboring stream, even a solitary woman could get a sufficient catch for her family. Considering the relative ease of fishing activities and the stability of fish procurement, it is possible that fish is a greater contributor to the foraging life in the African tropical forest than what the numbers in this survey represent.

Our survey indicated that annual yam tubers were essential for the foraging life in the southeastern Cameroon forest. Although their contribution was very humble in the controlled foraging trips, perennial wild yam tubers also had potential as a key food. As mentioned before, in the distribution survey using the line-transect method, the tuber biomass of yams and yam-like plants was estimated at 5.3 to 8.7 kg per hectare in three sites, the vegetation of which was semi-deciduous forest similar to that of our survey area (Sato 2001). More than 70 percent of it was from perennial yam plants, such as *D. minutiflora*, *D. smilacifolia*, and *D. burkiliana*. Furthermore, these plants were ubiquitously distributed in each forest. When the clumps of annual yam plants were not around a camp, or too far from it, perennial yam tubers were expected to be alternatives to annual yams. In fact, they constituted a fair amount of tuber biomass procured in between camps during the nomadic period of molongo (Yasuoka 2006).

In addition to these foods, other diverse food items such as honey, insects, and snails were obtained during the survey periods. The more numerous and lengthy the foraging trips, the more the variety of food items increased. Although Hart et al. (1986) and Bailey et al. (1991) pointed out the high cost in accessing food resources in small, widely dispersed patches as one of the challenges facing foragers in tropical rainforests, we can cite the presence of large, densely distributed clumps of annual wild yam plants and the tremendous resource diversity regardless of the sparse distribution of each food resource as factors that enhance the possibility of a foraging lifestyle in this area.

Cost for Foraging Lifestyle

The activity pattern commonly followed during the first two trips involved each couple going out to collect wild yam tubers every two days, while on the other days they engaged in fishing and collecting nuts, honey, termites, and so on. Such an activity pattern likely resulted from the fact that (1) yam tubers spoiled in a couple of days after being dug out and (2) the yam collecting places were not close enough—one hour on foot from the camp—to be easily visited every day. During the initial two trips, most of the wire snares were set around the campsite, and men made rounds to check on them every day before breakfast or on their way back from other foraging activities. Husband and wife couples were often able to engage in yam and nut collecting together because snaring was very economical in terms of work, allowing men more time to assist the women. On the third trip, the cooperators went out from camp to seek yam tubers four or five days a week. This was most likely because the main yam-collecting place was relatively close to the camp, about half an hour on foot, and the daily amount of tubers collected in the accessible forest, which probably was under the hunting and gathering pressure, was small.

Table 6.6 shows the number of steps walked by cooperators and the time they spent outside the camp on the days when their primary activity was searching for and digging out yam tubers, or yam-collecting days (YCD). On the Mt. Bek trips, an average of 12,000–15,000 steps were taken over the course of 400–500 minutes; both figures are significantly greater than those recorded on non-yam-collecting days (NYCD). Obviously, the cooperators tended to engage in comparatively light work on NYCD. The number of daily steps walked on YCD during the third trip was around 8,000. Unlike the previous two trips, the number of steps on NYCD was equivalent to that on YCD. This was probably the result of the cooperators having to seek diverse food resources other than game in the accessible forest.

An important consideration of a foraging lifestyle in tropical rainforests is that it likely requires a large energy expenditure. We estimated the daily energy intake at 2,528–2,865 kcal per person in the first trip, 2,479–2,777 kcal in the second trip, and 2,000–2,184 kcal in the third. Because during the first two trips the cooperators' body weight did not fluctuate, the daily energy intake is

Table 6.6. Steps walked and time spent on YCD and NYCD.

Survey Season	Male				Female	
	Steps	t-test M vs F	Time minutes	t-test M vs F	Steps	Time minutes
Aug. '03 YCD	14823±3160	ns	501±60.2	*	14512±3483	482±54.8
YCD vs NYCD	***		***		***	***
Aug. '03 NYCD	8418±3398	***	412±72.9	*	5398±3168	377±57.9
Oct. '05 YCD	13085±4523	ns	441±122.6	ns	12767±5035	426±89.4
YCD vs NYCD	***		***		***	***
Oct. '05 NYCD	6550±3225	*	313±109.1	ns	5392±3074	275±107.9
Apr. '10 YCD	8035±3337	ns	485±75.5	ns	7907±3095	489±72.6
YCD vs NYCD	ns		***		ns	***
Apr. '10 NYCD	8058±4540	ns	419±90.4	ns	8707±4346	412±103.2

YCD: Yam collecting day (YCD) means the day when the cooperators engaged in seeking and digging out yam tubers.
NYCD: The days other than YCD during the survey period.
t-test: Student's t-test. * $p < 0.5$, *** $p < 0.001$

essentially equivalent to the daily energy expenditure; likewise, the total energy intake (TEI) is equal to the total energy expenditure (TEE). We can estimate the mean basal metabolic rate (BMR) from the body weight of all the cooperators at 1,277 kcal/day in the first trip and at 1,308 kcal/day in the second trip (based on the BMR calculations of the Food and Agricultural Organization of the United Nations [FAO], 2001). If the physical activity level (PAL: TEE/BMR) was the same for both sexes, we can calculate the approximate PAL value at 1.98–2.24 in the first trip and 1.90–2.12 in the second, substituting total energy intake for total energy expenditure. In the same manner, the PAL value for the third trip was calculated at 1.54–1.68. According to the FAO (2001), a PAL value of 1.70–1.99 is consistent with an active or moderately active lifestyle and that of 2.00–2.40 with a vigorously active lifestyle. If the substitute PAL values indicated for the first two trips are legitimate, those high values were likely caused by the hard work on YCD, which included walking long distances, digging out tubers, and carrying them back to the camp. On the other hand, although the yam-collecting places were relatively close to the camp on the third trip, it was difficult to evaluate the lower substitute PAL value because that trip was conducted under different circumstances than the Mt. Bek trips (wire snare restriction and heavily accessed forest). Based on the findings from the first two trips, it is likely that the PAL value of the foraging lifestyle in the African tropical forest depended on the cost of yam collection, which was particularly impacted by the distance from the camp to yam–collecting places. If this is the case, then to decrease the cost, the camp should be built as close to yam–collecting places as possible. However, this is not easy for forest foragers, who must also secure water and other types of food. It was in fact for this reason that the cooperators did not build their camp close to the yam clump on the first two trips. As there was no watering place along the mountain roads between the camp and the yam clump, they set their camp on the bank of the river where they could easily get both water and fish. In addition, if men were compelled to dedicate themselves solely to hunting activity, each woman would have had to seek, dig out, and carry back about 10 kg of yam tubers for her family, every two days, by herself. Even if yam clumps were close, this work would still be hard for women. In this sense, Bailey and Headland's (1991) assertion is true: the costs of searching and traveling for resources in small, widely dispersed patches are high in the tropical rainforests, particularly for centrally placed foragers staying in camps for an extended period.

Leonard and Robertson (1992) described the PAL values of two hunter-gatherer groups as 1.71 for males and 1.51 for females among the !Kung and 2.15 for males and 1.88 for females among the Ache. The PAL value recorded on our first two trips is higher than that of the !Kung in the Kalahari Desert and seems to be comparable to that of the Ache in the Amazon forest. Yamauchi (2000) reported on the daily physical activity level of African populations, including the Baka leading a village life in their sedentary settlement. Among more than ten populations, a farmer group in Gambia had the highest PAL value for both sexes: 2.02 for males and 1.97 for females, which are equivalent to the values from our Mt. Bek trips. In contrast, the Baka leading a village life had a PAL value of 1.41 for males and 1.56 for females, corresponding to a sedentary, or lightly active lifestyle, according to FAO classifications (2001). The data gathered from our surveys suggest that a very high level of daily activity is necessary for a pure hunter-gatherer way of life to be viable in the African tropical rainforest.

What Was the Lifestyle of Paleolithic Foragers and Who Were They?

The surveys from our controlled foraging trips and Yasuoka's molongo (2006) indicate the possibility of a foraging lifestyle in the northwest margin of the Congo Basin. A future challenge will be to find out whether it is possible in other forest areas, especially in the northeastern part of Congo Basin. Our survey also suggests that Paleolithic foragers could have potentially lived in this survey area. This final section will discuss the lifestyle of Paleolithic hunter-gatherers in the tropical rainforest of the Congo Basin, what they may have been like, and who they were.

Based on our observation, it is highly likely that Paleolithic foragers in the African tropical rainforest depended on yam tubers, especially those of annual yam plants, for the bulk of their energy requirements. The rest of their diet consisted of diverse foods, including game, nuts, fish, and honey, as well as others. They led a nomadic life and moved along a circuit of clumps of annual yam plants. Even if they were not able to use these clumps for brief periods, they had reliable access to perennial yam tubers. Tuber procurement, however, was almost certainly a much more difficult process for Paleolithic foragers, who had only stone and wood tools (Dounias 1993), than for the Baka cooperators, who had the benefit of tools made from iron, such as machetes and iron fittings that attach to the ends of digging sticks (Hurtado and Hill 1989). Perhaps no change has been more substantial than the increase in hunting efficiency brought about by the wire snare, which made it possible for male cooperators to spend time collecting yam tubers and nuts together with their wives instead of hunting. In contrast, the use of far less efficient stone spears by male Paleolithic foragers meant that women would have been left without any help in the already difficult task of collecting tubers. Thus, the daily work of Paleolithic women foragers was probably much harder than that of the women cooperators in our foraging trips. Simply stated, a Paleolithic foraging life in the African tropical rainforest could not have been easy.

Who were the Paleolithic hunter-gatherers in the tropical rainforest of the Congo Basin? The strongest hypothesis identifies the ancestors of present-day Pygmy groups and other inhabitants of the tropical rainforest as the leading candidates. One basis for this idea is the short physical stature of Pygmies, which is likely an evolutionary adaptation to the environmental pressures of the African tropical rainforest environment (see chapter 3). Various hypotheses about this evolutionary process have been proposed, some of which are that their short stature is an adaptation to ecological and biological constraints (Hiernaux 1975), to the hot and humid climate (Cavalli-Sforza 1986), to food scarcity (Bailey et al. 1989), to the physical conditions that reduce mobility (Diamond 1991), or to the high mortality rate (Migliano et al. 2007) in the African tropical rainforests. None of these, however, have been proven definitively. Our results may contribute to the enhancement of the food scarcity hypothesis. According to the very recent findings of Becker et al. (2012), divergent selection may have occurred in two genes related to growth—the growth hormone receptor gene (GHR) and the insulin-like growth factor 1 (IGF1)—during the evolution of Pygmies and non-Pygmies, who had common ancestors 60,000–70,000 years ago (see chapter 3). This supports the evolutionary hypotheses concerning Pygmies' short stature described above. However, the single-nucleotide polymorphisms (SNPs) of GHR and IGF1 identified in their studies also suggest the possibility of the evolution of taller stature in non-Pygmies as well as shorter stature in Pygmies. Further population genetics studies, palaeoenvironmental studies, and ecological studies of foraging lifestyles in the palaeoenvironment in the Congo Basin are needed to solve the question of who the Paleolithic hunter-gatherers in the tropical rainforest of the Congo Basin were. Our observations on the controlled foraging trips revealed that the foraging life in the tropical rainforest of the western Congo Basin was possible, but costly. At present, hunter-gatherers who lead a purely foraging lifestyle, independent of agricultural and commercial food, cannot be found anywhere. Therefore, in addition to archaeology, a variety of methodologies such as the surveys we conducted should be utilized so that the details of the Paleolithic hunter-gatherer lifestyle can be revealed.

References

Bahuchet, S., D. Mckey, and I. De Carine. 1991. "Wild Yams Revisited: Is Independence from Agriculture Possible for Rain Forest Hunter-gatherers?" *Human Ecology* 19(2):213–43.

Bailey, R. C., G. Head, M. Jenike, B. Owen, R. Rechtman, and E. Zechenter. 1989. "Hunting and Gathering in Tropical Forest: Is It Possible?" *American Anthropologist* 91(1):59–82.

Bailey R. C., and T. N. Headland. 1991. "The Tropical Rain Forest: Is It a Productive Environment for Human Foragers?" *Human Ecology* 19(2):261–85.

Becker, N. S. A., P. Verdu, M. Georges, P. Duquesnoy, A. Froment, S. Amselem, Y. L. Bouc, and E. Heyer. 2012. "The Role of GHR and IF1 Genes in the Genetic Determination of African Pygmies' Short Stature." *European Journal of Human Genetics*, 1–6.

Brosius, J. P. 1991. "Foraging in Tropical Forests: The Case of the Penan of Sarawak, East Malaysia (Borneo)." *Human Ecology* 19(2):123–50.

Cavalli-Sforza, L. 1986. "African Pygmies: An Evaluation of the State of Research." In: *African Pygmies*, edited by L. Cavalli-Sforza. Orlando: Academic Press, 361–426.

Cavalli-Sforza, L., P. Menozzi, and A. Piazza. 1994. *The History and Geography of Human Genes*. Princeton: Princeton University Press.

Diamond, J. M. 1991. "Why Are Pygmies Small?" *Nature* 354:111–2.

Dounias, E. 1993. "Perception and Use of Wild Yams by the Baka Hunter-gatherers in South Cameroon." In *Food and Nutrition in the Tropical Forest: Biocultural Interactions*, edited by C. M. Hladik, A. Hladik, O. F. Linares, H. Pagezy, A. Semple, and M. Hadley. The Man and the Biosphere Series Vol. 15, UNESCO, 621–32.

———. 2001. "The Management of Wild Yam Tubers by the Baka Pygmies in Southern Cameroon." *African Study Monographs*, supplement 26:135–56.

Dwyer, P. D., and M. Minnegal. 1991. Hunting in lowland tropical rain forest: towards a model of non-agricultural subsistence. *Human Ecology* 19(2):187–212.

Endicott, K., and P. Bellwood. 1991. "The Possibility of Independent Foraging in the Rain Forest of Peninsular Malaysia." *Human Ecology* 19(2):151–85.

FAO. 2001. Human Energy Requirements. Report of a joint FAO/WHO/UNU expert consultation, Rome.

Hamon, P., R. Dumont, J. Zoundjihekpon, B. Tio-Touré, and S. Hamon. 1995. *Wild Yams in West Africa: Morphological Characteristics*. Paris: ORSTOM.

Hart, T. B., and J. A. Hart. 1986. The ecological basis of hunter-gatherer subsistence in African rain forests: the Mbuti of eastern Zaire. *Human Ecology* 14(1):29–55.

Headland, T. N. 1987. The wild yam questions: how well could independent hunter-gatherers live in a tropical rain forest environment? *Human Ecology* 15(4):463–91.

———. 1997. Revisionism in ecological anthropology. *Current Anthropology* 38(4):605–30.
Headland, T., and R. Bailey. 1991. Introduction: have hunter-gatherers ever lived in tropical rain forest independently of agriculture? *Human Ecology* 19(2):115–22.
Hiernaux, J. 1975. *The People of Africa*. London: Weidenfeld and Nicolson.
Hladik, A. S., S. Bahuchet, C. Ducatillon, and C. M. Hladik. 1984. Les plantes a tubercules de la foret dense d'Afrique centrale. *Revue d'Ecologie* (Terre et Vie) 39:249–90.
Hladik, A., and E. Dounias. 1993. Wild yams of the African forest as potential food resources. In: Hladik C. M., Hladik A., Linares O. F., Pagezy H., Semple A., and Hadley M. (eds.) Food and Nutrition in the Tropical Forest: Biocultural Interactions. The Man and the Biosphere Series (Volume 15), UNESCO, 163–76.
Hurtado, A. M., and K. Hill. 1989. Experimental studies of tool efficiency among Machiguenga women and implications for root-digging foragers. *Journal of Anthropological Research* 45(2):207–17.
Kitanishi, K. 1995. Seasonal changes in the subsistence activities and food intake of the Aka hunter-gatherers in northeastern Congo. *African Study Monographs* 16(2):73–118.
———. 2003. Cultivation by Baka hunter-gatherers in the tropical rain forest of central Africa. *African Study Monographs* 28:143–57.
Leonard, W., and M. L. Robertson. 1992. Nutritional requirements and human evolution: a bioenergetics model. *American Journal of Human Biology* 4:179–95.
Letouzey, R. 1985. Notice de la Carte Phytogeographique du Cameoun au 1:500000 Institute de la Recherche Agronomique (Herbier National) Toulouse, France.
Leung, W. W. 1968. Food composition table for use in Africa. FAO Nutrition Division: US Department of Health Education, and Welfare, Public Health Service, Bethesda.
Mercader, J. 2002. Forest people: the role of African rainforests in human evolution and dispersal. *Evolutionary Anthropology* 11: 117–24.
Mercader, J. 2003b. Foragers of the Congo: the early settlement of the Ituri Forest. In: Mercader J. (ed.) *Under the Canopy*. New Brunswick, NJ: Rutgers University Press, 93–118.
Mercader, J., and A. S. Brooks. 2001b. Across forests and savannas: later Stone Age assemblages from Ituri and Semliki, Democratic Republic of Congo. *Journal Anthropological Research* 57:197–217.
Mercader, J., M. D. Garralda, O. M. Pearson, and R. C. Bailey. 2001a. Eight hundred-year-old human remains from the Ituri tropical forest, Democratic Republic of Congo: the rock shelter site of Matangai Turu Northwest. *American Journal of Physical Anthropology* 115:24–37.
Mercader, J., and R. Martí. 2003a. The Middle Stone Age occupation of Atlantic central Africa: new evidence from Equatorial Guinea and Cameroon. In: Mercader J. (ed.) *Under the Canopy*. New Brunswick, NJ: Rutgers University Press, 64–92.
Mercader, J., F. Runge, L. Vrydaghs, H. Doutrelepont, C. E. N. Ewango, and J. Juan-Tresseras. 2000. Phytoliths from archaeological sites in the tropical forest of Ituri, Democratic Republic of Congo. *Quaternary Research* 54:102–12.
Migliano, A. B., L. Vinicius, and M. M. Lahr. 2007. Life history trade-offs explain the evolution of human Pygmies. *Proceedings of the National Academy of Sciences of USA*, 104:216–9.

Ministry of Education, Culture, Sports, Science and Technology, Japan. 2005. *Standard Tables of Food Composition in Japan*. Fifth revised and enlarged edition. Tokyo (in Japanese).
Sato, H. 2001. The potential of edible wild yams and yam-like plants as a staple food resource in the African rain forest. *African Study Monographs*, Suppl. 26, March:123–34.
———. 2006. A brief report on a large mountain-top community of Dioscorea praehensilis in the tropical rainforest of southeastern Cameroon. *African Study Monographs*, Suppl. 33:21–8.
Sato, H., K. Kawamuea, K. Hayashi, H. Inai, and T. Yamauchi. 2012. Addressing the wild yam question: how the Baka hunter-gatherers acted and lived during two controlled foraging trips in the tropical rainforest of southeastern Cameroon. *Anthropological Science* (Japan) 120(2):129–49.
Sato, H., K. Kawamura, H. Inai, and T. Yamauchi. 2006. A "pure" foraging lifestyle in a tropical rainforest of southeastern Cameroon: observations on a twenty-day trip of the Baka hunter-gatherers in the short dry season. *Journal of African Studies* (Japan) 77:1–14 (in Japanese with English abstract).
Sigaha-Nkamdiou. 1993. Caractérisation et Fonctionnement Hydrochimique d'un Bassin Versant en Milieu Forestier Équatorial Humide: L'exemple de la Ngoko Moloundou (Sud-est du Cameroun). Thèse de docteur de l'Université Paris XI Orsay.
Stearman, A. M. 1991. Making a living in the tropical forest: Yuqui foragers in the Bolivian Amazon. *Human Ecology* 19(2):245–60.
Tanno, T. 1981. Plant utilization of the Mbuti Pygmies: with special reference to their material culture and use of wild vegetable food. *African Study Monographs* 1:1–53.
Vivien, J., and J. J. Faure. 1996. Fruitiers Sauvages d'Afrique (Espèces du Cameroun). Ministère Francais de la Coopération et du CTA, Paris.
White, F. 1983. *The Vegetation of Africa*. Paris: UNESCO.
Yamauchi, T., H. Sato, and K. Kawamura. 2000. Nutrition status, activity pattern, and dietary intake among the Baka hunter-gatherers in the village camps in Cameroon. *African Study Monographs* 21(2):67–82.
Yasuoka, H. 2006. Long-term foraging expeditions (Molongo) among the Baka hunter-gatherers in the northwestern Congo basin, with special reference to the "Wild Yam Question." *Human Ecology* 34(2) April:275–96
———. 2009. Concentrated distribution of wild yam patches: Historical ecology and the subsistence of African rainforest hunter-gatherers. *Human Ecology* 37(5):577–587.
———. 2013. Dense wild yam patches established by hunter-gatherer camps: Beyond the wild yam question, toward the historical ecology of rainforests. *Human Ecology* (Published online: 16 March 2013).

7

Diversity in Pygmy Music: A Family Portrait[1]

Susanne Fürniss

Introduction

Like the term "Pygmy" itself, the concept of "Pygmy music" proves rather elusive to define. Still, despite the abundance of musical diversity found among Pygmy cultures, there does seem to be a particular style common to all. In every group, singing is the principle means of musical expression, and there are relatively few melodic instruments. The use of vocal polyphony is perhaps the most salient feature, while other prominent characteristics include the use of counterpoint[2] and yodeling[3] and singing without words. Pygmy music is also marked by a high degree of complexity that often distinguishes it from the music of their neighbors.

Music plays an outstanding role in the symbolic and religious systems of all Pygmy societies. Thus, the contexts for music making underline important spiritual and economic activities. Because of their vast knowledge of the forest, they are conferred by neighboring peoples with undisputed superiority in matters of hunting and manipulating spiritual forces for therapeutic or magical purposes. Each group marks these contexts with their own specific songs, rhythms, and dances. Music making is an essential activity, both for the society as a whole and for the individuals that are its foundation, each serving in their own way as intermediaries between the natural and supernatural worlds. The fundamental importance of listening to the forest soundscape as a condition and cause for the Pygmies' musicality is described with much sensitivity by Lewis and Abram (2006).

These musical and extramusical features contribute to the general fascination with Pygmies, which operates not only on Westerners but also on neighboring Africans. As a consequence, in any context, Pygmies are considered excellent musicians on which the neighbors and state representatives call upon in order to "animate" local or national ceremonies.

The concept of music is not limited to performance, the enactment of social relationships, or bearing symbolic charges; it is also a formal system, a language, with rules and specificities. The latter identify and distinguish different cultures and the musical repertoires they contain. Terminology focuses on the strong link between the music system and its symbolic meaning: it is quite common in African cultures that one term refers to a spirit, a ritual, the song repertoire that is sung during the ritual, the main instrument, and the polyrhythmic formula[4] that accompanies the dance (Arom 1991a). Music also has a material aspect that is represented in the instruments that often support and accompany singing. Thus, in the following overview, I will present each of the major Pygmy musical cultures in regard to three main aspects: performance context, instruments, and musical features (following the method of categorization used by Arom et al. (2008).

The first field studies on Pygmy music were conducted in the 1940s and 1950s by Colin Turnbull among the Mbuti (Kango-Sua) in the northern Belgian Congo district of Ituri and by Gilbert Rouget among the Babinga (Bambènzèlè and Bangombe) on the border region between the Central African Republic (CAR) and Cameroon. Simha Arom then contributed to the field in the 1960s with his studies of the Aka in the CAR. These authors published the first recordings and writings about Pygmy music and hence set the reference for all later research.

I shall present the groups that have been studied thus far, taking into account not only written documentation but also published CDs that are currently available. Thus, the interested reader may listen to the music discussed in this chapter. The order of presentation follows both a certain chronology of research and the general direction of the supposed migration of the various populations, originating in the east of the DRC and heading westward. To establish clear links between the different publications on Pygmy music, I list the equivalent designations in the title of each chapter, but have not unified the ethnonyms and spellings throughout the text.

(Ba)Mbuti, Efe, Kango-Sua, Asua

The music of the Mbuti became a famous anthropological object thanks to the research of Colin M. Turnbull (1961, 1965). Addressing music as a social practice, but also supplying a great deal of precious data on musical style and performance techniques, his writings fed the anthropological question concerning the relationship between social structure and musical style (Lomax 1968). Together with Francis S. Chapman, Turnbull recorded Mbuti music in ritual context in 1954, which was published in 1957 and 1961 (reedition Turnbull and Chapman 1992). During the same period, Paul Schebesta (1957) summed up his long experience with Pygmy music since 1929. At least in anglophone countries, this was the first scientific reference on the subject.

It was not until the 1980s that ethnomusicological research was specifically undertaken. Such work was first done by Didier Demolin (Demolin 1990, 1993;

Demolin and Bahuchet 1991) and Masato Sawada (1990), who completed Turnbull's work. Demolin was notably the first scholar to differentiate the subgroups of Mbuti—Kango-Sua, Efe, and Asua—on the basis of their distinct languages as well as their spatial distribution, each of which map onto different interethnic constellations.[5] Masato Sawada addressed the concepts of utterance, song, and silence as social interactions in ritual and nonritual contexts. Historical recordings have only been made available quite recently and may now be taken into account for coming research on musical permanences and changes: Gustaaf Hulstaert recorded the Efe in 1910 (Gansemans et al. 2000) and Hugh Tracey the Efe and Sua in 1952. Other recordings have been inventoried in the *Berlin Phonogramm-Archive*: the Batwa by Hulstaert in the 1930s, various Ituri Pygmies by Schebesta between 1929 and 1935, and the Bakango by Jean Claessens in 1935 (Ziegler 2006). These recordings are not yet digitized (Ziegler, personal communication, November 21, 2012), and the sound contents and accompanying documentation are still awaiting discovery.

Performance Context

Turnbull revealed a strong difference between the music played by the Mbuti while living near their villager associates and the music played in the forest, where they live about five months a year. In the village, they perform recreational music and dance as well as music for the villagers' initiation rite (*nkumbi*). While in the forest, they practice the "real" Mbuti repertoires, music related to hunting, honey gathering, and spiritual activities.

The Mbuti hunting repertoires distinguish the elephant hunt as the most important of all hunts. As Demolin specifies, elephant hunt songs are performed before and after the hunting expedition. The preparation for an elephant hunt requires performance of the *tore* ritual, a ceremony dedicated to the ancestors and to the Spirit of the Forest (1993). After a successful hunt, music is performed either on the way home or in the camp after the evening meal, where it expresses the Mbuti's satisfaction and joy at having good and abundant food. When a young man makes his first big kill, the repertoire associated with the initiation ritual *molimo* is performed to celebrate his new status as a hunter (Turnbull 1961). Some songs accompany tales or dances in which animals are imitated.

With the exception of honey gathering, gathering does not provide occasion for special musical repertoires. Honey gathering, which takes place about two months a year, is a dangerous activity because honey is mainly found in hives high up in the trees. Therefore, this activity is surrounded by a set of ritual acts associated with music. Demolin gives examples of Efe songs sung before, during, and after the gathering in different musical constellations. The Asua have two types of songs following the collection that express the Asua's joy on one hand and their profound gratitude toward the ancestors on the other.

All researchers recognize the Mbuti's close relationship with the villagers and their participation in the villagers' musical and ritual activities. This is reflected in particular by their shared initiation repertoires. Turnbull documents primarily two Mbuti initiations: *molimo* for the boys and *(el) ima* for the girls.[6] Demolin describes two distinct repertoires for the girls' initiation, one that is sung every night in a reclusion hut to the sound of a musical bow and another that accompanies a women's dance that takes place occasionally outside the hut. Whereas female initiation is determined by a girl's first menstruation, male initiation takes place after puberty, when the young man is able to hunt enough game to feed a family. As part of molimo, boys learn a specific song repertoire. As the songs are charged with the power to rectify evil and disorder, this repertoire and its dance are also performed in times of general misfortune. Thus, the performance of initiation music and dance restores cohesion when a state of imbalance has developed in the social body. Turnbull insisted that molimo and elima are purely Mbuti initiations and are the only institutions able to confer adult status to the candidates. However, according to Demolin's research in the 1980s, these initiations are also practiced together with the associated villagers.

Already in Turnbull's time, Mbuti boys participated in the BaNdaka villagers' initiation rite, *nkumbi*, which takes place at puberty and entails circumcision. Thus, the Mbuti practice the villagers' initiation music as well as their own. The same is true for the shared funeral ceremonies that occur at certain intervals for a period of time after the burial. A week after the death, but still in the mourning period, the Efe play music of the *tore* ritual. The Asua close the mourning period with songs of the *molinga* repertoire. The role of these songs is to reestablish the harmony that has been disturbed by death, not only between the deceased and their family but also between the Asua and the forest world.

In addition to these collective musical practices, which are often accompanied by recreational dances involving the entire community,[7] musicians also engage in individual musical activities meant for their own entertainment, as well as some lullabies. Sawada (1990) explains the Efe's categorization of songs: ò*bé* are songs that accompany the dance, *òwá* are songs without dance, and è*mu* is children's music.[8] Demolin (1993) specified that òbé comprises the majority of musical performances, òwá concerns only solo singing, and èmu are children's songs and dances.

Instruments and Vocal Technique

The predominance of vocal music has to be relativized in the Upper Congo and differentiated through time and probably also space. In fact, unlike in the western Pygmies' musical cultures, the Mbuti have orchestras composed of either flutes or horns (Demolin and Bahuchet 1991). Throughout the last century, the Mbuti played their neighbors' instruments in the villages, including drums, slit-gongs, lamellaphones,[9] and harps, but very few instruments were played in the forest. However, by the century's end, the Mbuti had taken to playing musical instruments while in the forest as well (Demolin 1990).

Wooden concussion sticks (*ngbenbe)* provide the main accompaniment to singing and are played in all kinds of music. Turnbull describes "percussive brushes" (*banza)*, which are split wooden sticks that are struck against a log and used especially in the context of the elephant hunt (Turnbull and Chapman 1992). Later researchers have not mentioned banza "brushes," but another split idiophone,[10] that is, a split struck bamboo used in the 1980s in songs for the Forest Spirit, *tore*, among the Efe and for feast music by the Asua (Demolin 1990; Demolin and Bahuchet 1991). This brings up the question of cultural evolution: is there one split idiophone in Mbuti culture that has evolved from one form to another and may be used in both hunting and spirit rituals? Or have the two researchers only observed one of two different instruments and not the same one?

Turnbull notes that a hand-clapping rhythm may be superimposed over the concussion sticks, but this fact is not documented in any of recordings by either him or Tracey in the 1950s. Demolin's recordings from the 1980s demonstrate hand clapping, but exclusively in the honey-gathering repertoires. Other idiophones may be used occasionally, such as basket rattles for entertainment dances, a bunch of pellet bells worn around the dancers' ankles as they enter the village, and a bell attached to the dance skirt of female candidates during initiation rites. An instrument used particularly for entertainment is the water drum, described in detail in the section dedicated to Baka music.

One or two one-headed conical drums, called *kuce*, are part of all music shared with the Pygmies' neighbors, whether for entertainment, dance, or ritual. The drum is struck both on the skin and on the side of the wooden body.

Two orchestras of aerophones[11] are among the most spectacular constellations of the region and characterize Mbuti music among their neighbors. The first one is a set of eight to ten, one-tone flutes (*luma*) played by men for feasts and at the end of the male initiation. Near the villages, the flutes are made of bamboo or papaya leaf stalks, whereas in the forest, the Kango use stalks of the *Olyra latifolia* plant. This plant is also called *luma*; it has passed its name on to the flutes and their musical repertoire (Demolin and Bahuchet 1991).

The second orchestra is composed of approximately seven horns, called *mai*, that are made of rolled bark (0.3–1.5 m in length). They are constructed and played exclusively for the *tore* rituals and are destroyed after the ceremony. According to the neighbors, these instruments have been borrowed from them by the Efe. However, although they occur in a shared ritual, no neighboring culture currently uses these horns, and the Efe are the only musicians to play them (Demolin 1990).

The girls' initiation, *ima*, integrates a specific chordophone,[12] the two-stringed musical bow, *kitingbi*. The presence of this instrument is a strong indicator of Pygmy culture. It is somewhat particular in that while her hands strike and pluck both strings, the musician uses her chin to produce the melody on the upper string. The bottom of the bow is placed on a drum, pot, or bowl that amplifies the sound. As the instrumentalist cannot sing while she is playing, she is accompanied by the singing of other girls and women. This instrument is only played in the initiation hut, and it is accompanied by a friction-string, a free-hanging rope fixed at the roof and rubbed with wet hands. The vibrations are transmitted to the roof, which functions as a resonator. The low sound symbolizes the forest's breath and materializes the ancestors' and the spirits' voices. In this it is comparable to the sound of the *àìmò*, a bull-roarer used during preparation for the elephant hunt that produces a sound that supposed to be the voice of *tore* (Demolin 1993).

Two other chordophones specific to the Mbuti of the Ituri should be mentioned: a mouth-bow played by men to attract the spirits of animals before a hunt or to accompany the tale songs and an earth-zither played by youngsters for their entertainment.

Except for the orchestras, all instruments accompany singing. Men, women, and children sing together for most of the performances and separate only when music accompanies gender-specific activities. Only a few songs have words, which allows for yodeling. However, Mbuti yodeling is not always based on a real change between chest voice and falsetto, but rather on an imitation of this change within the chest voice. Words are generally important in tale songs and, among the Asua, in the honey repertoires, where they are sung both by the soloist and the choir. In some Efe hunting repertoires, only the soloist pronounces the words (Demolin and Bahuchet 1991; Demolin 1990).

Musical Features

What is striking about the Mbuti in comparison with the western Pygmies is the relatively slow pace of the music. Also, all Mbuti music is based on a binary metricity[13] that is not marked by hand clapping. Musical periods are very short, usually based on two or four beats, but sometimes eight or sixteen. According to Demolin, the shorter periods are characteristic of Kango music (Demolin and Bahuchet 1991). Rhythms are played either by instruments or by hand clapping.

Melodies are mainly built on an anhemitonic pentatonic scale.[14] The Efe may expand this system in lamellaphone music (Demolin 1993), and the Asua also use scales with joint minor thirds for singing.

The Mbuti sing in polyphony to which they also adapt music borrowed from their neighbors (Demolin and Bahuchet 1991); singing in unison[15] is extremely rare (Tracey 1952). Counterpoint of very short intermingled melodic segments is a typical feature of Mbuti singing. The segments gradually fall in steps, which produces the impression of an Escher-esque circular movement of a perpetually falling melody (see transcriptions by Rose Brandel 1961). Although detailed work on Mbuti musical systematics has not been done yet, Demolin (1993) uses the example of the Efe girls' initiation songs, which requires at least four singers (the principal line being sung by an initiate in a high tessitura) to point out the four-part structure of the polyphony.

According to Turnbull, there are two song styles: the canon[16] (*rondisa*), which is adapted according to the number of singers, and solo singing over a polyphonic choir (Turnbull and Chapman 1957). In the latter situation, the soloist launches the song before blending into the choir, which provides an ostinato[17] that is meant to resemble the humming of bees (Demolin 1993). Kango music seems less elaborated than the Efe's, as the Kango play more with polyphonic densities through the alternation between soloist and choir. Another structural specificity that distinguishes Mbuti music from the western Pygmies' music is that in several repertoires, the circular flowing of sound is regularly interrupted by a break before starting to spin around again. This echoes Sawada's research on the antagonism between density of utterances and moments of silence in the construction of both song and speech, as well as in the conception of harmony.

The one-tone flute and horn orchestras are played according to the hocket technique,[18] also used with two whistles in a song celebrating the successful outcome of an antelope hunt (Tracey 1952). The term "hocket" is quite often used to characterize the Mbuti's vocal style, based on short melodic segments (Kisliuk 1991). However, true vocal hocketing as described by Turnbull (Turnbull and Chapman 1991) seems to be quite rare. Voco-instrumental hocketing[19] exists in some flute repertoires, such as *luma*, a repertoire the Kango play to close the initiation ritual (Demolin and Bahuchet 1991).

Finally, some ritual repertoires have a responsorial structure between a soloist and a choir singing, sometimes in parallel fourths or, among the Asua, in parallel thirds.

Aka, BaAka, Bayaka, Biaka, (Ba)Mbènzèlè, Mbendjele, Babinga

The bibliography shows clearly that Aka music is the most studied of all Pygmy music. Gilbert Rouget's first testimonies (Rouget and Didiet 1949, 1959; Rouget 1959, 2011) of the music of the western Pygmies contain both music from the Babinga Bambènzèlè (Aka) and the Babinga Bangombe (Baka) living in the Sangha region close to the Cameroonian border. The research Simha Arom carried out in the Sangha and in the Lobaye region from 1965 onward (Arom 1973, 1978, 1991a, 1991b, 1994; Arom and Dournon 1968; Arom and Dehoux 1978; Arom and Martin 1992; Arom and Fürniss 1993; Arom and Pahaut 1993) has enabled us to discover the musical features and the structural principles that underlie and characterize this music. There seem to be significant variations in the music of the Sangha and the Lobaye. Scholars have recently worked in both regions of the CAR: Louis Sarno (1993, 1995) in the Sangha and Michelle Kisliuk (1997, 1998, 2000; Kisliuk and Gross 2004) and Susanne Fürniss (1991a, b, 1992, 1998, 1999, 2006, 2008, 2012b; Fürniss and Bahuchet 1995) in the Lobaye. Jerome Lewis (2006b, 2009, 2013) is the first scholar to work on the music of the Mbendjele of the Republic of Congo, joined recently by Nathalie Fernando (2011). The great variety of approaches gives a very differentiated image both of music as a social practice and of musical categories and systematics as formal identity markers.

Performance Context

The Aka's musical tradition is deeply structured by hunting and, similar to the Mbuti, music making is closely linked to spirituality: "The Aka present an extreme example where religion is nearly exclusively expressed through music and dance, without officiant, without prayer and without offerings, that is without any perceptible religious gesture" (Bahuchet 1995). Singing and dancing are performed in offerance to the Forest Spirit and to the spirits of the ancestors, both of whom are responsible for humans' well-being. The verb *kàmuz*, meaning[20] "to be happy," "to agree," or "to give the response in a song" (Bahuchet 1995, translated by the author), denotes the musical aspect of the concept of happiness. This explains why the majority of musical repertoires call upon polyphonic singing. The significance of vocal polyphony and collective music-making for the society and interindividual relationships has been studied extensively by Kisliuk (1998, 2000) and also by Lewis (2013). The religious component of music and dance is not always visible. Still, through the social dynamics that operate in a performance, even what appears as a mere "Saturday night" event has multiple layers of meaning for the camp as a whole, as is demonstrated throughout Kisliuk (1998).

Musical variability is quite strong in this culture. Many songs and dances circulate with the people as they move frequently throughout the area (Bahuchet 1995; Kisliuk 1998). As these movements partially reflect the dynamics of family links between lineages, the musical and choreographical heritage is never identical from one place to another, although the main repertoires and rhythms are the same.

In the Lobaye region near Mongoumba, twenty-four musical repertoires have been identified[21]: for instance, *kólí*, a repertoire of collective lamentations, or other repertoires sung to accompany certain rituals, such as *bòndó,* an important healing ritual; *mòbándì*, which opens the honey-collecting season (Arom 1978); and the *mòkóndí* ritual, which is addressed to the Forest Spirit, *Edzengi*, but which is also performed to close the mourning period after a death and to confirm male initiation.

The majority of the repertoires, however, are related, either directly or indirectly, to hunting. In addition to the collective ritual repertoires preceding or following a hunt, there are others, reserved originally for either men or for women, that were played during the periods of the year when the men were out following game in the forest for weeks while the women stayed at home. An anthropological reading of the relationship between music and hunting context (Fürniss 2012b) reveals a network of technical, symbolic, and social aspects that are expressed through specific musical activities and appear in different musical repertoires (indicated by parentheses in the following list):

- Hunting techniques: collective net hunt (*zòbòkò*), collective spear or rifle hunt (*ndàmbò*), individual spear or rifle hunt (*nzòmbì*), trapping (*mbèlà*).
- Ritual functions: propitiatory, praying for success (*zòbòkò, ndàmbò, èsà, mbèlà*); expiratory, reparation for taking the lives of certain animals (*mònzòlì, kóbá*); celebration of a young hunter's first capture of a big animal (*mòpóndí*).
- Particular animals: the elephant (*mònzòlì*), a fundamental animal in Aka religion and the only one supposed to have its own spirit (Thomas et al. 2005); the white-bellied duiker, which bears the colors of the spirits (*kóbá*).
- Moments following a successful hunt: announcement (*mòbìó, nzòmbì*) and rejoicement (*mòmbènzèlè* or *léndè*).
- The indispensable spiritual cooperation between husband and wife: recall of the men who are on a hunting expedition for too long (*sàpá*). Indirectly, *èngbítí* and *bògóngó* also contribute to a successful hunt. The first, played by women, is linked to group cohesion and contact with far off husbands; the second is played by men and concerns marital love.

These last two repertoires, as well as the song celebrating the young hunter, are linked with another fundamental trait of Aka culture, that is, the attention paid to the individual. A young man's first capture of a significant animal indicates that he has acquired the skill necessary to nourish a family and is thus ready to get married. Nowadays, this music may also be performed when a boy has taken a degree from elementary school (Alain Epelboin, personal communication, November 2012). *Ndósí,* a song sung to an unweaned infant whose mother is pregnant again (Arom 1978), is another example of the care given to a single person by the whole community.

Sexual satisfaction and harmony are also crucial to the well-being of Aka couples. Some of the songs played with the *bògóngó,* a harp-zither, express the men's desire to take up sexual relationships again once a hunting expedition has come to a successful end (Fürniss 1998). This topic can also be found in the words and the choreography of the women's dance, called *sàpá*, performed while the men are away (Guillaume and Surugue 1982). The men's love songs are addressed to their spouses. This is quite exceptional in African music and underlines the importance of the individual's well-being as the foundation for the whole society's social and cultural perpetuation. The association between sexuality and success in hunting is conditioned by the Aka concept of happiness, which is central to their philosophy. It recognizes the individual and collective happiness that is brought about when there is an abundance of children and meat (Bahuchet and Thomas 1981). These songs—which seem intimate and individual by their lyrics—are nevertheless conceived for group singing, as they are based on two or three vocal parts (Fürniss 1998).

Instruments and Vocal Technique

The Aka have more melodic instruments than the Mbuti. Their inventory consists of four types of string instruments and two types of flutes. The string instruments include a harp-zither (*bògóngó*), one- and two-stringed bows (*mbèlà* and *èngbítí*), and a ground-harp (*dìgòmbé*).[22] There is a notched flute (*mòbìó*) and a pair of whistles (*mòbéké*, or *hìndèhú* in the Mbènzèlè dialect). The *mòbìó* seems to have fallen out of use in the 1980s, at which time a six- or seven-stringed harp, called a *kùndé*, began to appear. This harp along with a lamellaphone (*sànzé*) played frequently in Aka camps (Epelboin 1995b) were only recently borrowed from other populations and hence have neither their own song repertoire nor specific social functions other than individual entertainment.

Most of these melodic instruments accompany songs that are mainly played in intimate contexts. Thus, their presence may not be immediately evident to strangers, and the songs played on them may not be perceived as representative of Aka music. Nonetheless, as demonstrated by the songs of married couples discussed above, their functions within the society reveal a great deal about the concepts central to the Aka cultural way of life. The data sets collected from ethnological, terminological, technological, morphological, and musical research perspectives all confirm the deep-rootedness of these instruments in Aka culture, as each of them are used for particular purposes in specific social contexts (Bahuchet 1992; Fürniss and Bahuchet 1995; Fürniss 1998). For example, the type of harp-zither played by the Aka—found throughout the western part of central Africa—is the only one to have a wooden body and three strings made from a single aerial root of a Vanilla liana plant (Thomas et al. 2008).

The *èngbítí*, a two-stringed bow played by Aka women, is an important link to other Pygmy cultures, both eastward (Mbuti) and westward (Baka). Guillaume and Dehoux's (1995) study qualifies this instrument as emblematic of what is generally considered Pygmy culture. The bow is played with the same particular technique as the Mbuti's. Like the harp-zither, the *èngbítí* also has a string made from vanilla, with which the instrument shares its name (vanilla has excellent acoustical properties [Dehoux and Guillaume 1995] and is used exclusively for musical instruments). The *èngbítí* is specific to the Aka in that it bears a double resonator, formed by attaching a Marantacea leaf to the bow, which is then placed into a pot held on the knees of the player. Dehoux and Guillaume showed that the èngbítí repertoire initially functioned as a spiritual bridge between women and their husbands while the latter were on hunting expeditions. Nowadays, it is more commonly played for entertainment (Epelboin 1995a). Its songs also contribute to social harmony by commenting on deviant behaviors (Fürniss 1998).

The Aka, like the Mbuti, play a mouthbow. However, a closer relationship is apparent between the Aka and their neighbors, the Ngbaka, through the fact that, in addition to their shared use of this instrument, they also call it by the same name—*mbèlà* (Arom 1974). Furthermore, both cultures use it exclusively in the context of trapping.

Several instruments provide the rhythmic basis for singing and dancing. In the first place, there is the nailed one-headed drum (*mòkíndá*)[23] played in pairs (cf. infra *Baakaa). One of them is used in two ways, being struck on its head with the hands by one player and with two wooden sticks on its flank by a second player. This rhythmical part is called *dìkpàkpà*. The rhythm section is completed by the *dìkétò*, a pair of metal concussion blades.[24] For the divination ritual *bòndó*, a rattle, called *sòkò* or *zékézéké*, is shaken by the healer, and *mbòlà*, a pair of suspension rattles, are attached to his legs. The ritual *zòbòkò* is accompanied by a percussion log (*mòkóngò*), struck in unison by a group of men (Arom 1991a).

Many repertoires are sung with a simple meter-defining accompaniment provided either by hand clapping (*màkpòkpò* and *dìkáké*) or by the flagellation of the body with bunches of leaves (*lókò*). The main function of flagellation is to purify the body of evil forces (Bahuchet 1995). In several ritual dances that call for the purification of humans, this symbolic action is integrated into the musical activity: the legs are regularly struck with bunches of leaves that produce a thumping but clearly audible sound that provides the meter for the singing of the dancers.

Most songs are sung in chest voice, but falsetto is also used for three different vocal expressions that all have large melodies with large intervals (Fürniss 1991a). These include:

- yodeling: alternation between chest voice and falsetto
- whistling (*mòbéké*): regularly alternating voice and whistle sounds
- hunting calls (*mòngómbí*): falsetto only

Yodeling is characteristic of a polyphonic part, called *dìyèí*. It is also a variation technique for the other parts, and many of the dance songs are launched by a brilliant solo yodel. It is also a common way of soothing babies (Epelboin and Gaulier 1987).

As the hunting calls are not regularly measured, they are not "danceable," according to the Aka (Arom 1991). Therefore, they do not belong to the realm of "song" (*lémbò*) but are part of the net-hunting technique (Thomas et al. 2008).

Musical Features

There seem to be significant regional differences in the Aka's conceptualization of music. A discussion on this phenomenon has been raised by Victor Grauer (2009); comparing the writings of Kisliuk and Fürniss on Aka music, he notices different results that seem contradictory. However, knowing both the authors and the field situation, I would rather suppose that the variance seems to be caused by differences in methodological approaches of the authors who focus on different features of music and music making. Furthermore, there also appear to be regional variations in the fundamental conceptualization of music.

The Aka from the Lobaye region distinguish terminologically between various aspects of singing (*lémb-* "to sing") and "song" (*lémbó*), such as

- names of the polyphonic parts: *mòtángòlè* ("the one who counts"); *ngúé-wà-lémbò* ("the mother of the song"); *òsêsê* ("underneath"); and *dìyèí* ("yodeling")
- realization modalities of these parts: *kpókpó* ("straight on") designates the minimal version of each part; *kètè bányè* ("to take a shortcut") is a type of melodic and rhythmical variation; *kùkà ngó dìkùké* ("just cut it!") is a specific rhythmical variation procedure that splits up the melodic line in small segments
(Arom 1994)

The various combinations of these parameters give rise to significantly different musical realities, which range from sung or chanted monody to polyphony (Fürniss 1999). Two polyphonic techniques can be observed: a four-part counterpoint in which only one voice sings the words, and homorhythmic singing[25] in parallel intervals in which all singers pronounce the words (Arom and Pahaut 1993; Fürniss 1999). The latter takes place in a responsorial structure. As is the case with the Mbuti, the contrapuntal repertoire provides the key distinctions between Aka music and their neighbors.[26]

Melodic and rhythmic variation are two of the foremost concerns of good Aka musicians. Thus, the four-part structure is veiled by the creative interaction between singers and may not be recognized or even known by either the singers or the researcher. Arom's and Fürniss' inquiries have revealed that in the context of an adequate performance, the names of the parts are never mentioned. This happens only if there is an error (Arom 1994) or in the context of musical analysis in collaboration with a researcher.

Counterpoint singing always takes place in the context of a dance performance. From a musical point of view, each dance can be identified by its own polyrhythmic formula, which is a combination of different rhythms (cf. infra *Baakaa). The generic name of a dance refers to this formula as well as to the specific choreography and the song repertoire that goes with it. Metricity and the polyrhythmic formulae for all Aka music are analyzed at length by Simha Arom (1991).

The structure of Aka music is based on repeating cycles of either four, eight, twelve, or sixteen ternary beats split into three minimal values, or subdivisions. Some rare repertoires are based on a binary subdivision of the beat. Melodies are based on an anhemitonic pentatonic scale (Fürniss 1991b; Arom and Fürniss 1993).

Baka, Bibayak, Bangombé

Baka music was nearly unknown before the end of the last century. Gilbert Rouget's Bangombé recordings (Rouget and Didier 1949, 1959) were not particularly identified as Baka in origin. Neither were Pierre Sallée's two records with music recorded in the 1960s and 1970s among the Bibayak of Gabon (Sallée [1967/73]; De Fraysseix and Sallée 1976). The first record identified as "Baka music" was recorded by Simha Arom in Cameroon in 1975 (Arom and Renaud 1977). But no detailed studies were available until very recently. Daisuke Bundo (2001) wrote on the anthropological implications of dancing. Susanne Fürniss (2005a, b, 2008, 2011, 2012b, 2013; Fürniss and Lussiaa-Berdou 2004; Fürniss and Joiris 2011) researched girls' dances, musical structure, and ritual circulation, and Luis Devin (2009, 2012a, 2012b) discussed initiation and musical instruments. Nathalie Fernando (2011) provided some first recordings of the Baka in the Republic of Congo, and Sylvie Le Bomin presents some musical specifities of the Gabonese Baka (Le Bomin and Mbot 2012b).

Performance Context

Compared to the Aka, the Baka have a much more vivid ritual practice, in terms of both frequency and number. In spite of certain common cultural resources throughout the Baka territory, one may observe an important variability on a regional level because of different modalities of innovation. This leads to a fragmentation of the musical heritage, mainly through the creation of new local rituals or the borrowing from non-Baka neighbors. These diversification processes have been described in the domain of religious anthropology (Joiris 1996; Tsuru 1998, 2001)[27] and of music (Fürniss and Joiris 2011). The borrowing of the circumcision ritual (*bèkà*), as well as the circulation of several rituals and their music, has been discussed in detail by Fürniss (2008, 2011; Fürniss and Lussiaa-Berdou 2004). She has worked in different regions of the Baka area, mainly in the western part, where she collected over thirty repertoires, twenty of which are associated with rituals.

Musical practice is part of the activity associated with rituals and takes place during what the Baka call *bè*, ("dance, song, ceremony"), a public ceremony in which all members of the encampment participate by singing and dancing. Each ritual corresponds to a particular musical repertoire with specific traits distinguishing it from others (Fürniss and Joiris 2011). In the same way as for the Aka, these musical traits are identifying markers of the whole ritual complex and therefore reflect the signification of the ritual as a whole.

Rituals often have multiple meanings and purposes, and there is often a great deal of overlap between them. Most are mainly related to hunting and to first and second funerals (*mangélebò, èbùmà*), in addition to circumcision (*bèkà*) and the twins cult (*múkó*). There are three healing rituals with specific therapeutic indications: *ngàngà* corresponds to the Aka's *bòndó,* performed in case of misfortune and serious illness, combined with fire divination; *màwòso* heals "mystical" diseases; and *èdíò* fixes complicated fractures. However, most rituals are associated with hunting (*yéli, àbàlè, ndembà, gbelé yéyi, mòkìlà*), even if they do not always occur during a collective hunting party. These rituals "fulfill several functions at a time: healing, conflict resolution, witchcraft, hunting, etc." (Joiris 1997–1998). They are connected through the notion of "the balance of chances," that is, the relational equilibrium between the living and the spirits of the deceased that walk side by side with game and thus give access to the forest (Joiris 1997–1998). Thus, it is no surprise to find that symbolic categories overlap, accounting for the polyvalent character of some musical categories. Therefore, one musical repertoire may serve in several circumstances when it carries a specific symbolic function required in different contexts; as an example, *èbùmà* is linked to both sorcery and the end of mourning, but it is also considered the most representative of Baka dances. It is performed at any occasion to which the Baka have been invited for their ability to "animate" official ceremonies.

On the other hand, one circumstance may call for more than one musical repertoire when specific music is needed for a precise phase of a complex ritual, as is the case with circumcision (*bèkà, bè a sojà, gàlo, nkìyàà yàká*) and the ritual for the Forest Spirit (*ejengì, sàlèsálé*).

Sound plays an essential role in maintaining strong relations with the spirits, which "hunting power" is largely predicated upon. Some ways of singing, especially the yodeling technique (*yéli*), bear strong spiritual power. Similarly, the dancer of *èbùmà* reaches the climax of his sorcery power when his movements make the rattles suspended on his ankles and waist sound harmonically with the drums and the women's singing (Joiris 1997–1998).

Among the nonritual repertoires,[28] the water drum and the girls' dances have been the most documented. Devin (2012b) provides a very detailed acoustic and kinetic study of the water drum, played by women and girls while bathing or washing clothes at a river (see "Instruments" section).

The girls' dances, or "play songs" (*bè na sòlo*), provide a most interesting insight into the role of intragenerational learning that takes place within the group of prepubescent girls. Baka play songs show that in spite of discourse and music—neither of which present any particular features indicating the symbolic meaning of these games—there is an eloquent convergence of choreography and the topics of the songs. The songs' expression of feminine views on seduction and desire familiarizes the girls with women's concerns just as they are about to enter adulthood. What is special about the repertoire of the *bè na sòlo* is not its melodies and rhythms, nor is it the theme of love and seduction, but the fact that the choreography and the text join together in their explicitly sexual connotations. Indeed, through joyful and artistic expression, the young Baka girls' game participates fully in their sensual awakening, all the while remaining a corporal metaphor. This is a significant part in the psychological and physical development of future Baka women (Fürniss 2005a, 2005b, 2013).

Instruments and Vocal Technique

Baka melodic instruments are roughly the same as the Aka's; they include a harp-zither (*ngòmbi*), a two-stringed bow (*língbidi*), a harp (*àìtà*), a ground-harp (*g'bendi*),[29] and a pair of whistles (*èlehú* or *èlepú*). Here as well, a notched flute called *mòbìo* seems to have been abandoned about a generation ago.

The Baka harp-zither, *ngòmbi,* has a much different morphology than that of the Aka. Made of raffia, as are all other known harp-zithers in central Africa, it has an amplifying table of raffia sticks attached to the bridge and its four strings are nowadays made of cable. In Baka mythology, the *ngòmbi* is the demiurge Komba's instrument (Joiris 1997–1998; Brisson 1999). Thus, it accompanies those tale-songs (*lìkànò*) that evoke the achievements of the civilization hero Waïto and the first master-hunter, Tibola.

The *língbidi,* a metal-stringed bow played by women, exhibits a particular relationship between instrument and resonator. Whereas the Aka place the bow into a pot on the player's knees, the Baka turn the pot over on the floor and place the bow on its base. Thus, although it is "the same" instrument as the Aka's, it is clearly distinguishable by its playing position.

Dances are accompanied by two or three wedged one-headed drums (*ndùmù*) and, depending on the dance, occasionally by concussion blades (*màkeke*) and vessel or string rattles (*lìgbegbe, mangisa, màyoyo* or *màngota*). Some of these rattles are attached to the dancer's body. Sometimes, a percussion log (*mbànda*) is added on which two men double the rhythms of the drums. The *mbànda* is supposed to be the "original" rhythm instrument played by the Baka in the forest. As in Aka music, hand clapping marks the pulse nearly all the time.

Devin (2012b) made a detailed study of the Baka techniques for playing the water drum, called *likuende* or *likuindi* in Cameroon and *èkonda* or *mòkunda* in Gabon. They consist of hitting, squeezing, and stirring the water to produce different rhythmic patterns. He established a typology of sound production based on the reciprocal interaction between body parts and water. Some of these sounds have a definite pitch that the players "tune" by varying the strength, speed, and depth of the percussive moves to obtain two or more different pitches.

For all important musical events, singing is performed by a women's choir led by elder women who are repertoire specialists. In some rituals, the solo part is held by the male ritual leader. Yodeling (*yéyi* or *yéli*) is a variation technique that may appear in most of the dance songs. In Baka culture, yodeling is explicitly charged with a strong spiritual power that is essential to the group's survival (Joiris 1997–1998; Rouget 2011): the *yéli* ritual performed by elderly women to create magical protection for the men off on elephant hunt consists of a purely yodeled polyphony sung without any words.

Musical Features

Melodies are built on an anhemitonic pentatonic scale. Songs are all contrapuntal and contain two or three parts. In the large majority, counterpoint is integrated in an alternation between a solo part (*kpó njàmba,* meaning "to pick" or "intone") and a choir part (*na ja,* meaning "to take"). The latter is performed in two simultaneous tessituras; the higher one is named "little voice" (*líè na tè*) and the lower one "big voice" (*ngbè líè*). The responsorial structure can be quite difficult to perceive, as each part has multiple variations that often overlap and are sung simultaneously.

Each Baka dance is accompanied by a repertoire with a specific polyrhythmic formula. Rhythms are integrated in a regular metric structure of four or eight beats, each with a ternary subdivision of three minimal values. Still, some rhythms are based on a different metricity, which is an enduring testimony to the exogenous origin of the dance or the ritual to which they belong (Fürniss 2008).

*Baakaa: The Hypothetical Original Common Society of the Aka and Baka

Among the known Pygmy cultures, the Aka and the Baka are clearly the most closely related. The great proximity between their musical heritages supports Bahuchet and Thomas's (1986) hypothesis of these two cultures' common history. Bahuchet (1992) later gave the name "*Baakaa" to this hypothetical historical population. Fürniss initiated her research on Baka music from this perspective.[30]

Although the repertoires within their musical heritages are organized in quite different ways, the strongest affinities between Aka and Baka music are found in the instruments, musical terminology, and formal musical features.

The words for yodeling are nearly identical: *dìyèí* (Aka)/*yéyi* or *yéli* (Baka). The women's two-stringed musical bow is the same in its morphology and playing technique as well as in its use and name: *èngbítí* (Aka)/*língbidi (*Baka) [and *kitingbi* in Efe] (Bahuchet 1992; Demolin and Bahuchet 1991). Bahuchet (1992) reports two names for the harp-zither: ngòmbí or *bògóngó* (Aka)/*ngòmbi* or *bògongo* (Baka). The one-tone flute is called *hìndèhú* (Aka-Mbènzèlè), according to Arom and Dournon (1968), and *èlehú* (Baka).

In the domain of rhythm instruments, the name *mòkíndá* (Aka)/*mòkinda* (Baka) for the drum is specific to the Aka and Baka languages (Bahuchet 1992). In Aka, it designates a one-headed drum, which differs from the two-headed *ndùmù.*[31] In Baka, *ndùmù* is the generic name for drums, which are always one-headed. Thus, rather than the general name of the drum or its morphology, the link between the two cultures in this case is the term *mokinda* itself; whereas in Aka the name signifies the material object, in Baka it refers to the musical part played by one of the drums (Fürniss 2012b). In both cultures, drums are usually played in pairs, with each tuned to a slightly different pitch. The lower one is called *ngùè-wá-mòkíndá* ("mother of the drum") in Aka and the higher one *èndòmbà-wá-mòkíndá* ("the small one of the drum") or *ìkúbù*. In Baka, they are also called "mother" *nyéè ndùmù* (low) and "child" *lè ndùmù* (high) for all repertoires except the repertoire played during the Forest Spirit ritual, *(e)jéngì*. In this case, the two drums have their own names, the low one being called *mòkinda* and the high one *kubú*. This terminological convergence between the two musical cultures occurs in connection with the most important spirit for both societies, the Forest Spirit *èzéngì* (Aka)/*(e)jéngì* (Baka) (Bahuchet 1992), and is therefore an extremely strong indicator of a historical link between the two societies and the drum's fundamental role in *Baakaa music.

The most obvious convergences, though, are to be found in musical features, namely, in the domain of meter and rhythm. Except for repertoires borrowed from or shared with neighbors, all Aka and Baka music is based on a ternary metricity, the beat being subdivided into three minimal values (Arom 1991a). The repertoires dedicated to fire divination and the search for life-threatening diseases are based in both cultures on a four-beat periodicity. The extension to eight beats is found in the repertoires associated with hunting, establishing contact with the Forest Spirit, and marking the end of the mourning period. Although they are not exclusive, these correlations clearly indicate in both groups a symbolic convergence between musical grammar and meaning.

A close look at rhythmics shows an even closer proximity. In both cultures, only the low drum plays a unique rhythm for each repertoire. The other instruments share a common stock of rhythms that are combined in various ways according to the repertoire. This stock of rhythms is exactly the same in both cultures (figure 7.1).

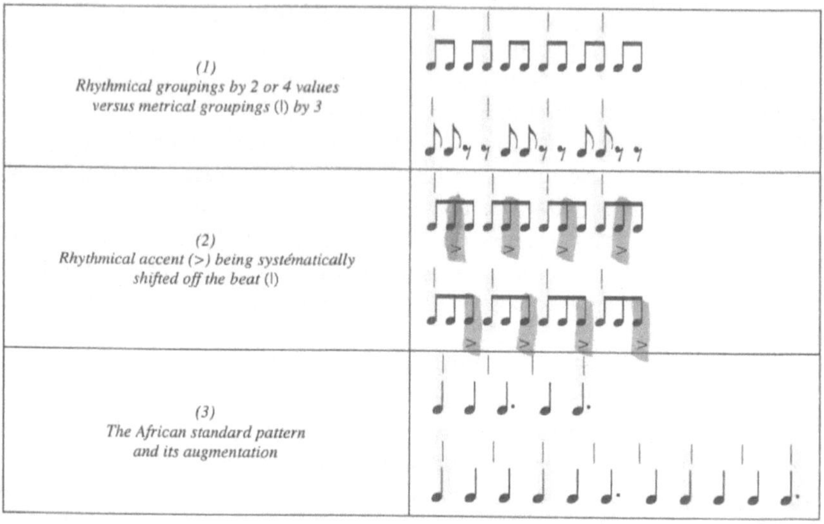

Figure 7.1. Common stock of rhythms in Aka and Baka music.

One of the characteristics of these rhythms is the metrical ambiguity, caused either through a two-against-three relationship between rhythm and meter (1) or by contrametricity that systematically places the rhythmical accent off the beat (2). The African standard pattern (3) is used as such, but mainly appears in an augmented version extending the period to eight beats (Arom 1991a). This rhythmical figure seems to be uniquely representative of the *Baakaa, as is currently unknown in any of the neighboring cultures.

Other Groups in Cameroon and Gabon

The other Pygmy groups are not as well-known as the Mbuti, Aka, and Baka. Musicological research was initiated only in the late 1990s among the Bedzan in Cameroon, at the beginning of 2000 among the Bongo and Koya in Gabon and in 2010 among the Gyeli in Cameroon. Therefore the data are still very heterogeneous in quality and aspect and overall quite incomplete.

Bedzan (Cameroon)

The music of the Bedzan has been studied by Fabrice Marandola and Nathalie Fernando. Their research is a methodological contribution to the exploration of musical scales (Marandola 2003; Fernando-Marandola 2002, 2007). Though their work has not concerned Bedzan musical heritage per se, they do offer a short overview of musical repertoires and contexts (Marandola and Fernando 2007).

The Bedzan are closely linked to the Tikar and share aspects of their social organization and musical practice. The Bedzan are essential intermediaries at key moments in the life of a Tikar chief, such as the presentation of his newborn child to the Tikar community. During this rite, the Bedzan sing the *melwœn* repertoire, passing the baby from one person to another while dancing. Their music and dance also play a crucial role in Tikar funeral ceremonies, which take place after forty days have passed since the death.

The Bedzan's song *lengbu*, which welcomes men returning from a successful big game hunt, is the only link with the other Pygmies' hunting repertoires. Two mask societies each have their own musical repertoires. The male society, which is linked to death, uses the *Win*, a type of "sound mask." It is made of a "set of aerophones and an instrument with a strange sound which the Bedzan carefully keep secret" (Marandola and Fernando 2007); for this reason it can only come out at night. In contrast, the women's mask, called *mgbènyè*, is associated with maternity and fertility and is worn during daytime dancing. Its dance is accompanied by women's songs that mock male sexuality. Finally, the children have their own mask, *mgba*, for public amusement.

The most representative Bedzan repertoire is *nan*, which may also function as an opening for any ritual occasion. It is accompanied by two one-headed drums, the *nkè meku* or *ngwin ndu*, which is struck with a stick and one hand, and the *mben* or *ngwin sedi*, played with the hands. A basket rattle (*shisha*) may be played in several repertoires, occasionally coupled with hand clapping.

The melodic instruments, for which the Bedzan have the generic name *mbe*, are borrowed from the Tikar or the neighboring Vute. The lamellaphone, *mbe pèrè* ("melodic instrument—thin and long"), has from eight to fourteen bamboo lamellas and is tuned in either a tetratonic[32] or pentatonic scale. The harp-zither, *mbe kelon* ("melodic instrument—chest"), has five metal strings and a raffia body. Its half-calabash resonator is moved back and forth from the player's chest to vary the timbre. Finally, a three-string pluriarc bears a morphological specificity: some corn grains are placed in the sound box to embellish the sound by their sizzle. None of these instruments has a specific song repertoire; they are played alone or within a small group of singers and listeners (Marandola 2003).

The melodies of the vocal polyphony are based on tritonic, tetratonic, or pentatonic scales (Marandola 2003). Songs are started by a soloist who is then joined by the other singers in counterpoint or, for more recent songs, in responsorial alternation. In the same way as the Aka, the counterpoint is based on four parts: *nkwo bunkin* (voice of big [men]), *nkwo bembeban* (voice of young men), *nkwo beyi* (voice of big [women]), and *nkwo bwèso* (voice of little ones) (Marandola and Fernando 2007). Here again, only the solo part bears words.[33]

(Ba)Gyeli, (Ba)Kola (Cameroon)

An initial overview of Gyeli music has recently been supplied by Camille Oloa-Biloa in her master's thesis (2011). She presents not only contemporary performances, but also abandoned repertoires that are still in some Bagyelis' memory.[34]

Of the polyvalent dance repertoires most currently practiced by the Bagyeli, *bàpéà* is the most representative: two or three masked male dancers compete in turns with acrobatics for the public's acknowledgement. As a dance, bàpéà is performed for funerals, for general entertainment, or for any official ceremony. Its songs may also be sung by women after particularly good fishing parties. This is not the case for the songs of the *bìsámbà* repertoire that accompany the solo dance of an elderly woman. These are performed for marriages, funerals, or fun. Both dances are accompanied by a struck bamboo and two drums. Basket rattles are attached to the *bàpéà*-dancers' legs.

Oloa-Biloa mentions four other recreational repertoires[35] and a set of women's songs related to birth and marriage. Oloa-Biloa qualifies the latter as probable relics of a female ritual. This is plausible in the specific interethnic context, where the immediate Ngumba neighbors have equivalent female repertoires.

Among the original Bagyeli ritual repertoires, only the therapeutic ritual *mángà* seems to be regularly practiced. As this ritual is efficient in fighting sorcery and other supernatural discords, Bagyeli healers are seen both by Pygmies and neighbors. As patients, the latter participate in the performance by singing and marking the beat. A ceremony contains different phases, each of which has its own song repertoire.

The Bagyeli have been net hunters in the past but rarely practice this kind of collective hunt anymore. As a result, the corresponding ritual repertoire, *sángà*, is currently falling into disuse. Outside the camp, hidden from the women, it featured *Maslé*, the master of all forest spirits who was sometimes joined by *Nzángálá*, the spirit of the net hunt. The nonaccompanied songs were sung in a very low register, wishing good luck to the hunters. When leaving, the hunters would make *mbèlèwá mbwímò* calls that, like the Aka's hunting calls, are musically unmeasured. In the forest, they would communicate with a one-tone bamboo whistle, *lìbóusá*. Also rare, but still remembered, is the trapping song *màlámbò*.

The most representative instrument in Bagyeli music is the bamboo log (*ngyèn*), struck by a group of women with two different rhythms: *sálò* is played by a single woman and *pàálé* by the others. It is noteworthy that in the case of the healing ritual, the log is named *bákà*, bearing a strong resemblance to the name of the Bongo's struck log, *baka* (Le Bomin and Mbot 2012a:7).

Another instrument is both called by a different name and played with a different technique when used in the healing ritual. The basket rattle is called *bégèlè* when it is attached to the legs of the *bàpéà* dancers. Shaken with the hand, the same object is the healer's essential tool to call the spirits' aid before a ceremony. In this instance, it is named *mèndjiébá*.

The sticks (*bàpáà*) used to strike the bamboo log may also be used as concussion sticks. Together with hand clapping, which may be performed in three different ways, the sticks simply mark the beat or perform a counterrhythm to the bamboo. The drums are morphologically identical to the Baka's and are played in pairs: a tall one (*mbè*) and a very short one (*téngè* or *ngòmò*).

Among the solo instruments, one finds a set generally equivalent to those found elsewhere, but with slight specificities: a lamellophone (*tàmè-támè*), played "wrong way round," with the lamellas pointed away from the player's body; a six-stringed pluriarc (*dúm*) with a specific song repertoire; and a one-string harp (*támìntúbà*) that is played with a technique resembling the way some other Pygmy groups play the ground harp (Fürniss 2012a). Finally, there is a women's bow that is no longer played but still existed two generations ago.

Like the Bedzan, the Bagyeli do not yodel nor do they use the falsetto. The Bagyeli singing technique is nonetheless recognizably Pygmy in style, with a particular tension that distinguishes their vocal projection from that of their neighbors (Oloa-Biloa 2011).

Melodies are based on a heptatonic scale[36] and take place in the formal framework of an alternation between a solo singer, *tándò* ("the one who entones"), and a choir, *bolègè* or *gyáyé* ("sing!"). The choir is divided in three parts that are superimposed according to the tessitura: *kínyà piú* ("high voice"), sung by young girls; *kínyà tétèmò* ("middle voice"); and *kínyà néní* ("low voice"), sung by elder women. They move in parallel movements of thirds/fifths or fourths/sixths, or in counterpoint. Again, only the soloist sings the words, and some songs are chanted without a melody.

The majority of music is based on a ternary beat, with at least two percussion instruments playing primarily polyrhythms. *Bàpéà* contains the same rhythm with systematically shifted accents that is found in *Baakaa music.

(Ba)Bongo, Moghama, and Murimba (Gabon)

Because of the Bongo's wide distribution over the Gabonese territory and the doubt that remains as to whether all groups named Bongo can or should be considered as one people, their music is most difficult to characterize. Magali de Ruyter established the first study in 2003 with one of the Bongo groups, followed by Sylvie Le Bomin, who has compared the main features of six different Bongo groups (Le Bomin and Mbot 2011, 2012a, 2012b). Both ethnomusicologists take into account the intensive interaction with neighboring cultures, but their research perspective is slightly different. De Ruyter focused on the common regional ritual corpus, considering the Bongo group and their respective neighbors, Tsogho and Masango, as two distinct ethnic groups within a single society. Le Bomin has conducted investigations on common origin, migration, and kinship of Pygmy societies and considers its dynamics in the context of three different scales: among Bongo groups, Gabonese Pygmies, and other central African Pygmies.

The Bongo share with their neighboring cultures several ritual institutions (*mwiri, nyembe, mungala, lisimbu, bwiti disumba* and *misoko, nzobi*). In the Tsogho-region, the *bwiti misoko* ritual has incorporated the Bongo's propitiatory hunting invocation, *buluma* (Bonhomme et al. 2012) or *boluma*, a repertoire linked to the elephant and bushpig hunt (Le Bomin and Mbot 2012a).

Le Bomin's synthesis of the Bongo's musical repertoires shows that "the common substrate of most Bongo groups is not a musical repertoire, a musical instrument, or a music technique, but rather a performance context, namely the hunt" (Le Bomin and Mbot 2012a). She lists six hunting repertoires (*ibwema, ikoro, boluma, mudimu, muri, mapaji*) that are differentiated according to the hunting technique (net or spear) or particular animals (elephant, antelope, bushpig, buffalo). But they are not practiced by all Bongo: only two repertoires are shared, and each of these are shared by only two groups (Le Bomin and Mbot 2012a).

Such is the case with the multipurpose *mudimu* ritual and its music; while it acts as propitiation for the bushpig hunt (Le Bomin and Mbot 2012a), it also functions as a rite of passage when a young man has killed his first big game and as a rite of affliction in the event of a bad hunt (Bonhomme et al. 2012).

All musical instruments—with the exception of three types of whistles, made of papaya leaf stalks, bamboo, and earth, respectively—are commonly played by both the Bongo and their neighbors. This includes the struck bamboo, *baka (bake, bakoko)*. No Bongo group uses the yodel technique (Le Bomin and Mbot 2012a), though one may hear a "pseudo-yodel" with large melodic intervals (Le Bomin and Mbot 2012b). Only one subgroup sings in counterpoint (Le Bomin and Mbot 2012a), whereas the others mainly sing in homorhythmic movement.

Some groups from southwest of Gabon are also called by the generic name "Bongo," but call themselves Moghama and Murimba. They have little in common with the other Bongo. A very short overview (Le Bomin and Mbot 2012b) highlights a divination dance, *diboka*, that has the same ritual features as the Aka's *bòndó*.[37] The Moghama seem not to have hunting repertoires, whereas the Murimba play *ditsatsa* music during the ceremonies linked to the bushpig hunt. They also play a papaya whistle (*mubembi*) (Le Bomin and Mbot 2012b).

(Ba)Koya (Gabon)

Le Bomin and Mbot (2012a, 2012b) compare the Bongo to three other Gabonese Pygmy groups—the Baka, Koya, and Moghama—with whom they share hunting rituals, whistles, and contrapuntal polyphony. The Bongo of the Nzebi area and the Koya have the same hunting repertoire *ibwema/abwema* (Le Bomin and Mbot 2012a, 2013). The Koya's own repertoires are all related to the hunt of different animals, but here as well, all Koya do not have the same rituals: only the Koya of Ekata practice the elephant ritual *nzokou* (Le Bomin and Mbot 2012b).

Koya are the musicians for their neighbors' girls' initiation (*lisimbu*) and the main actors for *mungala*, a circumcision, and twins ritual. They are the guardians of the *ngi* repertoire, which had formerly been performed during the Fang's rite for their ancestors. It has been passed on to the Koya via the Kwele, who do not practice it anymore. Thus, they sing in Fang and use a homorhythmic polyphony with melodic and harmonic intervals that are typical for Fang music.

The proximity with the Kwele is also visible in the musical instruments. The Koya have borrowed from them the harp-zither (*engue*) that has the same features as the Baka's. The Koya and Kwele have the same name for the struck bamboo (*kolo*/*kóló*).[38] However, the Koya's instrument deserves a special mention. It consists of four segments in a square configuration. As among the other Pygmies, this instrument is struck by several musicians (Soengas 2010).

Other Koya instruments are a two-headed conical drum (*andum*) struck with two sticks and a one-headed cylindrical drum-type played with the hands. The latter exists in three sizes: very short (*abela*), about one meter long (*ngom*), and about 1.6 meters long (also *ngom*). The long drum is used exclusively for the circumcision rite (Le Bomin and Mbot 2012b).

When clapping hands as accompaniment to their singing, the proper Koya repertoires can be identified by a special manner of suspending the rhythmic figure and of introducing a series of regular beats (Le Bomin and Mbot 2012b).

The Phantasm Gives Way to a Family Portrait

Once the myriad musical features of the various Pygmy cultures have been cataloged and codified under a unified paradigm, how do we move beyond a list of properties and understand these musics as more than

> often wordless yodelling that results in disjunct melodies . . . densely textured multi-part singing . . . pentatonic or sub-pentatonic forms . . . high degree of overlap between parts . . . a considerable freedom to improvise . . . primarily vocal . . . solo instruments . . . harp . . . the *mbiti*, a women's musical bow; and the *hundewhu*, a three-hole disposable bamboo flute (*sic*). (Cooke and Kisliuk 2007–2013)[39]

A synthesis of actual knowledge about Pygmy music necessarily follows a comparative perspective that cannot bypass the question of kinship. It is important not only to look at actual practices but to adopt a diachronic perspective. Sources have to be contextualized and abandoned practices to be included as far as they are still in people's memories in order to take into account as far as possible the aspect of dynamics in cultural changes. Likewise, it is also necessary to individualize the various musical heritages to avoid hasty generalizations and to appreciate the individualities of each culture. In this, writing this chapter was a challenge not easy to take up.

What is new, and where do these comparative condensations lead us? It clearly appears that there is a "family resemblance" that consolidates the formerly generalized characterizations. It also corroborates, from a musical point of view, Paul Verdu's findings in genetics (see chapter 2). Comparison reveals substantial musical markers for the cultural kinship of the *Baakaa and thus solidly anchors ethnomusicology among those disciplines that provide answers to the question of whether there is a valid reason to group different societies under the exogenous and homogeneous label "Pygmy." The most recent studies help refine a set of criteria to establish the musical characteristics of the various groups as well as those of their neighbors. Indeed, as the concept of Pygmy is legitimized by the mutual differentiation by Pygmies and their neighbors, the study of music must also take into account the music of the latter to understand the multiple layers of inclusion, juxtaposition, and exclusion. This increases the task considerably, and it is only recently that a large-scale comparative study has become seriously imaginable. During the last two decades, the music of several little-known Pygmy groups has been studied following the same methodology, and the corpus is becoming more and more solid.

One of the traits that appears is the implication of Pygmy music in their neighbors' rituals. Culture contact and ongoing migrations create unsynchronized dynamics that continue to shape these societies' common substratum, the core aspect of which is the outstanding function of music as prayer. Contact with the spirits, fire divination, and treatment of misfortune or sorcery are Pygmies' specialties and call as much upon music and collective performances as on therapeutic knowledge (Rouget 2011). The prevalence of hunting repertoires is linked, in many cases, to differentiation according to the animal being hunted or the hunting technique; some cultures celebrate the first efficient hunt of a young man. Gathering, however, does not necessitate a specific music, except for honey gathering, which is sometimes attended by ritual.

The egalitarian view of Pygmy cultures is also supported by the musicological research. In most of these cultures, men, women, and children perform music and dance together in a complementary way, each of them indispensable to effective music making. In this sense, music is as much a domain for collaboration as ritual and economic tasks (Joiris 1997–1998; McCreedy 1994; Kisliuk 2000).

Though the most efficient music in terms of conveying symbolic meaning is vocal—for example, the yodeling or pseudo-yodeling performed by several groups—musical instruments and their names have turned out to be important markers for common characteristics that distinguish Pygmy patrimonies from those of their neighbors. They underline the ephemeral nature of Pygmies' material culture and their maximal use of natural resources. The papaya or bamboo whistle has become an icon in world music, known by the name of "Pygmy flute." Both this flute and the percussion log have, along with their names, a wide distribution throughout central Africa. Hand clapping also represents a distinct feature in several regions. Many instruments are constructed from components that were not originally intended for that use, for example, concussion blades and certain chordophone parts. The women's two-stringed bow is a salient example of the more complex instruments, and the harp-zither has had an enduring presence in some groups as well.

The historicity and complexity of various intergroup relations represented by the drums certainly call for a more extensive discussion than what is possible here. A particularly conspicuous example, however, is offered by the deep terminological and symbolic links between Aka and Baka drums. The strong affinity of form and function of the drums in the mutually practiced core ritual of the Forest Spirit amply demonstrates the age-old integration of the drums within the *Baakaa culture from which these two groups diverged.

Music systematics confirm the predominance of counterpoint, practically absent among all known neighboring cultures. Hocketing, however, is found only in instrumental music played with horns and whistles. Certain rhythmical principles, such as complex polyrhythmics, off-beat shifts, and a particular extension of the African standard pattern hint at other musical traits that still have to be analyzed for many groups.

Much work remains to be done, and the music of many groups has yet to be documented and described systematically. To this end, several general research programs are currently integrating the study of Pygmy music into larger contextual frameworks[40] and helping to expand our knowledge of individualized contemporary Pygmy cultures.

Notes

1. The author wants to thank Scott Calvert, Barry Hewlett, and Laurence Fayet for their attentive reading and the very constructive revision of this text.
2. Parts are melodically and rhythmically independent from each other.
3. This characteristic singing technique alternates chest voice and head voice (or falsetto) on specific vowels. It produces melodies with large intervals (Fürniss 1992).
4. Polyrhythm designates the interaction of two or more complementary rhythmic figures played at the same time. The specific pattern that results from this simultaneity is called a polyrhythmic formula.
5. As I am not a specialist of the Ituri, it is not possible for me to identify exactly the subgroups in the work of Turnbull and Tracey. According to Demolin (personal communication, November 14, 2012), Turnbull has mainly worked with Kango-Sua and Tracey has recorded Efe and Sua. Still, I continue to use the collective name Mbuti when I do not have any specification.
6. Demolin (1990) specifies that the Efe's name for this ritual is *ima*, whereas *elima* is the equivalent term in Swahili, the common trade language in the eastern part of the DRC.
7. Other spiritual activities based on music are documented by Tracey, pertaining mainly to music performed for healing and after the installation of

a new camp. Unfortunately, there is no ethnographic description for these contexts. Likewise, it is not very clear whether the songs sung after fishing and for marriages are Mbuti or the neighbors' songs (Tracey 1952).

8. Efe and other Pygmy languages are tone languages where the relative pitch on which a syllable is pronounced is linguistically relevant. Some recent authors use phonetic transcriptions to correctly represent the different language sounds. I transliterate the vowels and tones as follows: [ɔ] = o, [ɛ] = e, high tone = ó, low tone = ò, and middle tone = o.
9. Thin lamellae on a wooden box or a plate that are plucked with the thumbs, commonly known as *mbira*, *sanza*, or *likembe*.
10. "[A]ll instruments made of materials sufficiently rigid to sound of themselves without skins or strings" (*Oxford Music Online*, July 23, 2013).
11. Commonly called "wind instruments": flute and horn, but also bull-roarer, "a thin, flat piece of wood, swung to produce a whirring noise, rising or falling in pitch with changing speed of motion" (*Oxford Music Online*, July 23, 2013).
12. Musicological term for "string instrument."
13. Metricity sets up the framework in which a rhythm takes place. It is qualified by the number of (mainly) regular beats that form the musical cycle and the type of subdivision of the beat: it is binary when the beat is divided in two or four minimal values, and ternary when the beat is divided in three minimal values.
14. Five degrees within the range of an octave. The smallest interval is a major second, an interval of a tone, as between C and D.
15. All singers sing the same melody at the same time, thus producing monody.
16. All singers sing the same melody, but starting at different moments, thus producing polyphony.
17. A constantly repeated melodic or rhythmic pattern; it may be varied or not.
18. The alternated blowing of different instruments.
19. The rapid alternation between singing and blowing a whistle by the same musician.
20. For any of the Aka terms, the reader is invited to consult the *Encyclopédie des Pygmees Aka* (Thomas et al. 1981–2013) which gives a very detailed insight in the Aka's culture, society, environment, and religion.
21. Related to hunting, directly: *zòbòkò*, *ndàmbò*, *sàpâ*, *nzòmbì*, *mòbìó*, *mò-mbénzélé*, *kóbá*, *èsà*, *mbèlà*, *mòpóndí*; indirectly: *èngbítí*, *bògóngó*. Other repertoires: *bòndó*, *kólí*, *ngbòlù*, *mòkóndí*, *mòbòmá*, *ndósí*, *màásá*, *mòzó*, *dìsàò*, *mòbándì*, *yómbè* / *léndè*. For a schematic representation see Fürniss (2006).
22. For this particular instrument's distribution and organological denomination, see Fürniss (2012).
23. One of them may be replaced by a two-headed ndùmù.
24. The concussion blades' name, dìkétò, is worth a terminological detour. The Aka use two common machetes called *nzènzè* (sg.)/*mànzènzè* (pl.); by joining two of these cutting blades into one musical instrument, *dìkétò (sg.)/màkétò*, (pl.), they change their name, thus indicating clearly that this new object has a completely different function than its two individual components.
25. Although they sing different notes, all singers sing the same rhythm.
26. The functioning of Aka counterpoint is described in detail in Fürniss (2006). Some paragraphs of the present article are identical to this text.
27. Tsuru (1998) details the combination of natural material and cloth in the creation of spirits and their masks.

28. Repertoires investigated by Fürniss (2013a): entertainment dances with or without masks (*èmbòàmbòà, mbala, líkòbo, ampir, mèbàsì*), tale songs and instrumental music (*lìkànò, ngòmbí, língbidi, èlehú*), lullabies, and play songs (*wàndèlì, na gbo ngo, bè na sòlo*).
29. I have never seen a ground-harp in person, but one is featured on Cradick on Avis (1993, track 11).
30. This section contains translated excerpts of Fürniss (2013a).
31. This term is not specific for Pygmy languages, as it exists in other Oubanguian and Bantu languages (Bahuchet 1992).
32. Four degrees within the range of an octave.
33. Track 11 of the CD contains a very interesting breakdown, part by part, of a song of the nan repertoire.
34. Oloa-Biloa's work is actually being completed by Fürniss who began research on historical recordings of Gyeli music made in Cameroon in 1907 by the German colonial physician Hans Waldow (Ziegler 2006), mainly featuring healing and hunting songs.
35. The women's *lúngà* songs; the "bachelor's dance" *mbálà*, a mixed dance, during which couples dance in turn in the middle of a circle; the foot game *kúrò* and girls' play songs.
36. Seven degrees within the range of an octave, close to an untempered diatonic scale.
37. In Aka, *dìbókà* means "divination fire" and, by extension, "divination and healing rituals and their dances" (Thomas et al. 1993).
38. The information for this parallel comes from different places: from the east of Gabon concerning the Koya and from southeast Cameroon concerning the Kwele. Fürniss has investigated a Kwele group that has recently migrated from a region near the Gabonese border.
39. Last consulted January 25, 2013.
40. UCL Anthropology, London: *Hunter-gatherers' Resilience: Past, Present and Future Adaptations to a World in Transition* (Leverhulme Trust); University of Montreal, *Etude comparative de la musique des populations pygmées du Nord-Congo*; Museum National d'Histoire Naturelle, Paris, *Etude interdisciplinaire de la diversité culturelle et de son évolution* (Labex BCDiv).

References

NB. Audiovisual material is preceded by a *.

General

Abram, Dave, and Jerome Lewis. 2006a. "Pygmy Music: Forest Songs from the Congo Basin." In *The Rough Guide to World Music: Africa and the Middle East,* edited by S. Broughton et al. London and New York: Rough Guides, 304–12.

Arom, Simha. 1976. "The Use of Play-Back Techniques in the Study of Oral Polyphonies." *Ethnomusicology* 20(3):483–519.

———.1974. "Trapping Considered as Liturgy." *The World of Music* 14(4):3–19.

Arom, Simha, Nathalie Fernando, Susanne Fürniss, Sylvie Le Bomin, Fabrice Marandola, and Jean Molino. 2008. "La Catégorisation des Patrimoines Musicaux de Tradition Orale." In *Catégories et Catégorisations. Une Perspective Interdisciplinaire,* edited by F. Alvarez-Pereyre. Louvain: Peeters, 273–313.

Brandel, Rose. 1961. *The Music of Central Africa*. The Hague: Martinus Nijhoff.
Cooke, Peter, and Michelle Kisliuk. 2007–2013. "Pygmy Music." *Grove Music Online*. Oxford University Press. www.oxfordmusiconline.com.
Lomax, Alan. 1968. *Folk Song Style and Culture*. Washington DC: American Association for the Advancement of Science.
Ziegler, Susanne. 2006. Die Wachszylinder des Berliner Phonogramm-Archivs. Berlin: SMPK.

Mbuti

*Demolin, Didier. 1990. *Chants de l'orée de la Forêt. Polyphonies des Pygmées Efe*, CD, Fonti Musicali Fmd 185.
———.1993. "Les Rêveurs de la Forêt. Polyphonies des Pygmées Efe de l'Ituri (Zaïre)." *Cahiers de Musiques Traditionnelles* 6:139–51.
*Demolin, Didier, and Serge Bahuchet. 1991. *Pygmées du Haut-Zaïre*, CD, Fonti Musicali Fmd 190.
*Gansemans, Jos, Didier Demolin, and Jean-Baptiste Nkulikiyinka. 2000. *Archives 1910–1960*, CD, Tervuren, MRAC, Fonti Musicali Fmd 220.
Kisliuk, Michelle. 1991. "Introduction of the 1991 Edition" of Colin M. Turnbull and Francis S. Chapman, *Mbuti Pygmies Sof the Ituri Rainforest*, Washington DC, Smithsonian Folkways, SF 40401, 4.
Sawada, Masato. 1990. "Two Patterns of Choir among the Efe, Forest Hunter-gatherers in Northern Zaïre—Why Do They Love to Sing." *African Study Monographs* 10(4):159–95.
Schebesta, Paul. 1957. "Pygmy Music and Ceremonial." *Man* 62:57–78.
*Tracey, Hugh. [N.D. 1952]. *On the Edge of the Ituri Forest? Northeastern Belgian Congo 1952*. Int. Library of African Music, SWP 009/HT 03.
Turnbull, Colin M. 1961. *The Forest People: A Study of the Pygmies of the Congo*. New York: Simon and Schuster.
———. 1965. *Waywards Servants: The Two Worlds of the African Pygmies*. London: Eyre and Spottiswoode.
*Turnbull, Colin M., and Francis S. Chapman. 1992. *Mbuti Pygmies of the Ituri Forest*. Washington DC: Smithsonian Folkways, SF 40401 [New Edition of FE 4457 and FE 4483].

Aka

*Arom, Simha. 1973. *Musique des Pygmées Aka*, UNESCO, Philips 6586–016 [Reed. 1994, CD, Auvidis, 8054A090].
———. 1978. *Anthologie de la Musique des Pygmées Aka*, Ocora-Radio France 558 526–28 [Reed.1987, C559012/13].
———.1991a. *African Polyphony and Polyrhythm: Musical Structure and Methodology*. Cambridge: Cambridge University Press.
———. 1991b. "La Musique Omniprésente." In *Encyclopédie des Pygmées Aka*, Vol. 1, edited by S. Bahuchet and J. M. C. Thomas. Louvain: Peeters, 227–34.
———. 1994. "Intelligence in Traditional Music." In *What Is Intelligence?*, edited by J. Khalfa. Cambridge: Cambridge University Press, 137–60.
*Arom, Simha, and Geneviève Dournon-Taurelle. 1966. *La Musique de Pygmées Ba-Benzélé*, UNESCO, Bärenreiter-Musicaphon BM 30L2303.
*Arom, Simha, and Denis C. Martin. 1992. *Polyphonies Vocales des Pygmées Mbenzele*, CD, Inédit, W260 042.

Arom, Simha, and Serge Pahaut. 1993. "Une Voix Plurielle." *Cahiers de Musiques Traditionnelles* 6:185–96.
Arom, Simha, and Susanne Fürniss. 1993. "An Interactive Experimental Method for the Determination of Musical Scales in Oral Cultures. Application to the Vocal Music of the Aka Pygmies of Central Africa." *Contemporary Music Review* 9:7–12.
Arom, Simha, and Vincent Dehoux. 1978. "Puisque Personne Ne Sait à l'avance Ce Que Tout Autre Que Lui-Même Va Chanter Dans la Seconde Qui Suit. . . ." *Musique en Jeu* 32:67–71.
Bahuchet, Serge. 1995. "De la Musique Considérée Comme Une Philosophie (Chez les Pygmées Aka de Centrafrique)." In *Ndroje Balendro, Musiques. Terrains Et Disciplines*, edited by V. Dehoux et al. Louvain: Peeters, 57–65.
Dehoux, Vincent, and Henri Guillaume. 1995. "Chasse Sexualité et Musique. Un Arc Musical des Pygmées Aka." In *Ndroje Balendro, Musiques. Terrains et Disciplines*, edited by V. Dehoux et al. Louvain: Peeters, 67–86.
*Epelboin, Alain. 1995a. *Gbédélé, Femme, Fille et Mère. Mongoumba, RCA 1993*, 20 min., CNRS Audiovisuel Coprod. Lacito, La Cathode Vidéo. Online Reference.
———. 1995b. *Chronique Aka 1994. Akungu. 12 le Lémurien et l'araignée, Chantefable*, Film 2'45, SMM-Lacito-CNRS AV, video.rap.prd.fr/video/mnhn/smm/ak94juinzike.rm.
*Epelboin, Alain, and François Gaulier. 1987. *Berceuse" Aka. Akungu, RCA 1987*, 6 Min. 30, Cnrs Audiovisuel Coprod. Lacito, La Cathode Vidéo.
Fürniss, Susanne. 1991a. "The Falsetto in the Sound World of the Aka Pygmies." *Bulletin d'Audiophonologie* 4(5–6):587–606.
———. 1991b. "Recherches Scalaires Chez les Pygmées Aka." *Analyse Musicale* 23:31–5.
———. 1992. *Die Jodeltechnik der Aka-Pygmäen in Zentralafrika*. Berlin: Dietrich Reimer.
*———. 1998. *Centrafrique. Pygmées Aka. Chants de Chasse, d'amour et de Moquerie*, CD, Ocora-Radio-France C560139.
———. 1999. "La Conception de la Musique Vocale Chez les Aka: Terminologie et Combinatoires de Paramètres." *Journal des Africanistes* 69(2):147–62.
———. 2006. "Aka Polyphony: Music, Theory, Back and Forth." In *Analytical Studies in World Music*, edited by M. Tenzer. New York, Oxford: Oxford University Press, 163–204.
*———. 2008. *Arts Pygmées*, Film 30 Min., Camera S. Fürniss and Claire Lussiaa-Berdou, Prod. CNRS-LMS. Online: www.vjf.cnrs.fr/clt/html/audio/videotheque.htm.
Grauer, Victor a. 2009. "Concept, Style, and Structure in the Music of the African Pygmies and Bushmen: A Study in Cross-cultural Analysis." *Ethnomusicology* 53(3):396–424.
*Guillaume, Henri, and Bernard Surugue. 1982. *Chasseurs Pygmées. Pygmées Aka d'Afrique Centrale*, ORSTOM/SELAF CETO 795.
Kisliuk, Michelle. 1997. "Musical Life in the Central African Republic." In *Garland Encyclopedia of World Music*, Vol 1(Africa), edited by R. Stone. New York: Garland, 681–97.
———. 1998. *Seize the Dance! Baaka Musical Life and the Ethnography of Performance*. New York, Oxford: Oxford University Press. Print-on-Demand Edition, 2006.

———. 2000, "Performance and Modernity among Baaka Pygmies: A Closer Look at the Mystique of Egalitarian Foragers in the Rain Forest." In *Music and Gender*, edited by B. Diamond and P. Moisala. Chicago: University of Illinois Press, 25–50.

Kisliuk, Michelle, and Kelly Gross. 2004. "What's the 'It' That We Learn to Perform? Teaching Baaka Music and Dance." In *Performing Ethnomusicology: World Music Ensembles*, edited by T. Solis, Berkley: University of California Press, 249–60.

Lewis, Jerome. 2006b. "Pygmy Music: Songs from the Forests of the Congo Basin." In *The Rough Guide to World Music, Africa and the Middle East*, edited by S. Broughton et al. London and New York: Rough Guides, 304–12.

———. 2009. "As Well as Words: Congo Pygmy Hunting, Mimicry and Play." In *The Cradle of Language*. Vol. 2, *African Perspectives*, edited by R. Botha and C. Knight. Oxford: Oxford University Press, 232–52.

———. 2013 (forthcoming). "Why Do Bayaka Sing So Much? Musical Participation, Cultural Transmission and Egalitarianism." *Before Farming.*

*Lewis, Jerome, Ingrid Pfrang-Lewis, and Nicolas Lewis. 1998. *Massana: Moments in Yaka Play and Ritual*, 36 Mins.

McCreedy, Marion. 1994. "The Arms of the Dibouka." In *Key Issues in Hunter-Gatherer Research*, edited by E. S. Burch and L. J. Ellanna. Oxford: Berg, 15–34.

Sarno, Louis. 1993. *Song from the Forest.* New York: Houghton Mifflin.

———. 1995. *Bayaka: The Extraordinary Music of the Babenzele Pygmies.* Roslyn, NY: Ellipsis Arts.

Thomas, J. M. C., S. Bahuchet, and A. Epelboin (since 1993) and S. Fürniss (since 2003), eds. *Encyclopédie des Pygmées Aka. Techniques, Langage et Société des Chasseurs-Cueilleurs de la Forêt Centrafricaine (Sud-Centrafrique et Nord-Congo)*, Paris/Louvain, SELAF/Peeters: Vol. I.1—*Les Pygmées Aka: Introduction* (1983), Vol. I.2—*Le Monde Des Aka* (1990), Vol. I.3—*La Société* (1990), Vol. I.4—*La Langue* (1990), Vol. II. *Dictionnaire Ethnographique Aka-Français*: 1. P (1981, Reed. 2003), Vol. II.2. B (1993), Vol. II.3. MB-M-V (1993), Vol. II.4. T-D (1998), Vol. II.5. ND-N-L (2003), Vol. II.6. S (2004), Vol. II.7. Z-NZ-NY-Y (2005), Vol. II.8. K (2007), Vol. II.9. G-NG-H (2008), Vol. II.10. KP-GB-NGB-W (2011), Vol. II.11. Voyelles (2013).

Aka and Baka

Arom, Simha, Nathalie Fernando, Susanne Fürniss, Sylvie Le Bomin, Fabrice Marandola, and Jean Molino. 2008. "La Catégorisation des Patrimoines Musicaux de Tradition Orale." In *Catégories et Catégorisations. Une Perspective Interdisciplinaire*, edited by F. Alvarez-Pereyre. Louvain-Paris: Peeters, 273–313.

Bahuchet, Serge. 1992. *Histoire d'une Civilisation Forestière I. Dans la Forêt d'Afrique Centrale: Les Pygmées Aka et Baka*. Louvain-Paris: Peeters.

Bahuchet, Serge, and Jacqueline Thomas. 1986. "Linguistique et Histoire des Pygmées de l'Ouest du Bassin Congolais." *Sprache und Geschichte in Afrika* 72:73–103.

*Fernando, Nathalie. 2011. *Congo. Polyphonies Pygmées du Nord-Congo*, CD, Genève: AIMP VDE 1339.

Fürniss, Susanne. 2012a. "Morphologies et Usages: La Harpe-En-Terre d'Afrique Centrale Face à la Classification Universelle des Instruments de Musique."

In *Annual Meeting of the CIMCIM*, edited by I. De Keyser. Tervuren, MRAC. www.africamuseum.be/research/publications/rmca/online.
———. 2012b. "Musiques Aka et Baka: Une Parenté de Référence." *Journal des Africanistes* 82(1):103–31.
Fürniss, Susanne, and Serge Bahuchet. 1995. "Existe-t-il des Instruments de Musique Pygmées?" In *Ndroje Balendro, Musiques. Terrains et Disciplines*, edited by V. Dehoux et al. Louvain: Peeters, 87–109.
Rouget, Gilbert. 1952. "Note Sur les Travaux d'ethnographie Musicale de la Mission Ogooué-Congo." *Conferência Internacional Dos Africanistas Ocidentais Em Bissau 1947* V(2), Lisboa, 193–204.
———. 2011. "Musical Efficacy: Musicking to Survive—The Case of the Pygmies." *Yearbook for Traditional Music* 43:89–121.
*Rouget, Gilbert, and André Didier. 1949. *Music of Equatorial Africa.* Folkways FE 4402.
*———. 1959. *Musique Pygmée de la Haute Sangha—Mission Ogooué-Congo 1946*, CNRS-Musée De l'Homme, BAM LD 325.

Baka
*Arom, Simha, and Patrick Renaud. 1977. *Musique des Pygmées Baka*, UNESCO, Auvidis D 8029 AD 090 [Reed. CD 1990].
Brisson, Robert. 1999. *Mythologie des Pygmées Baka.* Paris-Louvain: Peeters.
Bundo, Daisuke. 2001. "Social Relationship Embodied in Singing and Dancing Performances among the Baka." In *African Study Monographs*, supplement 26:85–101.
*Cradick, Martin, and Jeremy Avis. 1993. *Heart of the Forest: The Music of the Baka Forest People of Southeast Cameroon*, Coll. Rykomusic HNCD 1378.
Devin, Luis. 2009. *Gli Strumenti Musicali Dei Pigmei Baka del Camerun E del Gabon.* PhD diss., Torino, University of Torino, Italy.
———. 2012a. *La Foresta Ti Ha. Storia Di Un'iniziazione.* Roma: Castelvecchi Editore.
———. 2012b. "Baka Water Drums: Playing Rivers and Streams in the Rainforest of Cameroon and Gabon." *Before Farming [Online Version]* 2012/1 Article 2.
*Fürniss, Susanne. 2005a. *Jeux Chantés des Filles Baka*, Film 24'07 Min., Caméra C. Lussiaa-Berdou, Prod. CNRS-LMS, SMM 0533/CA. video.rap.prd.fr/mnhn/smm/00bakafilles.ram
*———. 2005b. "Femmes, Maîtresses, Mères. Chants et Danses des Jeunes Filles Baka." *Cahiers de Musiques Traditionnelles* 18:217–36.
———. 2008. "The Adoption of the Circumcision Ritual Bèkà by the Baka-Pygmies in Southeast Cameroon." *African Music* 8(2):94–113.
———. 2011. "Partages et Emprunts de Musiques Rituelles au Sud-Est Cameroun." In *Territoires Musicaux Mis En Scène*, edited by M. Desroches et al. Montréal: PUM, 313–28.
———. 2013 (forthcoming). "Sexual Education through Singing and Dancing." In *Music and the Art of Seduction*, edited by F. Kouwenhoven and J. Kippen. Delft: Eburon Academic Press.
*Fürniss, Susanne, and Claire Lussiaa-Berdou. 2004. "Bèkà. Rituel de Circoncision Chez les Baka Occidentaux du Cameroun." Website, www.vjf.cnrs.fr/lms/sf/accueil.htm.

Fürniss, Susanne, and Véronique D. Joiris. 2011. "A Dynamic Culture: Ritual and Musical Creation in Baka Context." *Before Farming [Online Version]* 2011/4 Article 3.

Joiris, Daou Véronique. 1996. "A Comparative Approach to Hunting Rituals among the Baka Pygmies." In *Cultural Diversity among 20th Century Foragers*, edited by S. Kent. Cambridge: Cambridge University Press, 245–75.

———. 1997–1998. *La Chasse, la Chance, le Chant. Aspects du Système Rituel des Baka du Cameroun*, PhD diss., Bruxelles, Université Libre de Bruxelles.

*Sallée, Pierre. [N.D. 1967–1973]. *Gabon—Musique des Pygmées Bibayak*, OCORA/Radio France 558 504.

*Sallée, Pierre, and Hubert de Fraysseix. 1973. *Pygmées et Bochiman*, CBS 80212.

Tsuru, Daisaku. 1998. "Diversity of Ritual Spirit Performances among the Baka Pygmies in Southeastern Cameroon." *African Study Monographs*, supplement 25:47–84.

———. 2001. "Generation and Transaction Processes in the Spirit Ritual of the Baka Pygmies in Southeast Cameroon." *African Study Monographs*, supplement 27:103–23.

Gyeli-Bedzan

Fernando-Marandola, Nathalie. 2002. "New Perspectives on Interactive Field Experiments." *Yearbook for Traditional Music* 34:163–86.

———. 2007. "Study of African Scales: A New Experimental Approach for Cognitive Aspects." In *Trans. Revista Transcultural de Música.* www.sibetrans.com/trans.

Marandola, Fabrice. 2003. *Les Polyphonies Vocales des Pygmées Bedzan du Cameroun: Une Approche Expérimentale du Système Scalaire*, PhD diss., Paris, Université Paris IV.

*Marandola, Fabrice, and Nathalie Fernando. 2007. *Cameroun. Pygmees Bedzan de la Plaine Tikar.* Maison des Cultures du Monde, Inédit W 260095.

Oloa Biloa, Camille. 2011. *La Musique des Pygmées Bagyeli (Sud-Ouest Cameroun). Étude des Conséquences du Changement de Mode de Vie Sur le Patrimoine Musical.* Master's thesis, Nanterre, Université de Paris-Ouest Nanterre.

Bongo-Koya

Bonhomme, Julien, Magali de Ruyter, and Guy-Max Moussavou. 2012. "Blurring the Lines: Ritual and Relationships between Babongo Pygmies and Their Neighbours (Gabon)." *Anthropos* 107:387–405.

De Ruyter, Magali. 2003. *La Musique des Bongo du Gabon: Première Approche.* Master's thesis, Nanterre, Université de Paris X.

Le Bomin, Sylvie, and Jean-Émile Mbot. 2011. "Identité Musicale et Absence de Territoire: Le Cas des Pygmées du Gabon." In *Territoires Musicaux Mis en Scène*, edited by M. Desroches et al. Montréal: PUM, 279–93.

———. 2012a. "The Musical Heritage of the Gabon Bongo: Heritage under Influence." *Before Farming [Online Version]* 2012/1 Article 3.

———. 2012b. "Sur les Traces de l'histoire des Pygmées du Gabon: Résultats de Cinq Ans de Prospection." *Journal des Africanistes* 82(1).

Soengas, Beatriz. 2010. *La Subsistance des Pygmées Bakoya à l'épreuve de l'agriculture: Dynamique des Savoirs Ethnobotaniques et des Pratiques.* PhD diss., Muséum National d'Histoire Naturelle, Paris.

8

Egalitarian Social Organization: The Case of the Mbendjele BaYaka

Jerome Lewis

The Pygmies of Central Africa

The steady reduction in access to forest by Pygmy groups across central Africa has resulted in most being more accurately called "former hunter-gatherers" than hunter-gatherers. Today, the different Pygmy groups are characterized by great diversity (Bahuchet this volume, 2012). One small group in Cameroon, the Medzan, now occupies a savanna; many Twa groups in DR Congo, Burundi, Rwanda, and Uganda have sedentarized among farming communities; and increasing numbers of Baka in Cameroon and Gabon are becoming sedentarized and alcoholized, along roadsides (Agland 2012). All experience increasing pressure from rampant commercial hunting, artisanal and industrial mining and logging activities, protected areas encompassing good forest, and in some places warring militias, government forces, and refugees.

Despite the great diversity of situations that many Pygmy groups find themselves in today, they share some remarkable similarities. In particular, their egalitarian social organization is bound up in a matrix with other key cultural practices. Hewlett identified some of these as spending at least four months a year hunting and gathering in the forest; strongly identifying with and preferring forest life; contrasting the "forest world" to the "village world"; having economies based on demand-sharing; practicing important rituals associated with elephant hunting; having intimate parent-child relations; and diverse relationships with neighboring farming groups (1996).

Ethnomusicologists working among Pygmy groups across the Congo Basin remark on similarities in their unusual highly integrated choral yodeled (alternating between chest and head voice) and polyphonic (multiple overlapping melodies) singing style among groups living very far apart (Arom 1978, 1981, 1985 on western Pygmies; Cooke 1980; Demolin 1993 on the eastern Pygmies; Fürniss 1993, 1999, 2006, 2007; Fürniss and Bahuchet 1995 on western Pygmies; Kazadi 1981 on similarities across Tua or Twa groups in DRC; Merriman 1980 on similarities in DRC; Rouget 2004 on the Pygmy musical style; and others).

Bahuchet tabulated his observations of cultural similarities and differences between Kola, Bongo, Baka, Aka, Twa, Asua, Mbuti, and Efe Pygmies stretching from west to east across the Congo Basin (table 5.1, 1996). Across the region yodel and polyphony together are consistently associated with forest mobility, camps made of round leaf and liana huts, woven-handled axes, and an egalitarian political and economic social order. The greater the degree of acculturation to farmer and village lifestyles the less frequent is yodeled polyphonic music. Those groups Bahuchet identifies as no longer singing polyphonies (Kola and Bongo) are those that Verdu et al. (2009) show to be the most influenced by outsiders' genes (also see Verdu, chapter 2).

The different Pygmy groups have been isolated from one another long enough to develop different languages, genes, technologies and techniques for exploiting forest resources. But there are underlying structural and cultural similarities in music, a predatory and mimetic language style (Lewis 2009), ritual structures (Lewis 2002), identification with a forest hunter-gatherer lifestyle, a gendered division of labor based on the symbolism of blood (Ichikawa 1987; Lewis 2008), economies based on demand-sharing, egalitarian social organization, and their status as the "first people" of the region.

These elements are too specific to emerge from convergent evolution and with genetic evidence proving a shared past, appear to be key components of a highly resilient and effective adaptation to forest hunting and gathering. I will elaborate on the political aspects of this adaptation using the Mbendjele BaYaka as an example of this egalitarian social order.

Mbendjele BaYaka

Mbendjele living in the equatorial forests of northern Republic of Congo (RC) will be the focus of this paper since this is the group I know best[1]. Most Mbendjele spend about two-thirds of the year hunting and gathering in forest camps and some part of the year near agriculturalists' villages. Although continuing to hunt and gather, here they will also trade, labor or perform services for villagers in return for food, goods, alcohol or money. However, the situation varies. Some Mbendjele near the Central African Republic (CAR) are evangelized and although relatively sedentary do not farm. Those living near logging towns may spend long periods working outside the forest. Others further south spend most of the year in the forest, with some groups not coming out to villages for years at a time.

Just like people from other Pygmy groups that I have talked to, the Mbendjele say that they belong to a larger group of forest people generically referred to as "*bayaka*" people. Indeed, Mbendjele more often refer to themselves as *bayaka* than Mbendjele. "Mbendjele" is principally used to distinguish themselves from neighboring *bayaka* groups such as the Mikaya, Ngombe, or Baka. While not concerned about height, *bayaka* is equivalent to the academic term "Pygmy."

The *bayaka* groups I will be focusing on here occupy forest west of the Ubangi River, in CAR, RC, Cameroon, and Gabon. They are made up of Mbendjele (15–20,000), Baka (45–60,000), Aka (15–20,000), and several smaller groups such as the Mikaya, Luma, Kola, Gyeli, Bongo, and others (maybe 10–15,000). Many still largely depend on hunting and gathering in an immediate-return society, though others, such as the Bongo, Kola, Gyeli, Luma, and increasingly Baka too, are engaged in increasingly diversified economies. The term "*bayaka*" is contracted to different extents and used by Aka, Baka, Luma, Mbendjele, and Mikaya, typically as *baaka,* or *baka*. Because the Baka are speakers of an Ubangian language, whereas the Aka and Mbendjele speak Bantu languages, I write the ethnonym as "BaYaka" to emphasize this dual classification. I shall use BaYaka to encompass all these western groups, but their individual ethnonyms when providing specific examples.

BaYaka groups claim shared identity based on common descent from the first forest hunter-gatherers, a shared history, some shared oral traditions (e.g., *gano* fables) and taboo complexes (e.g., *ekila*), an economy based on forest hunting and gathering, ritual and singing styles, and the possibility of marriage relations, but not on trading goods. This contrasts with BaYaka peoples' relations with "village people" that are predominantly based on trading and exchanging goods. Most villagers refuse to marry BaYaka, many will not eat together with BaYaka nor allow them to stay in their homes or villages. Rivers divide the territories of different BaYaka groups so they do not overlap; however, villagers superimpose their land claims over parts of BaYaka land.

BaYaka and Bilo

The Mbendjele distinguish between themselves as "forest people" (*bisi ndima*) with neighboring farming groups who they call "village people" (*bisi mboka*). The Mbendjele clans with whom we lived have exchange relations with four different groups of farmers: the Bongili, Kabunga, Sangha-Sangha and recently with the Bodingo. In addition to those just mentioned, I came across Mbendjele in relations with many different farmer groups: the Kaka, Ndongu, Ngando, Enyelle, Pomo, Yekinga, and Yasua. All these various groups are referred to using the ethnonym "Bilo." Bilo makes a meaningful distinction between non-BaYaka Africans and BaYaka (Pygmy) people that is based on perceived racial, ideological, knowledge, political, and economic differences. Instead of "Bantu," farmer or villager, in this chapter, as elsewhere, I follow their lead and use "Bilo."

Mbendjele describe Bilo as recent arrivals in the forest who discriminate against them, attempt to exploit them, claim rights over their land and labor, and make aggressive claims to own farmland, rivers, forest, and even other people. While Mbendjele resent these claims, Mbendjele elders often emphasize that it is their transience in the forest that makes Bilo claims vacuous and therefore not to be taken too seriously. Bilo are useful for providing Mbendjele with access to goods from outside the forest (notably iron and salt), and appreciated for their role in judging disputes between Mbendjele that the community is unable to resolve.

There exists a developed oral tradition that elaborates and entrenches cultural stereotypes differentiating BaYaka forest people from Bilo village people through accounts of the past. These numerous and widely told stories (*gano*) attest to the enduring and elaborate nature of the opposition between them. The cultural significance of the contrast between forest people and village people has been commented on by other ethnographers in central Africa as one of the most fundamental markers of ethnic difference in forested regions (see for instance Turnbull 1966; Bahuchet and Guillaume 1982; Waehle 1986; Grinker 1994; Kenrick 2000; Lewis 2001; Kenrick and Lewis 2001; Köhler and Lewis 2002). Where forest people no longer have access to forest, they speak the same language and share many similar cultural practices and beliefs with their farmer neighbors, and these oppositions do not break down. Indeed, they can become more entrenched as segregation and discrimination increase, as has happened to the Twa Pygmies of the Great Lakes Region (Lewis 2000).

The Egalitarianism of Hunter-gatherer Societies

In the late 1970s, James Woodburn developed his comparative analysis of the ethnography of hunter-gatherers to show that they could be divided into "immediate-return" or "delayed-return" societies (1982). Although taking economic activity as the starting point, the implications of the difference between immediate- and delayed-return societies go well beyond economics to determine key aspects of social structure and political organization. So, for instance, the sharing and immediate consumption of whatever has been hunted or gathered has political consequences because it ensures that individuals do not accumulate more than others. Individuals are therefore unable to use accumulated goods to exert authority or to oblige or influence others to do their will. Neither do people invest in long-term production strategies that would involve long-term binding commitments between them. People in immediate-return societies do not depend on specific others for access to food, land, resources, or tools and so can move easily should they so wish. In these societies pressure is not put on people to produce, but on them to share whatever they have produced.

By contrast, people in delayed-return societies invest labor over long periods before a yield is obtained. Typical examples include farming, herding, or capitalist systems, but also certain hunter-gatherer societies that invest labor over time or store yields (such as the Kwakiutl and Inuit, and most Amazonian farmer-foragers, including the Ache). The requirement to manage labor during the period in which the yield is being produced results in relations of dependence and authority developing between people to assure that labor is put in at the right times and that those who contribute are recompensed when the yield is obtained, and so willingly provide their labor again. Control over the distribution of vital resources promotes political inequality and hierarchy through the emergence of elites. Whereas delayed-return societies are by necessity hierarchically organized with inequalities between peers, seniors and juniors and gender groups, immediate-return societies are politically and economically egalitarian. While both delayed and immediate return societies exist among hunter-gatherers, only delayed return societies exist among non-hunter-gatherers.

In this paper I will focus on an immediate-return hunter-gatherer group to avoid the distorting effect of mixing delayed return hunter-gatherers into discussions of egalitarianism and inequality. By definition, delayed return societies cannot be egalitarian; therefore, the analyses of "egalitarian" societies done by Alden Smith et al. (2010) that consider delayed return, non-egalitarian societies as comparable to immediate-return ones produces confused and ultimately arbitrary results, based on unreasonably forced categories of analysis such as "relational wealth," "grip strength," and so on that focus on individual variation rather than the social mechanisms (such as demand-sharing) that ensure individual variation does not result in inequality. Social mechanisms must override individual variation for a society to be egalitarian.

While individual variation in skill or ability will exist everywhere, immediate-return societies impose economically egalitarian relations through procedures that force sharing on anyone with more than they can immediately consume and so prevent saving and accumulation. A range of mechanisms, notably demand-sharing but also gambling (Hadza), ritual (Pygmies), or gifting (San), ensure that valued goods circulate without making people dependent on specific other people. People are systematically disengaged from property and therefore from the potential for property to be used to create dependency. As a consequence, each member of such a society can freely move where they want, has direct individual access to the resources on which they depend for survival, and to the means of coercion. Such societies are politically egalitarian because no one can force others to do their will. People who brag or try to assert their wishes or views on others are mercilessly teased, fought, avoided, and, if they persist, even exiled.

Such societies are indeed rare today, but include some Pygmy groups in central Africa (Aka, Baka, Efe, Mbendjele, Mbuti); Hadza in Tanzania, some San groups in Namibia and Botswana; several groups in India such as the Jarawa and Ongee Andaman Islanders, Hill Pandaram, and Nayaka; and in Southeast Asia, the Agta, Batek, Maniq, Penan, and others. Though numerically insignificant today, these societies are hugely significant for anthropology because their egalitarian immediate-return orientation represents such a radically different mode of social organization to the numerous hierarchically organized delayed-return systems that currently dominate human societies.

Reviewing how his typology had stood up to the evidence from thirty years of new ethnography Woodburn (2005) noted that immediate-return societies have shown remarkable resilience over time. They are stable and enduring systems, internally coherent and meaningful to those who live in them. Despite the combined forces of government sedentarization and assimilationist policies, agricultural expansion, industrial exploitation, and forest conservation all putting huge pressures on these societies, they tenaciously cling to their immediate-return lifestyle. Rwanda's Twa Pygmies, for instance, are mostly denied access to forest and farmland. The majority work as itinerant day-laborers, beggars, potters, bar musicians, and similar jobs that provide immediate remuneration (Lewis and Knight 1995; Lewis 2000).

In northern Congo local, Bilo perceptions of the differences between Mbendjele and Bilo people were once expressed to me with surprising similarities to Woodburn's distinction between "immediate-return systems" and "delayed-return systems." A young educated Kabunga Bilo expressed it, unprompted, like this:

> Despite production being for subsistence, the Kabunga organize themselves to make reserves of food for the future. Using elementary conservation techniques, they preserve food from the harvest, fishing and hunting. In contrast, the *moyaka* [singular of *bayaka*] will always consume all the food he has before going to look for more. The BaYaka are most sociable people, their whole lives, and all activities, are carried out in groups. Their lives are in eternal communion with each other.
>
> In contrast, the Kabunga, whose life has evolved, is inclined to a solitary existence: "Each man for himself, God for all!" While the Kabunga wastes his time making politics, organizing himself, seeking to uplift his land and village, the Pygmy is totally preoccupied with the politics of the bush. He searches to discover all the possible procedures to trap or capture wild animals. From the point of view of education, the Kabunga orientates his children towards schooling, the BaYaka, to the domain of the bush.
>
> There is much conflict between the BaYaka and Kabunga. Most quarrels are caused by the BaYaka refusing to work. There are also quarrels caused by capricious acts committed by the BaYaka, such as theft, abuse of confidence, and refusal to honor debts.[2]

Woodburn's distinction is valuable because of this correlation with academic and local perceptions. Moise presents the relationship between economic outlook, sociality, and political relations with impressive clarity. Similar observations of the radical difference between immediate-return hunter-gatherers and their neighbors in other places have resulted in anthropologists elaborating distinctions of their own. For example, Barnard's contrast between the "hunter- gatherer mode of thought" and "accumulation modes of thought" (2001), Biesele's hunter-gatherer "way of thinking" or "imaginative substrate" (1993) that persists even when hunting and gathering are no longer possible, or Lee's concept of "communal foraging relations of production" (1981). Ingold argues that hunter-gatherer sociality is such "a radically alternative mode of relatedness" (1990) that the term "society" is inappropriate.

Demand-sharing and Ekila

While I will outline the Mbendjele's system for distributing material property through demand-sharing, similar practices are well-known from the work of anthropologists such as Blurton-Jones 1987; Ichikawa 2005; Peterson 1993; and Woodburn 1982 and 1998. Demand-sharing is the core practice that ensures egalitarian economic relations. In contrast to the donor-organized sharing familiar to most people, where the person owning the resource dispenses it according to their whim, demand-sharing is recipient controlled. Potential recipients constantly demand shares of things they suspect may be around. It is the donor's duty to give whatever they are requested; refusal is impolite even offensive. This is crucial to prevent sharing being manipulated to the donor's advantage.

For most material items, need determines who can claim the item, especially when they are consumable (Lewis 2005 provides more detail). In this context, possessing something is more like a guardianship or caretaker role until someone else needs it. Certain personal possessions, such as a woman's basket, her cooking pots and machete, and a man's bag, his spear, knife and axe, are recognized as belonging to named individuals, often the person who made, found, took, or bought the item. These individuals have priority over others' claims to the item. But when not in use by them, any of these objects will be shared on demand with someone who needs it.

Mbendjele men and women share in different ways. This is related to gender roles and their different productive activities. Women's gathering activities are geared to exploiting labor-intensive but dependable food sources for the regular provisioning of food for the family. These commonly include various wild yams, edible leaves and insects, ground-growing vegetables such as mushrooms and certain fruit, small fish, and crustaceans. When more than can be immediately eaten is gathered, the food is shared among all present in the forest before returning to camp. Once in the camp, women prepare and cook the food and share it again by sending plates (*djalu*) to the men's area (*mbandjo*) and to their female friends and relatives at other hearths. Women's production is rarely shared out on arrival in camp as the men's is.

Men specialize in obtaining foods with potentially large yields, such as wild animals, honey, and occasionally large fish. If a hunter returns with a large animal, it is publically taken from him as he enters camp by other men. They supervise the butchering and ensure the sharing is done equitably before being cooked and further redistributed by the women as they do with other food. Because the meat of game animals, though unpredictable, may be obtained in large amounts, it must be carefully shared out among all present to avoid possible favoritism or manipulation.

Rules called *ekila* determine exactly how each species should be butchered and to whom different parts should go. The hunter's meat (called *ekila*) is the heart; the men get the liver and kidneys (*piko*); a dog that participated would get the lungs, and so on. The remaining meat must be fairly shared among all present or the hunter's luck will be ruined. If sharing is not conducted according to *ekila* rules, it jeopardizes future success and the well-being of the camp. *Ekila* instills an ideology of proper sharing that is the key to the safe enjoyment of forest resources and the guarantee of their continued abundance (Lewis 2008).

Ekila taboos serve to enforce and define proper sharing: By not sharing animals and meat properly among all present, a hunter's *ekila* is ruined so that he is unsuccessful. If parents of infants eat *ekila* animals, it can provoke illness and even death in their children. If either husband or wife inappropriately shares his or her sexuality with others outside their marriage, both partners have their *ekila* ruined. A menstruating woman is *ekila* and must share her menstrual blood (also *ekila*) with spirits so that her male relatives continue to find food. Even laughter should be shared properly. Laughter shared between people in camp during the evening makes the forest rejoice, whereas laughing at hunted animals ruins the hunter's *ekila*.

Consistent with an egalitarian ethic that seeks to avoid singling out individuals for praise or condemnation, *ekila* provides a neutral medium for discussing success and failure. Thus, difficulties in the food quest or procreation are discussed in relation to *ekila* rather than to inadequacies in human skill or the environment's ability to provide. People recognize each other's skills, but it is impolite to refer to them. Rather, success is talked about in terms of proper conduct in personal and mystical relationships as defined by *ekila* taboos.

Sharing creates and sustains social relations of equality and affection. The importance of affection in defining sharing relations has consequences for spatial organization. There are "circles" of sharing depending on the quantity and type of food or other good and the degree of affection between people. The Mbendjele household (*mongulu*) is the basic unit of sharing. Any household member consumes whatever enters the household freely and normally without restraint. A typical household might consist of a young married couple with small children, the wife's mother, and unmarried younger siblings, all sleeping in the same liana and leaf house. Another common household type is based around a mature married couple living with a wide age range of children and sometimes grandchildren. The children will often include some of their own but also some of their siblings' and others' children. Children, like adults, value mobility and change their residence freely and easily should they desire to. In this way, affection is vital to keeping members of a household together.

Normally only people who like each other live together in the same camp as avoidance is a common way of resolving disputes. In situations of temporary high population density (such as during dry season ceremonies, when feasting on large game, or when visiting Bilo villages) Mbendjele often group their houses together with those people they particularly like and will tend to share more with them than those farther away. When numbers get over about sixty people, Mbendjele often make several separate camps close to each other to respect peoples' differences.

Sharing between camps is less frequent, but will occur when big game is killed and during *massana* forest spirit performances. When an elephant is killed, Mbendjele in the area go rapidly to where the carcass is lying. Large camps grow, and feasting and dancing go on until the elephant has been consumed. During lifting-of-mourning ceremonies (*eboka*), many *massana* spirit plays are performed, especially the three-day dance of *Ejengi*. People come from all around to join in. In addition to sharing out the euphoria of forest spirits, spirit plays share out prized consumables—meat, honey, wild yams, alcohol, tobacco, marijuana, and other goods, including money. Initiations often take place, and initiation fees will have to be paid. These fees are immediately redistributed among all present. In the past, people paid with coils of metal, alcohol, and food. Now they also use money.

Everyone is encouraged to share according to ability, but if you are old, physically or mentally challenged in some way and only rarely contribute, your entitlement is *not* diminished. You have just as much right as anyone else to demand a share of whatever comes into camp. Living in such a society is like living in a place where goods are free. If you do not have what you need, you simply look around to see who might have it and ask them for it. If it is a tool or object, when you have finished with it, they or someone else may ask you for the item again, and so it continues traveling around the community. If it is food, people will politely help themselves to the meal that you are eating.

The principle is that if someone has something that you need just ask them for it; and, as Mbendjele often say, "Since we have easy hands we just give it." Mbendjele adults should epitomize this quality by being generous to a fault, so they will give away all that is asked of them, even when this results in their having nothing left for themselves. They contrast this behavior with the "hard hands" of Bilo villagers.

Demand-sharing is not a form of tolerated theft, indirect exchange, or of generalized reciprocity, as has been suggested (Blurton-Jones 1987; Peterson 1993). Hunters do not leave the carcass in the bush, but make the often major effort of bringing it back to camp so that the men's *piko* meat is shared so as to ensure future success. Additionally, it tends to be the same people provisioning the group most of the time (Woodburn 1998; and my personal observations). They are forced to share more than others, while being denied any recognition of their greater contribution.

The implicit valuation of equality between members of the group can result in some surprising behavior from an economic perspective. Men are very sensitive to who is provisioning the camp with meat. Individuals who hunt a lot will become a target for teasing and mockery, even cursing, if people perceive that the group is eating their production too often. In contrast to models of economic behavior that assume that good producers will get recognition, status, and fame, here it is not the case. They stop hunting for a while rather than be subjected to teasing, gossip, and jealousy. I have described a man who was an obsessive hunter (Lewis 2003). Despite repeated calls for him to stop hunting so much, he continued. Eventually the women of his camp formed a coalition that refused to cook any meat that he killed. This was so offensive that he left to live with neighboring Luma Pygmies where he remains to this day. In effect, the women exiled him for producing too much.

While I have traced out the ways that material goods are shared on demand between Mbendjele, their behavior toward certain types of knowledge is different. Knowledge, such as the rights to perform a particular ritual and certain medicinal knowledge or mystical techniques, are not shared on demand, but selectively traded. There is a cultural logic to this seeming incongruity between the way people transact intellectual goods and the demand-sharing of material goods.

Komba, the creator and guardian of the world (*konja yombo*), made creation for all creatures to share. This is set out in *ekila* rules that organize sharing and are said to originate from this time. No individual or species has any greater right than any other to the forest and its resources. Once, when roared at by a silverback gorilla for camping too close, my Mbendjele companions were so annoyed that they shouted terribly rude insults back. It was unacceptable that the gorilla should claim part of the forest as his own. Similarly, they resent villagers' claims to own forest and fields and often refer to villagers simply as gorillas because of this likeness. Because *Komba* created all material things for all creatures to share, anyone can take what they need or demand it from someone who already has it.

By contrast, certain products of our own deductions, inspirations, dreams, and discoveries can belong to us. They only exist because someone thought or dreamed them into being. While the material world that *Komba* brought into being is shared on demand, as *Komba* wanted, people's ideas can be subject to exchange, negotiation, and trade. It seems that because they are the product of a particular person's imagination, their creator can decide on how they should be distributed.

Many choose to share their herbal remedies freely on demand, others may only do so in exchange for something else. It seems to depend on the individual.

It is similar for certain mystical procedures. For instance, the obsessive hunter had been cursed by other men to meet gorillas when he went in the forest. It was surprising how often he was charged by silverbacks. The Mbendjele healer who knew the remedy to this curse began by demanding several thousand francs payment to provide it. In the end, he settled for a 1000 CFA and a handful of cigarettes on the day he made the special liana-string necklace to protect the hunter. Typically, once a payment has been made, it is subject to demand-sharing just like any other item. Only by quickly hiding the item will the recipient have any hope of keeping some for later.

Mobility, Disputes, and Moadjo

As I hope my discussion of sharing illustrates, egalitarianism is not a passive state. It is an assertive, dynamic process that depends on a complex of interdependent practices that constantly resist the emergence of hierarchy, dependency, and inequality. Like demand-sharing, two key mechanisms that assure egalitarianism are mobility and "avoidance strategy" (Woodburn 1982). Rather than confront someone who is trying to oblige you to do something, or seeks to exert authority over you, or with whom you have a dispute, move away from them. Because adults do not depend on others for access to vital resources, they can simply and easily move away. If mobility is a leveling mechanism, it depends on this absence of dependency. Mbendjele encourage mobility from an early age. Children that can walk can chose where they sleep, and some often spend the night with other kin or friends rather than with their parents. When people leave a camp, they give no public reason. As they leave, those remaining sometimes say "*duaké!*"—"Go!"

In an egalitarian society, no one can play the role of "judge" because this would imply status or authority. Occasionally people may discuss a recurring problem and collectively suggest a solution, but no one has the authority to impose it. Often one party simply moves away, without even acknowledging the dispute. While mobility and avoidance works well most of the time, sometimes, as when large ceremonies are called, individuals in conflict meet up again. Because both wish to remain for the ceremony, their latent conflict may reignite, especially if alcohol is consumed. Although fighting is publicly frowned upon, it is seen as a legitimate means of expressing indignation or resisting others. Mbendjele theatrically structure fighting to minimize the potential for injury, depending on the seriousness of the combatants and their gender (Lewis 2002).

Despite the availability of powerful weapons, including spears, crossbows with poisoned arrows, occasional guns, axes, and the ever-present machetes, Mbendjele strongly dislike and disapprove of combat that draws blood.[3] It is *ekila*, and despite having witnessed numerous fights, I have never seen an Mbendjele use such weapons against a person.[4] However, the theoretical possibility that someone could pick up a poisoned arrow and wound an aggressor when they sleep or are not paying attention is a powerful deterrent against pushing someone too far, no matter how weak they may physically be.

While arguing and fighting are immediate ways that people can deal with their differences, there is a more institutionalized process of shaming called *moadjo* that is the monopoly of elderly women.[5] Some time after an event in which someone behaved particularly stupidly or unacceptably, one or two women will rise and begin comically reenacting the event. They will not say who they are mimicking but repeat the scene many times as an audience collects around them. The audience, among much hilarity, will begin shouting out comments to accompany the action. Although all are able to guess who is being ridiculed, their name is never mentioned.

By comically mimicking the wrongdoer, the women elicit a moralistic commentary from their audience that, by the end of the show, has served to communally map out the moral high ground. *Moadjo* educates those present about Mbendjele values. Children and younger girls tend to be less vocal in their comments, but laugh loudly. Older women quickly become boisterous, supporting the actors by making jokes and offering explicit but humorous condemnation of mimicked behavior. Mbendjele men only tolerate such explicit criticism from women. If men do this, it easily leads to serious fights. Widows have a special place in this type of humorous but directed criticism and are expected to do this in front of the whole camp at moments of high tension or when someone has committed a grave error. A good performer will succeed in calming the atmosphere by allowing everyone to laugh at themselves. Indeed, if the person being criticized is present, the *moadjo* will only end when they laugh publicly too. However, on realizing that they are becoming the center of the camp's mirth, the wrongdoer often flees and hides in the forest until things calm down.

Camp Organization—*Mosambo*

When non-BaYaka strangers arrive at an Mbendjele camp, they are often presented with a male *kombeti*. The term means "elder," and this man often becomes the main interlocutor between the strangers and the camp. Outsiders often interpret this as a sign that the *kombeti* is a "chief." This is misleading. From an Mbendjele perspective, whoever is the oldest in a group is the *kombeti*, and their responsibility is to provide anything that younger members demand from them. This is as true for a group of playing children as it is for a group of women, men, or any other group. Every camp therefore has many *kombeti*—ones for the women, for the men, for the young men, for the young women, for the girls, and for the boys of the camp. Unaware of this plurality, outsiders often expect to make decisions with "the" *kombeti* and for these decisions to be respected by the rest of the group. This causes problems because no one has the right to decide things on someone else's behalf, nor any authority beyond their individual charisma and skill at establishing consensus.

While I was visiting in 2012, an elder (X)—a renowned healer, former elephant hunter and now a drunk—presented himself as the "chief" of the camp to a newly arrived Bilo man. As they walked up the camp, the newcomer politely engaged him in conversation. X began insisting that he be shown "chiefly" respect by being bought some alcohol (a common practice among Bilo). As his demands became increasingly insistent, a group of boys playing nearby started to call out in Lingala *"X a di djoba"* (X is an idiot), *"a di faux Mokondij!"* (He's a fake chief!). He became embarrassed and, ignoring the children, moved the newcomer away from them and closer to the alcohol seller. They continued playing. Mbendjele recognize specialists but attribute them no authority and like these children, instinctively undermine those claiming status or privilege, regardless of age or gender.

Rather than depending on a recognized individual to coordinate activity, the camp is organized in nonhierarchic way through a public-speaking protocol called *mosambo.* It is the means by which the camp communicates with itself, organizes activities, and resolves problems. It should be heard twice a day, in the morning and evening, but anytime somebody addresses the whole group, it is *mosambo*. Through *mosambo* camp members inform the camp of what they have done, express their opinions, advise camp members, share news of general interest, and seek a consensus, or not, about what the camp will do and who should do what. It also provides a forum for children to learn about social and moral values and about the etiquette of public discussion.

Although the prospective speaker can be any member of the camp, some people are better at *mosambo* than others and they may be asked by others to speak for them. A "good" speaker (*lipwete*) is not a persuasive speaker, but one who is able to express the main points of view in camp with eloquence and humor. Those who are too shy, unable, or risk provoking trouble if they speak, often approach such a person and tell them what they wish said on their behalf.

The person wishing to speak shouts, "*Oka, oka, oka*!" (Listen, listen, listen!) and only begins speaking when the camp is silent. Even toddlers and small children are expected to be quiet. During the full length of the speech, no one should interrupt the speaker. Speech during a *mosambo* has a particular style. Words are stretched slightly, shouted rather than spoken, and short intervals are left between subjects. Listeners use expletives to accompany key moments in the speech and express their reaction to what is said. Humor is an important component of a good *mosambo*, especially when the orator is angry or upset. Once finished, the speaker says, "*Angamu ncia*" (Mine is finished), and anyone who wishes to speak may now begin.

The essence of *mosambo* lies in its role to advise, criticize, and organize the camp. The individual speaking, especially during an evening *mosambo*, is expected to express what most people think or want to do anyway. Ideally, *mosambo* leads the group by consensus. People who do not agree may make this known by punctuating the speaker's speech with appropriate expletives. They may or may not choose to speak when the speaker finishes.

Nobody has the right to oblige others to do anything that they do not want to do. If individuals, families, or groups do not agree with a *mosambo,* they might not say so publicly. But the following day when everyone else leaves to a particular place, they go elsewhere. This will happen without remonstrations from others, and their right to do as they please is respected. Sometimes, although the men may agree, the women do not. If the women act in solidarity and refuse to do what the men proposed in their *mosambo,* the men are forced to follow the women's decision.

The process by which the content of the *mosambo* is decided is relevant here. In camp, men tend to congregate at the *mbandjo.* This is, at most, a simple lean-to where the young unmarried men (*ngendja*) of camp sleep. It is often only a couple of logs on the ground for seating where men congregate and take their communal evening meal. As men return from the forest, they sit down here to chat. It is rude to ask questions, so a man is left his own time before beginning to speak. This comes easily as the men's conversations are dominated by the events of the day, by accounts of what people heard, saw, and did. Those who went to different places give their accounts, and time is taken to allow women's experiences to be shared with the men.

A woman who wants to share something with the men walks close to the *mbandjo* and addresses another woman or sympathetic listener in a loud voice so all can hear. If she is angry, her *mosambo* may slowly increase in volume as she repeats her main point, often emphasizing it with sweeping downward arm movements and sung expletives. She may even begin moving around the central space in a parody of the recounted events. As different accounts and points of view on the day's affairs are heard, men discuss them and slowly arrive at a consensus.

As the men discuss what they hear, they add their own points of view. After some time, one will suggest the *mosambo* to the other men. They listen and express agreement by exclaiming "*bonaapɛ*," by repeating key themes and embellishing the details. If they disagree, they remain silent until the *mosambo* is over and, waiting for a suitable moment, propose their own ideas or version of events that will in turn be discussed by the men present. The men may call certain individuals to the *mbanjo* to explain something in particular. When there is agreement on the subject of the evening *mosambo,* one of the men, considered a good speaker (*lipwete),* will be proposed, or volunteer, to give the *mosambo.* In this way, the *mosambo* should ideally express the majority view of the camp. It is in effect the camp talking to its members.

While evening *mosambo* tend to focus on advising people, the day's activities, and what people should do the next day, morning *mosambo* often focuses on the camp's internal dynamics. This generally takes two forms, one *mosambo* that occurs very early in the morning and tends to be individual complaints, then another after dawn that reinforces the messages of the previous evening's *mosambo.*

Typically, in the twilight before dawn, a speaker with a problem paces up and down the middle of the camp calling out their *mosambo* to the sleeping and waking occupants of the huts. Such speech often features an individual's complaints directed at a particular person (or group), though never mentioning their name, not even indirectly. General phrases will be used, such as "people who do this sort of thing are bad," as the affair that caused conflict is explained, often with the talker becoming increasingly agitated. Sometimes such *mosambo* provoke arguments and even serious fights. But mostly, morning *mosambo* is used to make a point and get the annoyance off the speaker's chest.

After dawn, *mosambo* tends to focus on reinforcing the consensus expressed during the previous evening's *mosambo.* "The young men *(boka)* should cut palm nuts at that abandoned plantation so-and-so saw yesterday! You young unmarried girls *(bangondo)* must bring us back lots of *mea* (wild yams) from that place the hunters saw!" and so on. Rather than identifying individuals with authority or status, camp life is organized through the institution of *mosambo.*

Specialist Roles

This doesn't mean that Mbendjele do not recognize individual skill or expertise, but rather that such recognition is not associated with any special advantage or privilege. Specialist roles are held by both men and women, except for men's role as *tuma* (elephant hunter). While the titles *konja mokondi* (spirit guardian), *kombo* (song composer), *lipwete* (speaker), and *nganga* (healer) recognize that certain people are particularly skilled or knowledgeable about a particular activity, they get no privilege or special treatment from this recognition. Rather each role is recognized because the activities that they are associated with are potentially dangerous or stressful to the community as a whole. People described by one of these titles are expected to manage these stressful situations well for the benefit of all, so that they have a positive outcome.

During the lifting-of-mourning ceremonies (*eboka*) invited guests will expect daily high-quality spirit play performances. This places considerable pressure on the hosting community. On their behalf the spirit guardians must ensure that the appropriate clothing is obtained for the spirits, that proper procedures are followed, that singers are enthusiastic, and so on. While the spirit guardians gets no more alcohol or smoke than anyone else and are not able to oblige anyone to do anything, they are expected to humorously cajole, persuade, and encourage all to do what is necessary for a successful performance.

In contexts such as elephant hunting, childbirth, severe illness, or during rituals, a particular individual may be recognized as having more skill or knowledge than others and is consulted for their advice when that area of knowledge is useful. But they have no ability to oblige anyone to do as they suggest, no recognition beyond this activity, and no privilege or lasting benefits from the role.

Egalitarian Gender Relations

A society is not egalitarian if individual differences such as strength, gender, or age confer any lasting status or authority. Mbendjele recognize, cultivate, and celebrate gender differences, but value them equally. To understand egalitarian societies, it is necessary to understand that individual variation and equality coexist, so to understand gender egalitarian societies, it is necessary to recognize that gender difference and equality coexist (Endicott and Endicott 2008).

Mbendjele men and women spend most of their waking time apart; in the forest, women gather and fish together with other women and children, and men go looking for honey or hunting in smaller male-only groups. Sometimes couples go on romantic foraging trips, but it is not the norm unless it is a very small camp. In camp, men spend most of their time sitting at the *mbandjo*, talking to other men and caring for children, while women sit at their hearths and talk to each other in a particularly songlike speech-style (Lewis 2009).

Before today's society existed, the stories tell that women lived independently from men fishing and collecting wild yams. The men lived in another part of the forest, hunting and collecting honey (Lewis 2002). The gendered work roles in these stories are the actual work roles of men and women today. Their daily spatial separation into gendered groups and spaces reinforces the contemporary political significance of their original mythical autonomy.

I have mentioned how *ekila* defines proper sharing; other aspects of this important polysemic concept define the gendered division of labor as a natural consequence of gendered bodies (Lewis 2008 provides detail). For instance, menstruation (called *ekila*) is the focus of ideological elaboration that has important consequences on women's solidarity. The smell of menstruation (and pregnancy) is said to provoke dangerous animals to attack, so women walk in large noisy groups and often sing to warn the animals. This communalism in their daily lives cultivates intense solidarity, so women quickly support one another in situations of conflict with men and can resist men's decisions or demands should they wish.

An Mbendjele woman or man does not depend on anyone for direct and unrestricted access to food and their basic needs. Men cannot control women's labor nor the yield of women's labor. Neither can they control the destination of women in marriage because they cannot oblige a woman to marry anyone against her wishes, and a woman wishing to divorce a man simply leaves him without any requirement to justify herself. Such an absence of dependency is the necessary prerequisite for egalitarian relations (Woodburn 1982). A person can exert power over others only to the extent that he can withhold basic requirements such as food and shelter, access to key resources, or marriage partners.

This degree of autonomy could imperil communal life if it were not for other areas of *ekila*. A complex ideological balance of difference and interconnectedness between men and women resists the tendency to fragmentation that each sex's potential for autonomy could lead to. Mbendjele often discuss this in terms of "cutting" (*moena*) and "tying" (*mokata*). Thus during *Yele* singing sessions, certain women enter trance and "tie-up" the spirits of game animals in order for men to be able to find them. Men "cut" the life of the animal and butcher it. Women's cooking "*ties*" the meat back into the community. Men "cut the moon" in order for women to become pregnant. Women "tie-up" the man's semen to grow the fetus. Women "cut" the umbilical cord at birth to separate baby and mother; later the father provides the name that "*ties*" the baby into society and his clan. At death, women are ritually concerned with cutting the spirit from the body (sending the spirit to *Komba*) and the men with burying the body and ensuring the body is "*tied*" into the earth.[6]

These dialectics of cutting and tying negate claims to higher status by either men or women by attributing the valued production of one sex to actions by the other sex and emphasizes the equally important contribution each makes to valued social production. Thus, women grow men's children by turning semen into a fetus and men kill animals that women's mystical activities have made available. These ideological leveling mechanisms effectively cut each gender group off from the potential status derived from the high social value of the tasks they uniquely perform, while tying them back together for successful production and social life to occur.

Massana—The Importance of Play

This ideology of gendered complementarity and difference is learned and reinforced in activities that the Mbendjele call *massana*. *Massana* can be translated either as "ritual" or "play": Mbendjele do not make the distinction. Beginning with the casual play of children, *massana* activities develop as a person grows to involve a wide range of games including role-playing games and spirit play rituals (*mokondi massana*) where forest spirits are sung to, to attract them to dance in camp to share euphoria among all present.

Masssana activities are based on the principle that the better the participants coordinate with each other the more pleasure they experience. They are the major social arena for learning gendered forest skills, cooperation, and the group coordination that is crucial to the success of hunting and gathering. Mbendjele explicitly work to establish a certain quality of relations between participants during *massana*: no arguing or shouting, and all must contribute as best they can.

Doing *massana* educates and genders Mbendjele in particular ways that I cannot fully describe here, but Lewis 2002 and 2013 provide more detail. This is most explicit during initiation ceremonies into the ritual associations responsible for each of the spirit plays. Each association has its sacred path (*njanga*), secret lore, and defined group of initiates (*bangonja*) responsible for preparing the spirit play and calling the spirit out of the forest. Hidden knowledge is shared: among women this concerns catching the spirits of game animals so men can kill them, "telepathy," using "sexiness" to control and manage men, maintaining fertility, safe childbirth, and healthy child rearing; for men, this concerns hunting, honey collecting, traveling in the forest (night walking, high-speed displacement, invisibility, etc.), and making themselves "awesome"—impressive, handsome, and fearsome.

Spirit plays occur often, sometimes nightly, sometimes weekly. Regularly performing them inculcates an egalitarian aesthetic of gendered interaction. As an example, the widespread spirit play called *Ejɛngi* combines the men's and women's groups together in a typical way for a wide range of *massana* that involve the whole community. Women's beautiful singing and sexual attractiveness lure and excite *Ejɛngi*, men's strength, and fearlessness controls and manages *Ejɛngi* safely. By combining their differences, they succeed in bringing him out of the forest into the human group, so all are able to share *Ejɛngi*'s joy (*bisengo*). These gender roles structurally resemble those of hunting and cooking. Men bring in the raw meat. Women cook it and enable all to gain energy from it. When men are with *Ejɛngi*, he is raw. Women's dancing cooks *Ejɛngi*, getting him hotter and hotter so his wild energy is enjoyed by all.

The different but complementary roles each gender plays during such spirit play instills certain ways of coordinating and structuring group activities. Participating in spirit play also inculcates distinctive ways for individuals to coordinate themselves. There is no hierarchy among singers, no authority organizing participation, but all must be present and give their best to create a rich overlapping polyphony. Each singer must harmonize with others but avoid singing the same melody; if too many sing the same part, the polyphony dissolves. Thus each singer has to hold his or her own and resist being entrained into the melodies being sung around them. This cultivates a particular sense of personal autonomy that is not selfish or self-obsessed, but is keenly aware of what others are doing and seeks to complement this by doing something different.

This has organizational implications in a society where daily hunting and gathering activities are intuitively coordinated without someone telling people what to do. If too many do the same thing, there may be nothing to eat, so being musically primed to do something different but complementary to others improves the chances that the camp will eat well without explicit coordination. Similarly, knowing a sufficient range of melodies and when to insert them into a song structurally resembles the way environmental knowledge is used to identify and extract resources from the forest efficiently (Lewis 2013).

Developing musical skill and regularly participating in musical performances seems to prime participants to culturally appropriate gendered ways of interacting with human, plant, and animal others. People's everyday choices are rarely made explicit but are instinctively understood by others because they are based on this musically shared aesthetic sense of what one ought to do. This social aesthetic is the central dynamic silently organizing daily camp life in a society where no one, not even parents to their children, can oblige others to do their will.

Certain spirit plays such as *Ngoku* and *Yelle* for women or *Sho* and *Niabula* for men focus on celebrating and cultivating gender differences: *Ngoku*, fertility and child health; *Yelle*, abundance of meat; *Sho*, courage, strength, and awe; and *Niabula*, invisibility and elephant hunting. During these gendered spirit plays, especially in sacred areas, Mbendjele publicly elaborate on the particular qualities and strengths of the initiates' gender. Using song and dance, these gendered ritual coalitions communicate this with the rest of society.

During the women-only spirit play of *Ngoku*, for instance, the interlocked body of the singing women dances up and down the central area of camp. As more and more women join them, they take over the camp, and men politely retreat with their sons to bathe or rest in the shade. As they begin a new song, whoever stopped the last song sings out a line, such as "*baito wonda to njɔmbɔ; dɔtɔ ba die ebɛ!*" (Women chase young men; old men are no good!) or "*Mapindi ma mu pola!*" (Their testicles are empty!), so all know which melodies to draw on to create the appropriate polyphony. Dancing as one interwoven body, this is "Woman" speaking to men.

Men, in turn, speak as "Man" to the women during spirit plays, such as *Sho* or *Niabula*, as they stamp up and down the camp, bound together as one, they frighten but also attract, making themselves desired but respected. A pattern of assertion and counterassertion is a central dynamic maintaining egalitarian relations between the sexes, allowing each group to publicly define, celebrate, and express their value to the rest of society. Individuals passing through these institutions explore what these identities mean as they move through life. Crucially, they do so without requiring explicit teachers.

By singing together, each gender group reinforces its message to the other gender, and repetition strengthens the point rather than annoying or tiring listeners as it would if spoken. The corporate body can speak and be understood without needing to single out a leader. The singing and dancing group can say things that no individual in the group could say without fearing repercussions. Strong, provocative, insulting, or political statements can be made or enacted without giving the intended recipients space to respond or interrupt. Tensions inherent in gender relations are expressed and acknowledged, even if not resolved. Indeed, these public assertions and explorations of gendered uniqueness and value are a key force in maintaining egalitarian gender relationships as Finnegan (2009, 2013) explores in more detail.

Massana celebrate gender and emphasizes independence yet interdependence, antagonism yet desire, separation and unity, subversion and respect, and the management of gender relations by same-sex solidarity, by taunt and praise or shaming and loving. These seemingly contradictory emphases are a critical dynamic by which the Mbendjele maintain relatively egalitarian gender relations. Relatively because at different moments, most obviously during *massana,* but at other times too, either men or women may appear to dominate the other gender group. However, this appearance is transitory. Learning this begins early with the children's spirit play called *Bolu.* Here, the children dominate the camp, demanding, and mostly receiving, respect and obedience from the adults. The interplay between men's and women's groups is better represented as an egalitarianism that depends on each gender group asserting itself effectively in front of the other and in spite of the other.

Reproducing an Egalitarian Society

There is no explicit discourse on "equality." Rather the implicit valuation of equality crucially underpins the cultural logic of complex cultural concepts such as *ekila* and the modes of participation required of key activities such as *mossambo, moadjo,* and *massana.* Establishing consensus, witnessing humorous reenactments of unacceptable behavior, participating in spirit plays, or respecting *ekila* rules of sharing and behavior exert an anonymous but pervasive pedagogic action that prompts each Mbendjele person to understand egalitarianism as a valued political, moral, and economic orientation. These are pedagogic processes that do not depend on defining any individual as a focus for Mbendjele to learn. Rather, they teach piecemeal over a lifetime, and to different extents in each individual, not through verbal exhortation but by the communal experience of a series of bodily practices and proscriptions and the pleasure, curiosity, and satisfaction these provoke.

The cultural instability of egalitarian societies that Brunton (1989) claimed is not so. Observing the inability of egalitarian societies to judge new innovations, Brunton reasoned that such societies are inherently unstable, their practices haphazard or accidental assemblages, and their continued existence fortuitous.

But institutions such as *massana* or *ekila* show how values and meanings can be condensed to establish a cultural store that ensures internal continuity between generations without attributing special status or authority to individuals. Their basically nonlinguistic nature means that it is difficult to articulate them explicitly as a coherent body of ideas. This makes them difficult to manage by "authority."

The persistence of the cultural matrix remarked upon at the beginning of this chapter that connects up those Pygmy groups still able to hunt and gather in the forests of the Congo Basin is remarkable because certain groups have not had contact for many thousands of years. Yet their similar polyphonic yodeled singing style, their hunting-focused rituals, their preference for forest activities and living in forest camps of dome-shaped leaf and liana huts, an economy organized by demand-sharing, a division of labor based on the symbolism of blood, and an egalitarian social and political organization remain so strongly associated together because they constitute an exceptionally resilient cultural adaptation to forest living in this region.

The sophisticated integration of this matrix of adaptations to the forest is currently menaced by the immense power of capitalism and modern technology to appropriate, transform, and degrade the environment. Mbendjele were protected from Bilo accessing forest by their dependence on the Mbendjele to guide them. This was also true of other outsiders, such as loggers, until recently. However, sophisticated navigation and positioning technology now allows anyone independence from BaYaka. The ability of bulldozers to make roads and chainsaws to fell trees has accelerated turning the forest into cash. Large urban developments have mushroomed around the activities of logging and mining companies. Road networks now spread throughout the forest and have opened up of previously inaccessible areas to commercial exploitation, often by professional hunters supplying urban dwellers with bushmeat. Other outsiders, we call conservationists, seal-off large areas of good forest from local use and enforce seemingly arbitrary hunting restrictions in surrounding areas. The impact of these recent occupations of the forest is that local people—both Mbendjele and Bilo—see their autonomy and resource base diminishing.

Hunter-gatherers and former hunter-gatherers across central Africa face similar situations. Everywhere they have great difficulty resisting their dispossession because their numbers are small and because their egalitarian social organization undermine anyone who emerges as a leader, rejects their authority, and generally favors avoidance over confrontation, sharing over private property, individual freedom over organized representation, and immediate over delayed returns on labor.

While there is a growing international indigenous movement in central Africa, they remain mostly marginalized from those they seek to represent, and from mainstream society and politics (Lewis 2001). At the local and national levels, the severe discrimination "Pygmies" experience from many non-hunter-gatherers, their inexperience of the world outside the forest, and the huge profits to be made by commercializing forest resources makes these processes likely to continue without much regard for the future of forest hunter-gatherers. A major challenge is to find ways for their voices to have an influence on important decisions over their lives and resources while respecting their egalitarian principles.

Notes

1. PhD research was supported by the Wenner-Gren Foundation, an Emslie Horniman Scholarship and the Swan Fund (1994–1997). I return regularly to the Mbendjele area.
2. Moise Taito, Kabunga chief's son and second-year psychology university student, aged twenty-six, 1996.
3. The early French military explorer Captain Cottes was struck by this dislike or "fear of spilling human blood" (1911).
4. In 2011, an Mbendjele shot another man in the forest when out hunting. The circumstances are unclear, but he was accused of murder and has been languishing in jail ever since.
5. Men occasionally do this, but rarely as elaborately as women.
6. I use italics to show where I have inferred, rather than heard, the use of these terms.

References

Agland, P. 2012. *A Cry from the Rainforest.* BBC, River Films. 90 minutes.

Alden Smith, E., K. Hill, F. Marlowe, D. Nolind, P. Wiessner, M. Gurvenf, S. Bowles, M. Borgerhoff Mulder, T. Hertz, and A. Bell. 2010. "Wealth Transmission and Inequality among Hunter-gatherers." *Current Anthropology* 51(1):19–34. Doi:10.1086/648530.

Arom, S. 1978. "Puisque personne ne sait à l'avance ce que tout autre que lui-même va chanter dans la seconde qui suit…." Entretien avec Vincent Dehoux, *Musique en jeu* n 32:67–71.

———. 1981. "Un ethnomusicologue chez les Pygmées." Entretien avec A. Jaubert, *La Recherche 123*, 768–73.

———. 1985. *Polyphonies et polyrythmies d'Afrique centrale, structure et méthodologie.* Paris: SELAF, 2 vol., 906.

Bahuchet, S. 1996. "Fragments pour une histoire de la Forêt Africaine et de son peuplement: les données linguistiques et culturelles." In *L'alimentation en forêt tropicale: interactions bioculturelles et perspectives de développement*, edited by C. M. Hladik, A. Hladik, H. Pagezy, O. F. Linares, G. J. A. Koppert et A. Froment. Paris: Éditions UNESCO, 97–119.

———. 2012. "Changing Language, Remaining Pygmy." *Human Biology* 84(1):11–43. DOI: http://dx.doi.org/10.3378/027.084.0101.

Bahuchet, S., and H. Guillaume. 1982. "Aka-farmer Relations in the Northwest Congo Basin." In *Politics and History in Band Societies*, edited by Eleanor Leacock and Richard B. Lee. Cambridge: Cambridge University Press, 189–211.

Barnard, A. 2001. *The Hunter-gatherer Peoples: Three Lectures Presented in Argentina.* Buenos Aires: Fundacion Navarro Viola.

Biesele, M. 1993. *Women Like Meat: The Folklore and Foraging Ideology of the Kalahari Ju/'hoan,* Johannesburg: Witwatersrand University Press.

Blurton-Jones, N. 1987. "Tolerated Theft, Suggestions about the Evolution of Sharing, Hoarding and Scrounging." *Social Science Information* 26 1:31–54.

Brunton, R. 1989. "The Cultural Instability of Egalitarian Societies." *Man,* New Series, 24(4):637–81.

Cooke, P. 1980. "Pygmy Music." In *The New Grove Dictionary of Music and Musicians* 15:482–3.

Cottes, A., (Capitaine). 1911. *La Mission Cottes au Sud Cameroun (1905–1908).* Paris: Leroux.
Endicott, K. M., and K. L. Endicott. 2008. *The Headman Was a Woman.* Long Grove: Waveland Press Inc.
Demolin, D. 1993. "Les rêveurs de la forêt. Polyphonies des Pygmées Efe de l'Itouri (Zaïre)." *Cahiers de Musiques Traditionnelles* n 6, "Polyphonies," 139–51.
Finnegan, M., 2009. "Political Bodies:Some Thoughts on Women's Power among Central African Hunter-gatherers. *Radical Anthropology Group Journal* 3: 31–8.
———. Forthcoming 2013. "The Politics of Eros: Ritual Dialogue and Egalitarianism in Three Central African Hunting and Gathering Societies." *Journal of the Royal Anthropological Institute.*
Fürniss, S. 1993. "Rigueur et liberté: la polyphonie vocale des Pygmées Aka (Centrafrique)." In *Polyphonies de tradition orale. Histoire ettraditions vivantes,* edited by C. Meyer. Paris: Créaphis, 101–31.
———. 1999. "La conception de la musique vocale chez les Aka: terminologie et combinatoires de paramètres." *Journal des africanistes* 69(2), 147–62.
———. 2006, "Aka Polyphony: Music, Theory, Back and Forth." In *Analytical Studies in World Music* (chapter 5), edited by Michael Tenzer. Oxford University Press, 163–204.
———. 2007. *Approche interdisciplinaire des musiques pygmées,* HDR, dir. Serge Bahuchet, Université Paris X-Nanterre.
Fürniss, S., and S. Bahuchet. 1995. "Existe-t-il des instruments de musique Pygmées?" In *Ndroje balendro. Musiques, terrains et disciplines.* edited by Vincent Dehoux, Susanne Fürniss, Sylvie Le Bomin, Emmanuelle Olivier, Hervé Rivere, and Fréderic Voisin. Textes offerts à Simha Arom. Louvain-Paris: Peeters (Selaf 359), 67–86.
Grinker, R., 1994. *Houses in the Rainforest: Ethnicity and Inequality among Farmers and Foragers in Central Africa.* Berkeley: University of California Press.
Hewlett, B. 1996. "Cultural Diversity among African Pygmies." In *Cultural Diversity among Twentieth-century Foragers: An African Perspective,* edited by Susan Kent. Cambridge: Cambridge University Press, 215–44.
Ichikawa, M. 1987. "Food restrictions of the Mbuti Pygmies, eastern Zaire." *African Study Monographs,* suppl. 6:97–121.
———. 2005. "Food sharing and ownership among central African hunter-gatherers: an evolutionary perspective." In *Property and Equality,* Vol. 1, edited by Thomas Widlok and Wolde Tadesse, Berghahn Books, 151–64.
Ingold, T. 1990. Comment on "Foragers, Genuine or Spurious: Situating the Kalahari San in History," by J. Solway and R. Lee. *Current Anthropology* 31:130–1.
Kazadi, N. 1981. "Méprises et admires: l'ambivalance des relations entre les Bacwa (Pygmées) et les Bahemba (Bantu)." *Africa* 51(4): 836–47.
Kenrick, J., 2000. "The Forest Peoples of Africa in the 21st Century: Present Predicament of Hunter Gatherers and Former Hunter Gatherers of the Central African Rainforests." *Indigenous Affairs* 1/00: 10–24. Copenhagen: IWGIA.
Kenrick, J., and J. Lewis. 2001. "Evolving Discrimination against the Forest People ('Pygmies') of Central Africa." In *Racism against Indigenous Peoples,* edited by Suhas Chakma and Marianne Jensen, Copenhagen: Asian Indigenous and Tribal Peoples Network (AITPN) and IWGIA, 312–25.

Kisliuk, Michelle. 2001. *Seize the Dance: BaAka Musical Life and the Ethnography of Performance*. Oxford: Oxford University Press.
Köhler, A., and J. Lewis. 2002. "Putting Hunter-Gatherer and Farmer Relations in Perspective. A Commentary from Central Africa." In *Ethnicity, Hunter-gatherers, and the "Other": Association or Assimilation in Southern Africa?*, edited by Susan Kent. Washington: Smithsonian Institute, 276–305.
Lee, R. 1981. "Is There a Foraging Mode of Production?" *Canadian Journal of Anthropology* 2:13–9.
Lewis, J. 2000. *The Batwa of the Great Lakes Region*. Minority Rights Group Report, London: Minority Rights Group.
———. 2001. "Forest People or Village People: Whose voice will be heard?" In *Africa's Indigenous Peoples: "First Peoples" or "Marginalized Minorities"?*, edited by Alan Barnard and Justin Kenrick, Edinburgh: CAS, 61–78.
———. 2002. *Forest Hunter-gatherers and Their World: A Study of the Mbendjele Yaka Pygmies and Their Secular and Religious Activities and Representations*. PhD Thesis, University of London.
———. 2003. *The Hunter's Curse*. Film for the "What's going on?" video and document annotation tool. London School of Economics. 7 minutes. http://elearning.lse.ac.uk/dart/wgo/wgoLevel3.html
———. 2005. "Whose Forest Is It Anyway? Mbendjele Yaka Pygmies, the Ndoki Forest and the Wider World." In *Property and Equality*. Vol. 2, *Encapsulation, Commercialisation, Discrimination*, edited by Thomas Widlok and Wolde Tadesse, Berghahn Books, 56–78.
———. 2008. "Ekila: Blood, Bodies and Egalitarian Societies." *Journal of the Royal Anthropological Institute* 14:297–315.
———. 2009. "As Well as Words: Congo Pygmy Hunting, Mimicry and Play." In *The Cradle of Language*. Vol. 2, *African Perspectives*, edited by Rudolf Botha and Chris Knight. Oxford: OUP, 232–52.
———. 2013. "A Cross-cultural Perspective on the Significance of Music and Dance on Culture and Society, with Insight from BaYaka Pygmies. In *Language, Music and the Brain: A Mysterious Relationship*, edited by Michael Arbib. MIT Press.
Lewis, J., and J. Knight. 1995. *The Twa of Rwanda: Assessment of the Situation of the Twa and Promotion of Twa Rights in Post-war Rwanda*. IWGIA Document 78. London: World Rainforest Movement/Copenhagen: IWGIA.
Merriam, A. 1980. "Zaïre. 3. Pygmy Music." In *The New Grove Dictionary of Music and Musicians* 20, coordinated by Stanley Sadie. London: MacMillan, 623.
Peterson, N., 1993. "Demand Sharing: Reciprocity and the Pressure for Generosity among Foragers." *American Anthropologist* 95(4): 860–74.
Rouget, G. 2004. "L'efficacité musicale: musiquer pour survivre. Le cas des Pygmées." *L'Homme* n: 171–2, "Musique et anthropologie," 27–52.
Turnbull, C. 1966. *Wayward Servants: The Two Worlds of the African Pygmies*. London: Eyre and Spottiswoode.
Verdu, P., F. Austerlitz, A. Estoup, R. Vitalis, M. Georges, S. Théry, A. Froment, S. le Bomin, A. Gessain, J-M. Hombert, L. Van der Veen, L. Quintana-Murci, S. Bahuchet, and E. Heyer. 2009. "Origins and Genetic Diversity of Pygmy Hunter-Gatherers from Western Central Africa." *Current Biology* 19:312–8.

Waehle, E. 1986. "Efe (Mbuti Pygmy) Relations to the Lese Dese Villagers in the Ituri Forest, Zaire: Historical Changes During the Last 150 Years." *Sprache und Geschichte in Afrika* 7.2:375–411.

Woodburn, J. 1978. "Sex Roles and the Division of Labour in Hunting and Gathering Societies." Unpublished paper presented at the Conference on Hunting and Gathering Societies, Paris, June 1978.

———. 1982. "Egalitarian Societies." *Man, the Journal of the Royal Anthropological Institute* 17, no. 3:431–51.

———. 1998. "'Sharing is not a Form of Exchange': An Analysis of Property Sharing in Immediate-Return Hunter-Gatherer Societies." In *Property Relations: Renewing the Anthropological Tradition,* edited by Chris Hann. Cambridge: Cambridge University Press.

———. 2005. "Egalitarian Societies Revisited." In *Property and Equality.* Vol. 1, *Ritualisation, Sharing, Egalitarianism,* edited by Thomas Widlok and Wolde Gosse Tadesse. Oxford and New York: Berghahn, 18–31.

9

Hunter-Gatherer Childhoods in the Congo Basin

Barry S. Hewlett

Introduction

Children represent about 40 percent of Congo Basin hunter-gatherer communities, but the vast majority of studies are conducted with adults. Only 13 percent of the 345 entries in a bibliography on Congo Basin hunter-gatherers (Hewlett and Fancher 2010) concentrate on children. Bird-David (2005) and Hirschfeld (2002) reflect on some of the reasons why children are so invisible in hunter-gatherer and anthropological studies. They describe how the status and conceptions of children in the West as well the fact that researchers seldom have children of their own when they conduct their PhD research contribute to the limited research with children. Studies of Congo Basin forager children are important for a variety of reasons. How can one characterize a forager culture without talking with or attempting to understand how almost half of the population thinks and feels about particular issues? How do individuals acquire the egalitarianism, autonomy, and extensive giving described in other chapters in this book? At what age do individuals learn particular forest skills and knowledge, from whom do individuals learn, and how do they learn (e.g., observation, imitation, teaching, stories) the skills and knowledge essential to survive in the rainforest? What special skills and knowledge do children have that are essential to dealing with other children (i.e., children's culture)? These and other questions are addressed in this chapter.

The chapter is divided into three parts. The first part describes general features of Congo Basin forager childhood that are relatively distinct from their farming neighbors. The second part examines cultural diversity in childcare patterns, both between and within forager ethnic groups. The final part identifies and discusses the anthropological, developmental psychology and evolutionary biology theories that researchers used to guide their studies with Congo Basin hunter-gatherer children.

Studies in the Overview

The chapter reviews 77 studies of forager children; 46 from the Hewlett and Fancher bibliography mentioned above and 31 more recent child-focused publications. All of the child-focused research has been conducted with four ethnic groups—the Aka, Baka, Mbuti and Efe—so the discussion of intercultural and intracultural variability is limited to these groups. The Aka, Bofi, and Mbendjele of the Central African Republic (CAR) and Republic of Congo (ROC) are considered together. The Bofi foragers were Aka net hunters until about fifty years ago when they started to associate with Bofi farmers and over time adopted the Bofi language and began to view themselves as different from Aka foragers (e.g., have fewer forest spirit powers), although they seldom farm and continue to net hunt. The Mbendjele speak a dialect of the same C10 Bantu language as the Aka, regularly interact with Aka, and both groups refer to themselves BaYaka or BiAka.

Table 9.1 lists the first authors of the child-focused publications on Congo Basin forager children within particular ages. Several biases and limitations exist in the Congo Basin hunter-gatherer childhood literature. The vast majority of studies have been conducted with infants and young children (59 percent of all studies); relatively few studies exist on middle-aged children (12 percent) and adolescents (16 percent). The remaining studies (14 percent) included children from several age groups. The Efe do not have any studies on children over age three, and the Mbuti lack any child-focused studies until adolescence.

The published sample is also biased toward studies with the Aka/Bofi/Mbendjele (66 percent of studies); relatively few child-focused studies have been conducted with Baka (14 percent), Mbuti (4 percent), and Efe (16 percent). The ethnic group bias is due, in part, to the fact that Hewlett has conducted research with Aka infants for over forty years, has trained several graduate students to work with Aka/Bofi, and they in turn have published articles on the same ethnic group. Few studies exist with Mbuti and Efe because of political instability and the termination of field research in the Democratic Republic of Congo (DRC) since the mid-1980s. Japanese researchers started child-focused research with Baka children about ten years ago.

Most of the Congo Basin foragers in this review were relatively "traditional" (i.e., high mobility, temporary shelters, regular foraging, and limited use of guns) at the time the research was conducted. The Efe and Mbuti were the most mobile and traditional, in part, because data were collected in the 1950s (Mbuti studies) or 1980s (Efe studies). Aka experienced intermediate levels of acculturation, depending on the year of study and location, while the Baka encountered the greatest number of acculturation forces because of government sedentarization programs and formal schooling.

Table 9.1. First authors of published works on child-centered research on Congo Basin hunter-gatherers.

Ethnic Group	Infancy (0–1 year)	Early Childhood (2–5 years)	Middle Childhood (6–12 years)	Adolescence (13–18 years)
Aka/Bofi/ Mbendjele	B. S. Hewlett, Meehan	Fouts	Berry, Boyette, Neuwelt-Trunzler, van de Koppel	B. L. Hewlett, Lewis, Takeuchi
Baka	Hirasawa	Avis	Kamei,Sonoda	Hagino, Hayashi
Efe	Ivey Henry, Tronick, Winn,	Morelli		
Mbuti				Turnbull

Note: See references cited for a list of the authors' publications.

Common and Distinctive Features of Forager Childhood

This section of the chapter describes features of Congo Basin hunter-gatherer childhoods that are (1) common to ethnic groups with data at various ages and (2) relatively distinct from childhood in neighboring farmer groups. Although pronounced limitations and gaps exist in the literature, some consistent and distinct patterns have emerged from child-focused studies with Congo Basin foragers. Some of the features reflect common physical and social settings of hunter-gather life (e.g., camps with 10–35 individuals) or similar cultural values, such as respecting the autonomy of others. Brad Schore (1998) calls these core values foundational schema—ways of thinking and feeling that pervade and cut across many domains of social-emotional life. It is important to briefly describe some of the foundational schema of both the forest foragers and their farming neighbors because they profoundly influence features of childhood in both groups.

The four forager groups share foundational schema that are associated with hunters and gatherers in many parts of the world—egalitarianism, autonomy and giving/sharing (Lee and Daly 1979). Lewis (chapter 8) explains how forest-forager egalitarianism is transmitted and maintained, but in general it is a way of thinking where individuals respect each other's strengths and weaknesses and it is not appropriate to draw attention to oneself or judge someone as better or worse than another. Men and women, young and old, are viewed as relatively equal and have similar access to resources.

Respect for an individual's autonomy is also a foundational schema among foragers. One does not coerce others, including children. Men and women, young and old, are generally free to do what they want. If an infant wants to play with a machete, she is allowed to do so. Finally, a giving or sharing way of thinking also pervades Congo Basin forager life. Individuals often share 50–80 percent of foraged foods with others. They share their food with most everyone in camp and do this on a daily basis. This foundational schema also applies to childcare and information. Individuals regularly provide allomateral care and share forest information and knowledge with anyone who is interested.

Congo Basin farmers have foundational schema that are relatively distinct from those of the foragers: gender and age hierarchy, communalism, and material/economic dimensions to social relations (Turnbull 1965; Vansina 1990). Village women are expected to defer to the requests of men, and the young should be respectful of elders, whether they are older siblings or parents. Communalism refers to the cultural value placed on putting the needs of the group, generally clan members or the extended family, over the needs of an individual and the importance of relying upon and expecting support from these specific others. The third foundational schema refers to the thoughts and feelings that interpersonal relations have economic or material components. Material and economic dimensions of relationships are embedded within the social and emotional aspects of relationships.

The common and relatively distinct (i.e., statistically distinguishable from farmers) features of forager childhood described below are often based on systematic observational studies, with both forest forager and farmer children living in association with each other. For instance, studies with the following Congo Basin forager and farmer ethnic groups were conducted simultaneously: Efe forgers and Lese farmers, Bofi foragers and Bofi farmers, Aka foragers and Ngandu farmers, and Baka foragers and Bombong farmers. The forager and farmer communities often have similar fertility and mortality rates and are exposed to similar infectious and parasitic diseases.

The first five features occur across all ages—from infancy through adolescence—the sixth and seventh features are specific to infancy and early childhood, the eighth is specific to middle childhood, and the final feature is specific to adolescence.

Physical and Emotional Intimacy throughout Childhood

Physical and emotional proximity are particularly important to forest hunter-gatherers. Forager camps are generally very dense, often occupying a space the size of a large dining and living room in the United States or the space of one or two farmer houses. When hunter-gatherers sit down in the camp, they are usually touching someone. In terms of holding during infancy and early childhood, figure 9.1 shows that Aka three- and four-month-old infants were held 91 percent of the day and farmer infants were held substantially less frequently at 54 percent of the day (Hewlett et al. 2000). Bofi forager two-, three-, and four-year-olds were held 44 percent, 27 percent, and 8 percent of daylight hours, while farmer children of the same age were held only 18 percent, 2 percent and 0 percent of the day (Fouts et al. 2005). Longer breastfeeding among the Bofi foragers was not the sole factor that influenced holding in young childhood; fully weaned Bofi foragers were held 30 percent of daylight hours, which is ten times more than fully weaned young Bofi farmer children and more than Bofi farming children were held even while still breastfeeding (Fouts et al. 2005). During middle childhood, Boyette (2012) found that Aka forager children were significantly more likely than Ngandu farmer children to be in physical contact with (1) others, (2) more individuals, (3) a greater age range of individuals, and (4) children of the opposite sex.

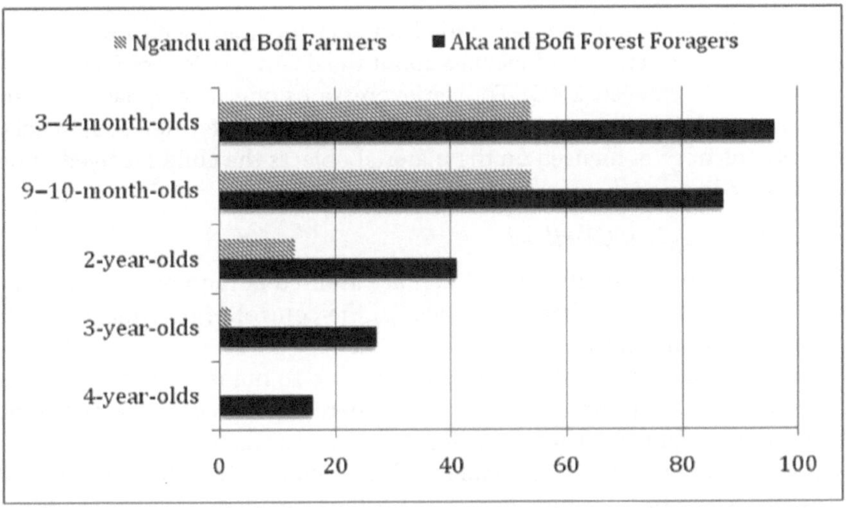

Figure 9.1. Percentage of time forager and farmer infants and young children are held.

Among the Efe, infants are constantly held (Tronick et al. 1987; Ivey 2000), but holding and touching decreases in early childhood, in part, because weaning occurs earlier among the Efe than it does among the Aka or Bofi (18–24 months for Efe versus 3–4 years of age for Aka and Bofi) foragers (Morelli 1987; Morelli and Tronick 1992). Among the more sedentary Baka, young infants are held 85 percent of the day, which was significantly more time than neighboring Bombong farmers held their young infants (Hirasawa 2005).

The proximal relations continue into the night. Aka forager children are less likely than Ngandu farmer children to sleep alone at night, especially in adolescence, and Aka beds are much smaller and have more people per bed than do neighboring farmers (Hewlett and Roulette, in press).

The importance of physical as well as emotional proximity to others is illustrated in two studies. In a study of conflicts between toddlers and older juveniles among Bofi hunter-gatherers and farmers, Fouts and Lamb (2009) found that Bofi forager toddlers were substantially more likely to have conflicts over staying close to juveniles (38 percent of conflicts among forager toddlers versus 2 percent of conflicts among farmer toddlers), while Bofi farmer toddlers were more likely to have conflicts with juveniles over competition for objects (48 percent of farmer toddler conflicts versus 14 percent of forager toddler conflicts) or over the juvenile hitting the toddler, which never occurred among the Bofi hunter-gatherer toddlers. This study illustrates early acquisition and manifestation of foundational schema—emotional proximity to others among the forest hunter-gatherers and the economic-material dimensions of social relations among the farmers.

In another study, Aka forager and Ngandu farmer adolescents were asked about their experiences and feelings about the death and loss of friends and relatives (B. L. Hewlett 2005). Forager expressions of grief emphasized their love and emotional connections to the person, while farmer adolescents' expressions of grief focused on the material objects the child received upon the death of a relative.

Autonomy throughout Childhood

Congo Basin hunter-gatherer children are granted autonomy in their daily lives, while farmer children are subject to the control of parents and older children. For instance, Hewlett found that Aka forager three- and four-month-old infants took the breast on their own to nurse during 58 percent of feeding bouts by comparison to only 2 percent of feeding bouts among farmers. Ngandu farmer mothers decided when to nurse, not the infant. At weaning, Bofi forager mothers said the child decided when she or he wanted to wean, while Bofi farmer mothers said they decided when to wean the child. The forager mothers said that if they initiated the weaning it would cause the child to get sick, whereas the farmers said nursing too long causes the child to become lazy (Fouts et al. 2001). In a study of cosleeping among the Aka foragers and Ngandu farmers (Hewlett and Roulette, in press), forager parents indicated their children slept wherever they wanted, whereas farmer parents said they told their children where to sleep.

Mixed-age Groups throughout Childhood

Adult-child groups. Time with parents and other adults, generally grandparents, gradually declines with age in most cultures, but by comparison to farmers, foragers spend considerably less time in child-only groups. Figure 9.2 summarizes Fouts's data on who is proximal (i.e., within arm's reach) to young Bofi forager and farmer children; forager children were much more likely to be proximal to more categories of people and parents and other adults than were farmers. By ages four to five, Bofi foragers are still proximal to parents and adults 33 percent of the time, while farmer children are proximal to them only 6 percent of the day. Bofi farmer children at this age spent 59 percent of their day in child-only groups, while Bofi hunter-gatherer children spent only 18 percent of their day in proximity to child-only groups. Boyette (2013) found that four- to sixteen-year-old Aka forager children spent considerable time in mixed-aged child groups, but they were still within visual range of an adult 64 percent of the day and within six meters of parents or other adults 45 percent of the day.

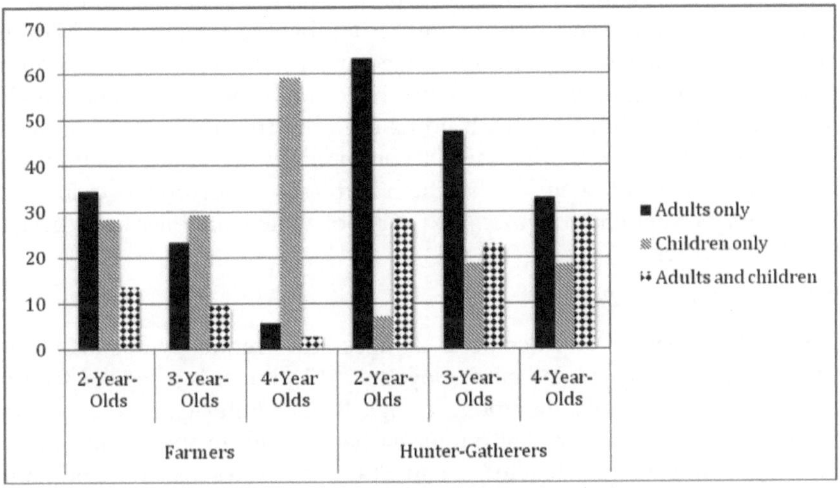

Figure 9.2. Percentage of time Bofi hunter-gatherer and young Bofi farmer children are in proximity of adults and other children.
Source: Data from Fouts, modified from Hewlett et al. 2011.

An early behavioral study of Aka children by Neuwelt-Trunzler (1981) found a similar pattern: middle-childhood Aka spent 40 percent of their day in mixed adult-child proximity groups (defined as the three closest individuals to child) and 30 percent of their day in child-only proximity groups while in a camp setting. Outside of the camp, these children spent 70 percent of their time with an adult social or work group and 30 percent of their time with a child-only social or work group. This is also consistent with Boyette's (2013) recent finding that children are more likely to spend time with adults in the forest and less likely to spend time with younger children in camp. Neuwelt-Truntzer (1981) found that this pattern continued into Aka adolescence.

The pattern of greater time near adults continues beyond daylight hours. The cosleeping study (Hewlett and Roulette, in press) found that Aka forager children and adolescents were more than three times more likely than Ngandu farmer children of similar age to sleep with parents or other adults.

Mixed-age child groups. Turnbull (1965a) described multiage play groups of Mbuti children long ago, and several studies since that time have confirmed this pattern among the Aka (Neuwelt-Trunzer 1981; Boyette 2013), Efe (Morelli 1987; Rogoff et al. 2010), and Baka (Kamei 2005). Adults are still relatively close-by, generally within visual or hearing distance. Boyette indicates that substantial social learning occurs in these mixed-age play groups, and Kamei shows how the different activities of children in mixed-age groups pertain to dealing and interacting with other children (i.e., child culture).

These patterns are generally different from those among farmer children, who are more likely to play in similar-aged groups. The differences are often a matter of demography; forager camps have about twenty-five individuals, half of which are children, which limits the opportunities for similar-aged children to interact. Likewise, farmer communities are larger, often having a population size in the hundreds, which increases opportunities to interact with peers. Farmer children often go to school, which amplifies age segregation even further.

Play throughout Childhood

Play is highly valued and occurs at all ages. The frequency declines from middle childhood to adolescence (Boyette 2013), but it is a regular feature of both child and adult life among Congo Basin hunter-gatherers. Several researchers have reported that hunter-gatherer children spend most of the day playing and are not expected to contribute much to subsistence or maintenance (Konner 2005, 2010; Kamei 2005). By comparison, children in farming communities are more likely to be given responsibilities for childcare and other tasks (Barry et al. 1959). Boyette found that forager four- to sixteen-year-old children spent a considerable amount of time playing (26 percent of the day) and lying around (resting 28 percent of day). Aka forager play is relatively equally divided between solitary play, social play, and imitation of work in play (Boyette 2013). Kamei's (2005) study of types of play among seven- to fifteen-year-old Baka hunter-gatherers identified eighty-five different types of games, the majority (61 percent) dealing with hunting and gathering, camp life (cooking and childcare), and singing and dancing. All of this play takes place in multiage child groups, and most of the play involves learning about making a living as a hunter and gatherer as well as learning about the modern world (e.g., making wooden trucks, playing soccer).

Quantitative studies of Aka (Hewlett et al. 1998), Bofi (Fouts 2009), and Efe (Morelli 1987) infants and young children demonstrate that play occurs early and regularly in Congo Basin forager life. Quantitative and qualitative studies of play in middle childhood and adolescence among the Mbuti (Turnbull 1965a), Mbendjele (Lewis 2002), and Aka (Boyette 2013; B. L. Hewlett 2013) provide rich descriptions of how social play contributes to learning subsistence skills, cooperation, sharing, and nonviolence. Research on children's play in all three forager groups shows that children often have an area next to or just outside camp (e.g., a place where they make a swing) where the multiage play group imitates adult dances, subsistence activities, and spirit-ritual activities. Boyette (ibid) indicates that social play is particularly important for learning foundational schema, such as sharing and cooperation.

Learning Environments throughout Childhood That Are Conducive to Early and Rapid Acquisition of Knowledge and Skills

Studies have shown that forest foragers learn how to share, take care of infants, and hunt and gather by age ten (Hewlett and Cavalli Sforza 1986). Features of the forager learning environment that enhance early and rapid learning include security and trust in others, tolerant instructors (i.e., adults do not push children away if they want to learn something), freedom to explore the natural and social environment, highly self-motivated learners, learning through highly valued play content, easy access to material artifacts and multiple skilled models, collaborative learning in multiage child groups from infancy through childhood, and plenty of time to practice and innovate.

Children in farming communities also have rich learning environments, but they are different from those of foragers in several ways. First, farming parents or anyone older than a child regularly gives commands to children, and children are expected to listen and obey parents, older siblings, and elders. Parents and older children have higher social status, and they frequently direct learning. The status differences contribute to less autonomy, less freedom to explore, and lower tolerance of potential instructors (parents and older children). The social-emotional environment of farming children is less sensitive and secure than that of foragers. As described below, children in farming communities are held and touched less frequently than forager children, and fussing and crying are responded to less rapidly or ignored by farming caregivers. Farmers often live in larger villages, but homes are father apart from each other than they are in forager camps, which means access to others with knowledge and skills may more difficult and costly. While knowledgeable others may not be close-by, children in farming communities have access to a larger number of individuals.

Giving and Responsiveness in Infancy and Early Childhood

Studies that have compared Congo Basin forager and farmer infant and early childhood consistently indicate that the foragers are more giving and responsive to their infants and young children than their neighboring farmers. The Western child development literature sometimes refers to this high responsiveness as "indulgence," which tends to imply that parents are "child-centered" and indulge (and potentially spoil) the child in whatever he or she wants. Congo Basin forager childcare is not child-focused, and parents are not trying to spoil the child. The term "giving" is used instead of "indulgence" because childcare is similar to giving and sharing food. Caregivers give to infants or young children who need or request care just as they would give and share food with others.

One measure of giving and responsiveness is frequency and nature of breastfeeding. Aka infants and young children were breastfed on demand about four times per hour, whereas farmers averaged about two times per hour (Hewlett et al. 2000). Fouts et al. (2011) examined breastfeeding among Aka and Bofi forager and Ngandu and Bofi farmer three- and four-month-olds, nine- and ten-month-olds, and one- to four-year-olds and found that at all ages forest foragers breastfed more frequently, had more breast-feeding bouts per hour, and were more likely to be holding infants when nursing than did neighboring farmers. Hirasawa (2005) found similar differences between Baka foragers and Bombong farmers. Forest foragers breastfed a greater percentage of the day, had more nursing bouts per hour, and breast-fed a shorter time per nursing bout than did farmers. This is particularly interesting because Baka are more sedentary and farm more frequently than Aka and Bofi foragers, but they maintain the giving and responsiveness of breastfeeding.

Aka and Baka forager caregivers are also significantly more likely than Ngandu and Bombong farmer caregivers to respond to infant crying and fussing. Farmer infants cried significantly longer and more frequently than did forager infants in both groups (Hewlett et al. 2000; Hirasawa 2005). Forager caregivers responded to fussy or crying infants much more quickly than farmer caregivers and responded to most every fuss and cry event, while farmer caregivers did not respond to about one-third of the fuss and cry events, and it took them much longer to respond when they did. Efe caregivers also respond to infants' cries or fussing rapidly—within ten seconds of a fuss over 85 percent of the time at three and seven weeks of age and over 75 percent of the time at eighteen weeks of age (Winn et al. 1987; Morelli et al. 2013). Foragers give and are responsive when infants are hungry, desire physical contact, want help walking, want attention, do not feel well, and so on. Foragers tend to feel secure and trust their social environments because responses to distress are rapid and come from many individuals in the camp (see below).

Allomaternal Care in Infancy and Early Childhood

Tronick et al. (1987) were the first to provide quantitative evidence of allomaternal care in Congo Basin forager early infancy (one to four months of age). They found that Efe mothers were not the first to nurse a newborn; four-month-olds spent only 40 percent of their time with their mothers, infants were transferred to alternative caregivers 8.3 times per hour on average, and infants were cared for by 14.2 different people on average during an eight-hour period. Hewlett (1989) examined his data on young Aka infants and identified a similar pattern. Meehan (2009) found that older (nine- to ten-month-old) Aka infants received some form of allocare (from holding to checking infant) from twenty different caregivers on average during nine hours of observation, while Ivey (2000) found that older (twelve- to fifteen-month-old) Efe infants had interactions with eleven different caregivers on average during two hours of observation. Morelli et al. (in press) calculate that Efe infants and young children spend, on average, a change in social partners every three minutes, regardless of age. Most of the allocaregivers in these studies are the infant's genetic kin (i.e., fathers, grandmothers, aunts, sisters).

Hewlett's (1991) research with one category of Aka allomothers—fathers—indicated that fathers were more involved with their infants than fathers in any other known culture. Aka fathers held or were within an arm's reach of their infants 51 percent of a twenty-four-hour period, did 22 percent of the caregiving with four-month-old infants in the camp setting, were the second most active caregivers after mothers, and were more likely than mothers to kiss and show affection to infants while holding. Some fathers offered their breast to fussy infants.

Considerable variability between the four forager ethnic groups in father involvement exists (next section), but the forager fathers consistently provided more direct care to infants and young children than did fathers among neighboring farmers (Hewlett 1991; Fouts et al. 2005; Fouts 2010). Farming fathers may help with caregiving on occasion, but they see their role as disciplinarians and providers of resources, such as paying for school fees or medical care.

Several researchers demonstrate that maternal care increases and allomaternal care decreases from early to late infancy (Hewlett 1991), which is consistent with attachment theory, but maternal care decreases and allomaternal care increases and remains steady in early childhood (Tronick et al. 1987; Meehan 2010; Morelli et al., in press).

Allomaternal nursing is also normative in the Congo Basin hunter-gatherer ethnic groups with data on early infancy—that is, among the Aka and Efe (Hewlett and Winn, in press). Young infants (one to four months) receive, on average, 15–25 percent of their total breastfeeding time from allomothers, and, in some cases, an infant spends up to half or more of his or her nursing time with an allomother. Allomaternal nursing in these groups disappears by late infancy, and it does not or rarely occurs among neighboring farming groups.

The studies mentioned above are based upon quantitative behavioral observations, but they are consistent with more qualitative accounts from Congo Basin hunter-gatherer ethnographers. For instance, Colin Turnbull (1978) describes allomaternal care among the Mbuti:

> The mother emerges and presents the child to the camp . . . and she hands the boy to a few of her closest friends and family, not just for them to look at but for them to hold him close to their bodies. . . . In this way an initial model of predictability and security becomes multiplied and so it is throughout the educational process: vital lessons, such as non-aggressivity, are learned through a plurality of models.

Few Gender Differences in Middle-Childhood Caregiving

Extensive research with farmers around the world indicates that middle-childhood-aged females are particularly important allomaternal caregivers of infants and young children (Weisner and Gallimore 1977; Whiting and Edwards 1992). Farming mothers assign middle-aged girls more childcare tasks than boys, and cultural models expect girls to help more with childcare than boys. This gender bias with middle-childhood caregivers does not exist with Congo Basin foragers. Systematic research with Efe (Ivey Henry et al. 2005) and Aka (Hewlett 1991; Boyette 2013) foragers indicates both males and females in middle childhood contribute similar amounts of caregiving to infants and young children.

Male and Female Initiation during Adolescence

Congo Basin hunter-gatherers are relatively distinct from neighboring farmers in that both male and female adolescents are initiated into forest spirit associations. Farmers, by contrast, often have adolescent initiation ceremonies for males, generally associated with circumcision, but seldom for females. Congo Basin forager boys and girls acquire knowledge about powerful forest spirits and appropriate forager conduct, especially about foundational schema such as sharing and giving, and learn about the mythical power and solidarity of their gender (i.e., how men or women used to be able to reproduce on their own, how they used to have all the spiritual powers of the forest) during initiation. Circumcision is generally not a part of Congo Basin forager initiations among the Aka, Efe, and Mbuti, but males are often circumcised in middle childhood, either as part of a farmer initiation ceremony (e.g., *Nkumbi* among Mbuti) or individually without any ceremony.

Among the Mbendjele, Lewis (2002) describes boys' initiation into *Ejengi* and girls' initiation into *Ngoku*. *Ejengi* initiation occurs when elders feel boys have the cognitive abilities to keep spirit secrets and the courage to go through the dangerous ceremony; the initiation often takes place after age seven or eight. A group of two to four boys are taken to a special location (*njanga* path) outside of camp where their bodies are painted red and they reside with other males for three days to learn about the mythical past, *Ejengi* forest spirit secrets, elephant hunting, and honey collecting. The *Ejengi* initiation also occurs among the Baka, Aka, and Kola.

Less is known about the Mbendjele female's *Ngoku* forest spirit initiation at puberty, but it is for girls only and focuses on learning female power and solidarity, female reproductive skills, and the mythical past when women owned all the forest spirits and obtained their babies from *Ejengi*. *Ngoku* dancers tease (we want young men, not old men!) and ridicule (not providing enough sex or food, chasing other women) men. Among the Aka (Takeuchi 2013), there is a spirit dance ritual, called *Waya*, that adult females organize when a girl reaches menarche.

Turnbull (1957, 1965a, 1965b) describes similar patterns among Mbuti adolescents; boys are initiated into the *Molimo* (called *Lusumba* in Turnbull's 1957 publication), and girls are initiated into the *Elima*. Wilkie and Morelli (1991) describe a similar girls' puberty ceremony, called *Ima*, among the Efe. In the *Elima*, a girl moves into a special hut with selected friends (*bamelima*) for a month or two shortly after she has her first menstruation. The special hut is in the care of an older female who is in charge of the instruction in *Elima* songs for the forest and the arts of motherhood and sexual life. Instruction takes place both in the hut and in the forest. The initiate invites boys to come visit her in the hut, but the boys must fight to get past women guarding the hut. The women have vines and pieces of wood and use them to whip or throw wood at boys trying to enter. The initiates are looking for potential spouses, but the process also emphasizes female power and solidarity.

With the *Molimo*, boys cannot be initiated until they have proved their prowess and strength as hunters. Turnbull (1965b) indicates most boys enter the forest spiritual association by age fourteen. As with *Ejengi*, boys learn songs and dances to the Spirit of the Forest as well as knowledge and skills about forest life. Turnbull (1965b) points out that Mbuti boys are circumcised in the neighboring farmers' *Nkumbi* adolescent ceremony, but that the entry into the *Molimo* is not contingent upon circumcision or participation in the *Nkumbi* initiation.

Discussion of Distinctive Features

The features of childcare described above are not unique to Congo Basin forest foragers; many exist in hunter-gatherers in other parts of the world (Konner 2005). The feature that appears to be particularly pronounced in Congo Basin hunter-gatherer childhoods is the nature and frequency of allomaternal care. Aka fathers provide more direct care to their infants than fathers in any known culture, whether farmers, hunter-gatherers, or urban industrialists. The number of different caregivers and frequency of allomaternal care in Aka and Efe infancy and early childhood has not been documented in other hunter-gatherers. Allomaternal nursing exists with Efe and Aka but does not occur among the Hadza or !Kung foragers in other parts of Africa.

Finally, before moving on to discuss diversity between and within groups, it is important to mention a few ways in which Congo Basin forager and farmer children are similar because childhood in foragers and farmers is more similar than is, for instance, Congo Basin foragers and Euro-American childhoods. First, table 9.2 summarizes the results of one of our studies (Hewlett et al. 1998) that compared Aka forager and Ngandu farmer infant development. Forager and farmer caregivers did not significantly differ from each other in a wide range of behaviors: frequency of face-to-face interaction with their infants, how often they watched their infants, showed affect to their infants, soothed their infants, and vocalized to their infants.

Table 9.2. Similarities and differences between Aka forager and Ngandu farmer infancy.

No Significant Differences between Aka Foragers and Ngandu Farmers	Aka Foragers Do Significantly More Often Than Do Ngandu Farmers	Ngandu Farmers Do Significantly More Often Than Do Aka Foragers
Caregiver and infant in face-to-face interaction	Caregiver holds infant	Caregiver grooms, dresses, cleans infant
Caregiver and infant watch something at same time	Caregiver is within arm's length of infant	Caregiver stimulates or arouses infant
Caregiver watches or checks on infant	Caregiver feeds infant	Infant fusses or cries
Caregiver shows physical affect toward infant	Infant sleeps	Infant is alone
Caregiver shows nonphysical affect toward infant		Infant smiles
Caregiver physically soothes infant		Infant vocalizes
Caregiver nonphysically soothes infant		Infant plays alone
Caregiver vocalizes to infant		Infant plays with objects
Infant looks at caregiver		

Source: Summarized from Hewlett et al. 1998.

Second, both foragers and farmers in the Congo Basin experience high mortality and divorce rates (Hewlett 1991, 1996). These two factors mean that most forager and farmer children in middle childhood and adolescence live with single parents or stepparents. Among the Aka, 42 percent of eleven- to fifteen-year-old children live with a stepparent or a single parent. Forager children decide which parent to live with, the other parent lives some distance away at his or her parents' camp, and the child often visits the other parent. At the time an Aka selects his or her spouse, he or she is living with both natural parents only 29 percent of the time. Ivey (2000) reports that 27 percent of Efe children do not have parents in camp because both parents have died, their parents divorced and live separately from either parent, or they are living temporarily away from their parents. Comparable demographic data do not exist among forest farmers, but observations and household data indicate that the frequency of stepparenting or single parenting is similar, if not higher.

Cultural Diversity between and within Ethnic Groups

Extensive diversity in childcare exists between ethnic groups. Table 9.3 summarizes some of the demographic, subsistence, and settlement variability between the four forager ethnic groups. Given the diversity in these limited parameters, it is not surprising that extensive diversity exists in how children grow up in these groups. In this section, we examine culture variability in two domains: father involvement and allomaternal care.

Table 9.3. Cultural diversity in demography, subsistence, and settlement patterns of the four forager ethnic groups with data on children.

	Efe[a]	Mbuti[b]	Aka[c]	Baka[d]
Estimated population	10,000	26,000	30–50,000	30–40,000
Total fertility rate	2.6	5.5	6.2	ND
Infant mortality rate	12.0	33.0	20.0	ND
Percentage of population under age fifteen	24.0	52.9	48.0	42.0–53.7
Polygyny rate	3.0	14.0	17.5	19.5
Frequency forager females marry village males	Common (13–28%)	Rare (<1%)	Rare (<1%)	Occurs (3–6%)
Number of farming ethnic groups living in association	3	4	15	19
Primary hunting technique	Bow	Net	Net	Spears and traps
Percentage of calories provided by meat or fish	13–27%	35–86%	36%	10%
Distance to most forest camps	4–5 km	5–15 km	8–40 km	4–8 km
Number of months/year in forest camps	5	8	7–8	4–5
Mean camp size	17.8	37.4	25.5	31.7

[a] Bailey 1991; Peacock 1985; Bailey and Peacock 1988; Hill 1982; Harako 1976; Tereshima 1987.
[b] Turnbull 1965a and b; Hart 1978; Ichikawa 1978; Hill 1982.
[c] Bahuchet 1985, 1988; Hewlett 1989, 1991.
[d] Vallois and Marquer 1976; Yamauchi et al. 2000; Dodd 1987; Takanori personal communication.

Father Involvement

Efe hunter-gatherer fathers held their infants 2.6 percent of the time in the camp setting (Winn 1989, personal communication) in comparison to the

22 percent of the time among the Aka fathers mentioned above. Bailey (1991) found that Efe men actively engaged in childcare only 0.7 percent (about five minutes per day) of daylight hours and indicated that "strong father-child attachments among the Efe were uncommon." Efe fathers were also not the secondary or even tertiary caregivers of their infants; several other females (older siblings, grandmother, mother's sister) provided more care than fathers. Efe are different from Aka in substantial ways: Efe do not cooperatively net hunt (men hunt with bows and arrows or small traps, while Aka men and women net hunt), which means Efe men and women go their separate ways most of the day. And as is evident in table 9.3, Efe have very high infertility rates, so several other adult women without children are potentially available to help with childcare.

Although precise data do not exist for the Mbendjele, the forager group just south of the Aka who speak a dialect of the Aka language, father involvement appears to be substantially lower than the Aka because Mbendjele men use traps, spears, and guns to hunt rather than nets, which limits Mbendele men's availability to their children during the day. Baka father involvement is likely to be similar to the Mbendjele because men also use traps to hunt, while among the net-hunting Mbuti, Turnbull (1965a) gives the impression fathers are very involved in childcare.

In terms of within cultural variability, Aka are generally very involved fathers, but remarkable variability exists between fathers. Some Aka fathers hold their infants 2 percent of the time, while others hold their infants about 20 percent of daylight hours. Hewlett's (1991) study found that highly involved fathers tended to have the following characteristics: no brothers, a wife from a distant clan, a small hunting net, greater reliance on individual hunting techniques rather than nets, and a close relationship with Ngandu farmers. Fathers with lower involvement had the opposite features. Hewlett suggested that kinship resources were important for understanding intracultural variability in paternal care.

Meehan (2005) and Fouts (2008) also demonstrate that fathers are more likely to be involved with infant care when the family is living patrilocally rather than matrilocally, and Fouts (2008) shows that the father holds the child less and proximity declines as the child gets older.

Allomaternal Care

Allomaternal care is common among Congo Basin hunter-gatherers, but how frequently it occurs, how many allomothers provide care, and who provides the care vary substantially between ethnic groups. Observational studies indicate that Aka infants receive allocare from an average of 20 different caregivers (Meehan 2005) while Baka infants receive allocare from 5.3 (Hirasawa 2005) and Efe infants receive care from 14.2 (Tronick et al. 1987) different caregivers. In terms of who provides allocare, Aka fathers are secondary and grandmothers the tertiary infant caregivers (Hewlett 1991; Meehan 2005), while among the Efe, adult females provide the majority of the allocare (Tronick et al. 1987; Ivey 2000). Among the Baka, older middle childhood-aged siblings, both male and female, are the second most important allomaternal caregivers, with fathers and grandmothers the secondary and tertiary caregivers (Hirasawa 2005). In terms of allomaternal nursing in early infancy, Efe infants receive more allonursing in comparison to Aka infants (60 percent of infants among the Aka versus 78 percent of Efe) (Hewlett and Winn, in press).

A variety of factors contribute to the observed diversity: demography, subsistence pattern, and sedentism. Table 9.3 indicates Efe have low total fertility rates, and a high percentage of Efe women (47 percent) have one child or less, primarily because of gonorrhea and other sexually transmitted infections (Bailey and Peacock 1988). As mentioned, Efe have more adult females without children in camp than do Aka or Baka to help provide allocare or allomaternal nursing. The Aka are net hunters where men, women, and children hunt together, while Baka and Efe men hunt and women gather and farm. Aka fathers are available to their infants for a longer part of the day than the Baka and Efe fathers. Carrying infants for long distances on the net hunt is also energetically demanding, which helps to explain why Aka brothers and sisters in middle childhood are not involved more in infant care. The Baka are relatively sedentary, hunt less, and more acculturated in comparison to the Aka and Efe.

In terms of cultural variability in allomaternal care, figures in the classic Tronick et al. (1987) paper illustrate pronounced variability between infants in the amount of time they spent in allomaternal care and the frequency they were transferred between caregivers. Of eight infants in the study, one received no allocare (was with mother all of the time) while another spent 80 percent of her time in allocare. Winn (1989) analyzed intracultural variability in Efe infant care and found that fussier infants spent more time with mothers and had fewer caregivers, and Efe infants in larger groups were less likely to spend less time with their mothers, were transferred more often, and encountered a greater number of different caretakers. Along similar lines, Fouts and Lamb (2005) found that young Bofi children were more likely to cry in the presence of their mothers and less likely to cry under the care of allomothers when their mothers were absent.

Meehan's (2005) research on allomaternal care among the Aka demonstrates pronounced intracultural variability in who provides the care, but not the amount of time an infant receives allomaternal care. When the family lives matrilocally, the mother's female relatives provide allocare, but when living in a patrilocal setting, the father and his relatives provide more care.

Extensive culture diversity also exists in the frequency infants receive allomaternal nursing (Hewlett and Winn, in press). In the study, twelve Aka three- and four-month-old infants received allomaternal nursing during observations, and eight Aka infants did not. The two groups were compared on a wide range of variables—from the frequency infants fussed or cried to the amount of time the mothers worked. Frequency of fussing and crying and frequency of the mothers' work did not matter, but how long an allomaternal caregiver held an infant during the day was a significant predictor of the allomaternal nursing and helped to explain intracultural diversity. The amount of time an allomother held an infant during the day was a good predictor of the frequency she allonursed the infant.

Theoretical Contributions

Research with Congo Basin hunter-gatherer children was often initiated to test theories and hypothesis in anthropology, developmental psychology, and evolutionary biology. This section briefly reviews a select few of the hypotheses and their results to demonstrate how research with Congo Basin hunter-gatherers has contributed significantly to our understanding of human nature, biases in Western conceptions of child development, how children learn, and the importance of fathers and others for child survival and healthy social-emotional and cognitive child development.

Development Psychology

Do differences in cognitive styles exist between hunter-gatherers and farmers? A multinational, multidisciplinary team conducted research in the Central African Republic for three years (1975–1977) to test Herman (Hy) Witkin's theory of psychological differentiation (Witkin et al. 1962) among Aka foragers and Ngandu farmers (Berry et al. 1986; van de Koppel 1983). The basic hypothesis was that the hunter-gatherer social ecology (mobility, flexibility, low population density) should contribute to the development of a "field independent" cognitive style (e.g., individuals perceive self and objects as separate and independent), while the farming lifestyle should lead to a "field dependent" way of thinking and perceiving the world (i.e., individuals view self and objects as embedded in a network). Parents in the different social ecologies should socialize their children in these different ways of thinking and perceiving the world. The psychologists developed a battery of tests for both adults and children (e.g., embedded pictures test, body adjustment test, draw a person test) to try and measure how well Aka and Ngandu could evaluate their position in space, how many embedded triangles they could identify in a picture, how parents socialized children, and how individuals viewed self with others. Overall, the tests supported the predictions of the theory; the Aka foragers were generally more field independent and autonomous, while the Ngandu were generally more field sensitive/dependent in their cognitive style. The results fit nicely with the forager and farmer foundational schema previously described.

Do infants become attached to their fathers through vigorous play? Hewlett's research (1991) with Aka infants focused on evaluating Lamb's (1981) hypothesis regarding the role of rough-and-tumble play in an infant's attachment to father. The prevailing hypothesis was that infants become attached to their fathers, who are not around their children as often as mothers, through their vigorous play and interactions, whereas infants become attached to their mothers via their regular and sensitive care. Studies in urban industrial societies in many parts of the world indicated that vigorous play was a distinctive feature of fathers' versus mothers' style of interaction with infants. Unlike fathers in urban-industrial cultures, Aka fathers were frequently with their infants and rarely engaged with them in vigorous play. Fathers engaged in physical play only once in 264 hours of systematic naturalistic father and infant focal observations. Fathers were also more likely to kiss and hug an infant while holding than were mothers.

Hewlett suggested that Aka fathers were not vigorous because they intimately knew their infants through their extensive care. Because Aka fathers knew their infants so well, they did not have to use vigorous play to initiate communication or interaction with their infants. Aka fathers are often around their infants because men, women, and children participate together in net hunting. Net hunting, in part, contributed to regular husband-wife cooperation and the fathers' intimate knowledge of their infants (1992a).

Do infants become attached to one primary caregiver? How important are mothers? John Bowlby (1980) hypothesized that infants become attached to a primary caregiver, usually the mother, because it enhanced their survival; infants that became attached to a primary caregiver and regularly reached for, fussed for, or crawled toward the primary attachment figure to maintain proximity were more likely to be protected from predators or aggressive others by the caregiver and therefore more likely to survive. Attachment theory (1980) is one of the leading psychological theories of social-emotional development. According to the theory, this dimension of human caretaking has biological component because they evolved in the long period of the hunting-gathering environment of evolutionary adaptation, and the infant had a primary attachment to the person who provided most of his or her care, usually the mother. The theory developed at a time when research showed that mothers were the primary caregivers in most primate species, and Konner's (2005) research with !Kung foragers supported the idea that mothers were primary. Bowlby actually indicated that infants could be attached to others, but the vast majority of research that followed focused on mothers and infants. Tronick et al. (1987) went to the Ituri Forest to test the continuous contact and care from the mother component of attachment theory with Efe foragers. Their study demonstrated that multiple caregivers provided more than 50 percent of the total early infant care and several other dimensions of care described above. More recent studies by Morelli et al. (in press) and Ivey (2000) provide greater detail on the multiple attachments of Efe infants and young children.

Meehan and her colleagues (Meehan 2005; Meehan and Hawks, in press) are also interested in the impact of allomaternal care on attachment and have found that Aka forager children demonstrate attachment behaviors to six different allomaternal caregivers, on average, and that allomaternal caregivers (both juveniles and adults) are just as responsive and sensitive to the fusses and cries of infants and young children as the mothers. They also report that young children with more allomaternal caregivers are, on average, taller and heavier than young children with fewer allomaternal caregivers, suggesting a reproductive advantage to having multiple caregivers.

Research with the Efe and Aka infants and young children demonstrate the importance of allomaternal care for child growth and survival and has contributed significantly to our understanding of humans as cooperative breeders (Hrdy 2009).

Is theory of mind a human universal? Theory of mind refers to the cognitive capacity of individuals to take on the perceptions and intentions of others. Many have suggested this is part of human nature and a critical cognitive ability that enables children to learn from others, feel empathy, and give and share extensively with others (Tomasello 1999). This emerges in late infancy and becomes clear in urban industrial children by four years of age. Avis and Harris (1991) evaluated this theory among Baka foragers in Cameroon and found support for the universal nature of theory of mind. This study is the only study of theory of mind in a hunter-gatherer community.

Evolutionary Biology

Who provides allomaternal care? Why do particular individuals provide allomaternal care? Tronick et al. (1987) pointed out the high frequency of allomaternal care among the Efe, and Paula Ivey Henry (2000) returned to their study site to test three evolutionary predications about who should provide allomaternal care: (a) genetic kin, (b) women who could share food or infant care (reciprocity), and (c) young adolescent girls wanting to learn how to parent. She found both males and females (males provided 46 percent of all allocare), young and old (juveniles provided 56 percent of allocare), provided allomaternal care but that genetic kin provided twice as much allocare as nonkin. Reciprocity did not explain much of the variability, nor did the learning to mother hypothesis because both male and prereproductive female individuals provided relatively equal allocare. This is also one of the few studies to demonstrate that allomaternal care enhances infant survivorship. She found that the number of allocaregivers an Efe infant had at one year of age was positively associated with survivorship of that child at three years of age.

Is the time of weaning characterized by parent-offspring conflict (fussing, crying, temper tantrums)? Evolutionary theory (Trivers 1974) indicates that reproductive interests of parents and children may conflict when parents want to have another baby and the existing child wants continued parental attention and resources. Mothers often wean their child because they want to have another child or are already pregnant, which leads to parent-offspring conflict. Fouts et al. (2005) evaluated this hypothesis among Bofi foragers and farmers and found that the farmers conformed to the predicted theory: mothers decided when to wean, they used dramatic steps to wean the child (e.g., use of hot pepper on nipples), and toddlers fussed and cried during this period. By contrast, parent-offspring conflicts among the Bofi foragers were minimal: the toddlers decided when to wean, and the toddlers did not fuss and cry during the process, in part because of an increase of involvement by allomaternal caregivers.

What are the modes and processes of cultural transmission? Luca Cavalli Sforza (1986) conducted some of the earliest systematic and comparative biomedical and demographic studies of Congo Basin hunter-gatherers. He and Marc Feldman (Cavalli Sforza and Feldman 1981) provided innovative theoretical contributions to evolutionary approaches to culture. They suggested that different modes of cultural transmission—that is, acquiring knowledge from parents (called vertical), similar-aged peers (called horizontal), or older children and adults (called oblique)—influenced how fast or slow a trait would change or the intracultural variability for a cultural trait. Hewlett wanted to understand the roles of parents versus others in a foraging community because some cultural anthropologists said parents were important while others suggested everyone in camp contributed equally. In an early interview-based test of these models, Aka adults and children (Hewlett and Cavalli Sforza 1986) indicated their parents (i.e., vertical transmission) were the primary transmitters of about 80 percent of the fifty cultural items they had learned.

Among the Baka, Hattori's interview-based study with adults (2006, in press) found that women learned about the uses of ninety plants from their mothers 80 percent of the time, fathers 15 percent of the time, and others 5 percent of the time; Baka men said they learned about the plants from their mothers 10 percent of the time, fathers 65 percent of the time, siblings 11 percent of the time, and others 13 percent of the time. Hayashi et al. (2012) also found that 61 percent of Baka men learned how to make metal spear points from their fathers (vertical transmission), and 11 percent learned from their grandfathers (oblique). B. L. Hewlett's interview-based qualitative study (2012) of Aka forager adolescents is replete with expressions of vertical transmission. One adolescent boy explained, "Father showed me how to care for younger brothers and sisters and to have a good character. He showed me how to hunt and find honey. [My mother] showed me how to guard the baby and how to wash and comfort the babies."

When asked, forager adults and children tend to identify parents as primary transmitters of culture, but Boyette's (2013) recent observational-based study of Aka cultural transmission indicates older children and adults (i.e., oblique transmission) are much more important at this age than the parents. This is consistent with Aunger's study (2000) of how Efe foragers acquire food taboos; he found that adults said that they learned food taboos from their parents, but when he systematically looked at who actually shared food taboos, peers and nonparental adults were more important than parents. Hewlett et al. (2011) suggest that vertical transmission is particularly important up to age four or five, but horizontal and oblique transmission are favored in middle childhood and adolescence.

Most anthropologists indicate that forest foragers primarily learn by observation and imitation (Turnbull 1965b; Hayashi et al. 2012). Boyette's study (2013) is one of the few to systematically examine several processes of social learning, such as imitation, observation, and teaching, among Aka forest foragers in middle childhood and finds that social play and some direct instruction (teaching) are particularly important for learning foundational schema, such as sharing and cooperation. Hewlett's (2013) research with Aka infants indicates various forms of teaching exist; e.g., parents use pointing, demonstration, and negative feedback to transmit a range of subsistence (e.g., how to dig for roots) and technology skills (e.g., how to use a machete). Sonada (2013) also describes how very subtle teaching (e.g., simple comment or body language) occurs with the transmission of knowledge to Baka children.

Sociocultural Anthropology

How do children learn about the giving environment? Hunter-gatherer researchers have tried to understand why foragers do not store food and tend to stop hunting and gathering as soon they have enough food. Bird-David (1991) hypothesized that forager economic behavior is linked to metaphors (i.e., cultural models) that contribute to a trusting, giving, and generous view of the environment. Foragers trust that the forest will provide and give just as other foragers share extensively with them. Hewlett et al. (2000) wondered how hunter-gatherers in diverse (i.e., from forests to deserts) environments develop trusting views of the environment. Using attachment theory, they hypothesize that children in responsive and sensitive caregiving environments develop internal working models (a type of cultural model) where the individual trusts self and others. The study compared Aka forager and Ngandu farmer patterns of infant care, and as described above, Aka caregivers are more responsive and sensitive than Ngandu caregivers, which according to attachment theory means Aka children should be more likely to trust others and the environment as giving. The study also looked at data from 10 other hunter-gatherer and 168 farmer groups and found that foragers were generally more likely to hold and respond to fussing and crying children than farmers. The study supported

Bird-David's hypothesis and identifies a mechanism by which "the giving environment" was transmitted to hunter-gatherer children in diverse natural environments.

How do forager children learn to be nonaggressive and nonviolent? Several studies indicate that hunter-gatherers are generally more peaceful, nonviolent or nonaggressive by comparison to farmers and peoples in other modes of production (see Fry and Soderberg 2013 for a recent review). Social-cultural anthropologists have been interested in how this way of life is transmitted to children. Turnbull (1978) examined how Mbuti children learn nonaggression and identifies four mechanisms as being particularly important. First, children are highly desired; they are brought into a joyous forest world where they are loved; many people provide children with food, support, and knowledge; children call several people mother and father; and children are indulged (e.g., breastfeeding on demand, immediate response to fussing, lots of physical contact, left to explore the environment when they want). According to Turnbull (ibid), children feel that life is predictable and secure. Young Mbuti children are secure and are not afraid of the world and therefore not suspicious or hostile to others. During middle childhood, children play in the *bopi* area next to camp, learn to play noncompetitive games, and develop trust with many other children of both sexes and several ages.

Second, children learn how to avoid conflict and aggression by moving away from contentious situations, at first by crawling away and later by staying away from places with lots of *akami* or noise (e.g., yelling at each other), which is associated with the farming way of life. Third, Mbuti tend to ostracize and stay away from aggressive and assertive individuals. Finally, Mbuti have an array of stories, primarily for youth, that describe interpersonal and human-nature conflicts and the negative consequences of violence and killing.

A quantitative study of aggression among Aka children (Hess et al. 2009) found no sex differences in physical aggression between adult men and women, but did find that male children and adolescents were generally more aggressive than females. Several researchers have indicated that Congo Basin forager parents rarely if ever use corporal punishment to discipline children, whereas it is common among farming populations in the Congo Basin (Hewlett 1991; Kamei 2005; Boyette 2013). Cross-cultural studies (Ember and Ember 2008) indicate a statistical relationship between corporal punishment and the frequency of violence and social stratification in a society.

How do forest forager children learn about religion? As mentioned in the introduction to this book, ecological issues have dominated Congo Basin forager research. Studies of forager religion are particularly limited, but Lewis (2002) provides one of the few ethnographic accounts of how Mbendjele forager children acquire an understanding of the forest spirit world. He indicates that a variety of *massana* activities—communal activities that involve plenty of fun and laughter—among children are particularly important. One type of children's *massana,* called *Bolu,* is used to call boys' forest spirits (*mokondi*). Boys sing *Bolu* songs, go to a special place (forbidden to girls and adults) in the forest near camp, and use tree branches to make clothes for the *Bolu* spirit. Girls sing and dance in the middle of the camp to entice the *Bolu* spirit to come out. Adults will give explicit instructions to the girls about how to improve their performance. The *Bolu* spirit enters the camp surrounded by the boys, and the singing and dancing of both boys and girls enable *Bolu* to empower everyone present. Lewis (ibid) indicates that *Bolu* is just one example of how children begin to learn about the much more complex and diverse forms of adult *mokondi massana.*

Children begin to learn about the spirit world in infancy and early childhood through lived experiences of dance and song, but Lewis (2008) demonstrates how physical and biological maturity contribute to a more complex understanding of the supernatural. *Ekila* is a complex set of beliefs, often associated with blood, that influence Mbendjele relations with animals and each other (the same term and similar beliefs also exist among Aka and Baka foragers). Children learn about some cultural practices associated with *ekila*—for example, food taboos, how menstruation impacts hunting and practices—in early childhood, but it is not until adolescence when girls start to menstruate and boys and girls start to have sex that Mbendjele become curious about and learn more extensively about the complex cosmological, supernatural, and political nature of *ekila.*

Summary

This chapter provided a brief overview of child-centered studies of Congo Basin hunter-gatherers. The overview is limited and biased because only 13 percent of all studies with Congo Basin hunter-gatherers have focused on children, only four out of the fifteen Congo Basin forager ethnic groups have child-centered research, and almost 70 percent of the published child studies have been conducted with one of those four ethnic groups. We have a long way to go, and substantially more research is needed before it is possible to provide a comprehensive and realistic understanding of hunter-gatherer children in the Congo Basin.

Given what is known, the chapter identifies some general features of Congo Basin forager childhoods that are somewhat distinct from their farming neighbors, describes some of the cultural diversity within and between ethnic groups, and discusses a limited number of theories that researchers have used to guide their child-centered research in the Congo Basin. Physical and emotional intimacy, autonomy, allomaternal care, play, mixed-aged groups, female initiation, and early and rapid social learning are characteristic features of Congo Basin forager childhoods. The commonalities exist because of similar foundational schema and the demography of mobile hunter-gatherers. The frequency and scope of allomaternal care seems to a particularly distinctive feature of Congo Basin hunter-gatherer childhoods. It exists in other hunter-gatherers, such as the Hadza, !Kung, and Australian Aborigines, but the number of different caregivers, the level of father involvement, and the high frequency of allomaternal care are much less pronounced among these foragers. Of course, enormous intercultural and intracultural variability exists in patterns of Congo Basin hunter-gatherer childhood because of differences in history, relations with farming neighbors, ecology, and demography. It is my impression that most researchers who conduct child-centered field research with forest foragers are more impressed by the pronounced diversity than they are with the commonalities. Finally, research with Congo Basin hunter-gatherer children has contributed substantially to a better understanding of human nature and of biases in Western ideas of children and appropriate ways for children to learn and develop.

References

Aunger, R. 1994. "Are Food Avoidances Maladaptive in the Ituri Forest of Zaire?" *Journal of Anthropological Research* 50:277–310.

Aunger, R. 2000. "The Life History of Culture Learning in a Face-to-face Society. *Ethos* 28:1–38.

Avis, J., and P. L. Harris. 1991. "Belief-desire Reasoning among Baka Children: Evidence for a Universal Conception of Mind." *Child Development* 62:460–7.

Bahuchet, S. 1985. *Les Pygmees Aka et la Foret Centrafricaine*. Paris: SELAF.

Bailey, R. C. 1991. *The Behavioral Ecology of Efe Pygmy Men in the Ituri Forest, Zaire*. University of Michigan Anthropological Papers No. 86. Ann Arbor: University of Michigan.

Bailey, R. C., and N. R. Peacock. 1988. "Efe Pygmies of Northeast Zaire: Subsistence Strategies in the Ituri Forest." In *Coping with Uncertainty in Food Supply*. Oxford: Oxford University Press, 88–117.

Barry, H., I. L. Child, and M. Bacon. 1959. "Relation of Child Training to Subsistence Economy." *American Anthropologist* 61:51–63.

Berry, J. W., S. Bahuchet, J. Van De Koppel, R. Annis, C. Sénéchal, L. L. Cavalli-Sforza, and H. Witkin. 1986. *On the Edge of the Forest: Cultural Adaptation and Cognitive Development in Central Africa*. Swets and Zeitlinger, Lisse.

Bourdieu, P. 1977. "Outline of a Theory of Practice." Cambridge: Cambridge University Press.

Bowlby, J. 1980. "Attachment." 2nd ed. *Attachment and Loss*. Vol 1. New York: Basic Books.

Boyette, A. H. 2012. "Physical Contact and Egalitarianism among Hunter-gatherers and Farmers in the Central African Republic: An Ontogenetic Perspective." Paper presented at Annual Meeting of the American Association of Physical Anthropologists, Portland, OR.

———. 2013. "Social Learning during Middle Childhood among Aka Foragers and Ngandu Farmers of the Central African Republic. PhD diss., Washington State University.

Caro, R. M., and M. D. Hauser. 1992. "Is There Teaching in Nonhuman Animals?" *Quarterly Review of Biology* 67:151–74.

Csibra, C., and G. Gergely. 2011 "Natural Pedagogy as Evolutionary Adaptation." *Phil. Trans. R. Soc. B.*
Ember, C. M., and M. Ember. 2008. "Explaining Corporal Punishment of Children: A Cross-cultural Study." *American Anthropologist* 107:609–19.
Fouts, Hillary N. 2002. *The Social and Emotional Contexts of Weaning among the Bofi Farmers and Foragers of Central Africa.* Unpublished PhD diss., Washington State University.
———. 2004. "The Social and Emotional Contexts of Weaning among Bofi Farmers and Foragers. *Ethnology* 43:65–81.
———. 2005. "Central African Families: A Comparison of Bofi Farmer and Forager Families. In *Families in Global Perspective*, edited by J. L. Roopnarine and U. P. Gielen, Boston: Allyn and Bacon, 347–62.
———. 2008. "Father Involvement with Young Children among the Aka and Bofi Foragers." *Cross-cultural Research* 42:290–312.
———. 2010. "Bofi Foragers and Farmers: Case Studies on the Determinants of Parenting Behavior and Early Childhood Experiences." In *Formative Experiences: The Interaction of Caregiving, Culture, and Developmental Psychobiology*, edited by C. Worthman, D. Schechter, and P. Plotsky. New York: Cambridge University Press.
Fouts, H. N., and R. A. Brookshire. 2009. "Who Feeds Children? A Child's-Eye-View of Caregiver Feeding Patterns among the Aka Foragers in Congo." *Social Science & Medicine* 69:285–92.
Fouts, Hillary N., Barry S. Hewlett, and Michael E. Lamb. 2001. "Weaning and the Nature of Early Childhood Interactions among Bofi Foragers in Central Africa. *Human Nature* 12:27–46.
———. 2005. "Parent-offspring Weaning Conflicts among the Bofi Farmers and Foragers of Central Africa. *Current Anthropology* 46:29–50.
———. 2011. "A Biocultural Approach to Breastfeeding Interactions in Central Africa." *American Anthropologist* 114:123–36.
Fouts, Hillary N., and Michael E. Lamb. 2005. "Weanling Emotional Patterns among the Bofi Foragers of Central Africa: The Role of Maternal Availability and Sensitivity." In *Hunter-Gatherer Childhoods*, edited by B. S. Hewlett and M. E. Lamb. New York: Aldine De Gruyter, 309–21.
———. 2009. "Cultural and Developmental in Toddlers' Interactions with Other Children in Two Small-scale Societies in Central Africa." *Journal of European Developmental Science* 3:259–77.
Fry, D. P., and P. Soderberg. 2013. "Lethal Aggression in Mobile Forager Bands and Implications for the Origins of War. *Science* 341:270–3.
Hagino, Izumi, and Taro Yamauchi. 2013. "Gener-age Differences in Daily Time-place and Activity Pattern of African Rainforest Forager's Children in Cameroon." Paper presented at the International Conference on Hunting-Gathering Societies, Liverpool, UK.
Harako, Reizo. 1976. "The Mbuti as Hunters: A Study of Ecological Anthropology of the Mbuti Pygmies (I)." *Kyoto University African Studies* 10:37–99.
Hart, J. A. 1978. "From Subsistence to Market: A Case Study of the Mbuti Net Hunters." *Human Ecology* 6:325–53.
Hattori, Shiho. 2006. "Utilization of Marantaceae Plants by Baka Hunter-gatherers in Southeastern Cameroon." In *African Study Monographs*, supplement 33:29–48.

———. 2006. "Diversity and Variability of Plant Knowledge among Adult Baka Hunter-Gatherers in Cameroon Rainforest." Proceedings of Kyoto Symposium: 229–35.

———. In Press. "'My Medicine (Ma a Le)': Variability of Medicinal Plant Knowledge among Adult Baka Hunter-gatherers of Southeast Cameroon." *African Study Monographs.*

Hayashi, K., R. Ishii, Y. Nakamura, H. Tereshima, and Y. Nishiaki. 2012. "Technical Transmission of Hunting Tool Manufacture: A Case of Spear Hunting among Modern Hunter-gatherers in Southeastern Cameroon." Paper presented at the International Conference on Replacement of Neanderthals by Modern Humans, Tokyo, Japan.

Hess, N., C. Helfrecht, E. Hagen, A. Sell, and B. Hewlett. 2010. "Interpersonal Aggression among Aka Hunter-gatherers of the Central African Republic." *Human Nature* 21:330–54.

Hewlett, B. L. 2001. "Adolescent Culture: An Exploration of the Socio-emotional Development of the Aka Adolescents of the Central African Republic." *The Oriental Anthropologist* (India) 1:84–96.

———. 2004. "Aka and Ngandu Adolescents of the Central African Republic: An Exploration of the Inter- and Intracultural Variability of Socio-emotional Development, Response to Loss and Health Status." Unpublished PhD diss., Washington State University.

———. 2005. "Vulnerable Lives: Death, Loss and Grief among Aka and Ngandu Adolescents of the Central African Republic. In *Hunter-Gatherer Childhoods: Evolutionary, Developmental and Cultural Perspectives,* edited by B. S. Hewlett and M. E. Lamb. New Brunswick: Aldine Transaction, 322–42.

———. 2012. *Listen, Here Is a Story: Ethnographic Life Basin.* Oxford: Oxford University Press.

Hewlett, B. L., and B. S. Hewlett. 2012. "Hunter-gatherer Adolescence." In *Adolescent Identity: Evolutionary, Cultural and Developmental Perspective,* edited by B. L. Hewlett. New York: Routledge.

Hewlett, Barry S. 1987. "Intimate Fathers: Paternal Patterns of Holding among Aka Pygmies." In *Father's Role: Cross-cultural Perspectives,* edited by M. Lamb. New York: Erlbaum, 76–124.

———. 1988. "Sexual Selection and Paternal Investment among Aka Pygmies." In *Human Reproductive Behavior: A Darwinian Perspective,* edited by L. Betzig, M. Borgerhoff Mulder, and P. Turke. Cambridge: Cambridge University Press, 263–76.

———. 1989. "Multiple Caretaking among African Pygmies." *American Anthropologist* 91:186–91.

———. 1991. *Intimate Fathers: The Nature and Context of Aka Pygmy Paternal Infant Care.* Ann Arbor: University of Michigan Press.

———. 1992a. "Husband-wife Reciprocity and the Father-infant Relationship among Aka Pygmies." In *Father-child Relations: Cultural and Biosocial Contexts,* edited by B. S. Hewlett. New York: Aldine De Gruyter, 153–76.

———. 1992b. "The Parent-infant Relationship and Social-emotional Development among Aka Pygmies." In *Parent-child Socialization in Diverse Cultures,* edited by J. L. Roopnarine and B. Segal. New York: Ablex Publishing, 223–43.

———. 1996a. "Cultural Diversity among African Pygmies." In *Cultural Diversity among Twentieth Century Foragers: An African Perspective*, edited by S. Kent. Cambridge: Cambridge University Press, 215–44.

———. 1997. "Cultural Nexus of Aka Fatherhood." In *Gender in Cross-cultural Perspective*, edited by C. B. Brettel, and C. Sargent. Upper Saddle River, NJ: Prentice Hall, 88–112.

Hewlett, B. S. 2013. "Teaching in Hunter-gatherers." Paper presented at the Conference on Hunting and Gathering Societies, Liverpool, UK.

Hewlett, B. S., and A. H. Boyette. 2012. "Play in Hunter-gatherers." In *Evolution, Early Experience and Human Development: From Research to Practice and Policy*, edited by D. Navarez, J. Panksepp, A. Schore, T. R. Gleason. New York: Oxford University Press.

Hewlett, B. S., H. N. Fouts, A. H. Boyette, and B. L. Hewlett. 2011. "Social Learning among Congo Basin Hunter-gatherers." *Philosophical Transactions of the Royal Society (B)* 366:1168–78.

Hewlett, Barry S., and L. L. Cavalli-Sforza. 1986. "Cultural Transmission among Aka Pygmies." *American Anthropologist* 88:922–34.

Hewlett, Barry S., and Michael E. Lamb. 2002. "Integrating Evolution, Culture and Developmental Psychology: Explaining Caregiver-infant Proximity and Responsiveness in Central Africa and the USA." In *Between Culture and Biology: Perspectives on Ontogenetic Development*, edited by H. Keller, Y. Portinga, and A. Schölmerich. Cambridge: Cambridge University Press, 241–69.

Hewlett, Barry S., Michael E. Lamb, Donald Shannon, Birgit Leyendecker, and Axel Schölmerich. 1998. "Culture and Early Infancy among Central African Foragers and Farmers." *Developmental Psychology* 34:653–61.

Hewlett, B. S., M. E. Lamb, B. Leyendecker, and A. Scholmerich. 2000. "Internal Working Models, Trust, and Sharing among Foragers." *Current Anthropology* 41:287–97.

Hewlett, B. S., and J. W. Roulette. In press. "Cosleeping beyond Infancy: Culture, Ecology and Evolutionary Biology of Bedsharing among Aka Foragers and Ngandu Farmers of Central Africa." In *Ancestral Landscapes in Human Evolution: Culture, Childrearing and Social Wellbeing*, edited by D. Narvaez, K. Valentino, A. Fuentes, J. Mckenna, and P. Gray. New York: Oxford University Press.

Hewlett, Barry S., Jan M. H. Van De Koppel, and Luigi Luca Cavalli-Sforza. 1982. "Exploration Ranges of Aka Pygmies of the Central African Republic." *Man* 17:418–30.

———. 1986. "Exploration and Mating Range of the Aka Pygmies of the Central African Republic." In *African Pygmies*, edited by L. L. Cavalli-Sforza, 65–79.

Hirasawa, Ayako. 2005. "Infant Care among the Sedentarized Baka Hunter-Gatherers in Southeastern Cameroon." In *Hunter-gatherer Childhoods: Evolutionary, Developmental and Cultural Perspectives*, edited by B. S. Hewlett and M. E. Lamb. New Brunswick: Aldine Transaction, 365–84.

Hrdy, Sarah B. 2009. *Mothers and Others: The Evolutionary Origins of Mutual Understanding*. Cambridge, MA: Belknap Press.

Ivey, Paula K. 2000. "Cooperative Reproduction in Ituri Forest Hunter-gatherers: Who Cares for Efe Infants?" *Current Anthropology* 41:856–66.

Ivey Henry, Paula, Gilda A. Morelli, and Edward Z. Tronick. "Child Caretakers among Efe Foragers of the Ituri Forest." In *Hunter-Gatherer Childhoods*, edited by Barry S. Hewlett and Michael E. Lamb. New Brunswick, NJ: Aldine Transaction, 191–213.

Kamei, N. 2001. "An Educational Project in the Forest: Schooling for the Baka Children in Cameroon." In *African Study Monographs*, supplement 26:185–95.

———. 2005. "Play among Baka Children in Cameroon." In *Hunter-gatherer Childhoods: Evolutionary, Developmental and Cultural Perspectives*, edited by B. S. Hewlett and M. E. Lamb. New Brunswick: Aldine Transaction, 343–62.

Konner, M. 2005. "Hunter-gatherer Infancy and Childhood: The !Kung and Others." In *Hunter-gatherer Childhoods: Evolutionary, Developmental and Cultural Perspectives*, edited by B. S. Hewlett and M. E. Lamb. New Brunswick: Aldine Transaction, 19–64.

———. 2010. *The Evolution of Childhood: Relationships, Emotion, Mind*. Cambridge, MA: Harvard University Press.

Koulaninga, A. 1987. "L'Education Chez les Pygmées de Centrafrique. Unpublished PhD diss., Université Paris V René Descartes, Paris.

Lancy, D. F., and M. A. Grove. 2010. "The Role of Adults in Children's Learning." In *The Anthropology of Learning in Childhood*, edited by D. F. Lancy, J. Bock, and S. Gaskins. Lanham, MD: Altamira Press, 145–80.

Lee, R. B., and R. Daly. 2004. *The Cambridge Encyclopedia of Hunters and Gatherers*. Cambridge, UK: Cambridge University Press.

Lewis, J. 2002a. "Forest Hunter-gatherers and Their World: A Study of the Mbendjele Yaka Pygmies of Congo-Brazzaville and Their Secular and Religious Activities and Representations." Unpublished PhD diss., University of London.

———. 2007. "Ekila: Blood, Bodies, and Egalitarian Societies." *Journal of the Royal Anthropological Institute* (N.S.) 14:297–335.

Meehan, C. 2005. "The Effects of Maternal Locality on Alloparental Behavior and Frequency of Caregiving among the Aka Foragers of the Central African Republic." *Human Nature* 16:58–80.

———. 2008. "Allomaternal Investment and Relational Uncertainty among Ngandu Farmers of the Central African Republic." *Human Nature* 19:211–26.

———. 2009. "Maternal Time Allocation in Two Cooperative Childrearing Societies." *Human Nature* 20:375–393.

———. In Press. "All Mothers and Child Well-being." In *Handbook of Child Well-being: Theories, Methods and Policies in Global Perspective*, edited by A. Ben-Arieh, F. Casas, I. Frones, and J. Korbin. New York: Springer.

Meehan, C. L., and S. Hawks. In Press. "Cooperative Breeding and Attachment among Aka Foragers." In *Attachment Reconsidered: Cultural Perspectives on a Western Theory*, edited by N. Quinn and J. Mageo. New York: Palgrave.

Meehan, C. L., R. Quinlan, and C. M. Malcom. 2013. "Cooperative Breeding and Maternal Energy Expenditure among Aka Foragers." *American Journal of Human Biology* 25:42–57.

Meehan, C. L., J. W. Roulette. In press. "Early Supplementary Feeding among Central African Foragers and Farmers: A Biocultural Approach." *Social Science and Medicine*.

Morelli, G. A. 1987. "A Comparative Study of Efe (Pygmy) and Lese One-, Two-, and Three-Year Olds in the Ituri Forest of Northeastern Zaire: The Influence of Subsistence-related Variables and Children's Age and Gender on Socio-emotional Development." Unpublished PhD diss., University of Massachusetts at Amherst.

Morelli, G. A., P. I. Henry, and S. Foerster. In press. "Relationships and Resource Uncertainty: Cooperative Development of Efe Hunter-gatherer Infants and Toddlers." In *Ancestral Landscapes in Human Evolution: Culture, Childrearing and Social Wellbeing,* edited by D. Narvaez, K. Valentino, A. Fuentes, J. Mckenna, and P. Gray. New York: Oxford University Press.

Morelli, G. A., and E. Z. Tronick. 1991. "Parenting and Child Development in the Efe Foragers and Lese Farmers of Zaire." In *Cultural Approaches to Parenting,* edited by M. Bornstein. Mahwah, NJ: Erlbaum.

———. 1992. "Male Care among Efe Foragers and Lese Farmers." In *Father-child Relations: Cultural and Biosocial Contexts,* edited by B. S. Hewlett. New York: Aldine De Gruyter, 231–61.

Neuwelt-Truntzer, S. 1981. "Ecological Influences on the Physical, Behavioral and Cognitive Development of Pygmy Children." PhD diss., University of Chicago.

Peacock, N. R. 1985. "Time Allocation, Work and Fertility among Efe Pygmy Women of Northeast Zaire. Unpublished PhD diss., Harvard University.

Rogoff, Barbara, G. A. Morelli, and P. Chavez. 2010. "Children's Integration in Communities and Segregation from People of Differing Ages. *Perspectives in Psychological Science* 5:431–40.

Saurois, Jacques. 1979. *Les Pygmées Aka de Basse Lobaye (Centrafrique): Approche Psychogénétique des Dynamiques de l'évolution Culturelle Aka.* These de Docorate de 3e Cycle.

Shore, Bradd. 1998. "Culture in Mind: Cognition, Culture, and the Problem of Meaning." New York: Oxford University Press.

Sonoda, K. 2013. "Teaching in Face-to-face Interactions among the Baka of Eastern Cameroon." Paper presented at Conference on Hunting and Gathering Societies, Liverpool, UK.

Takeuchi, Kiyoshi. 2013. "Food Restriction and Social Identity of Aka Forager Adolescents in the Republic of the Congo." In *Adolescent Identity,* edited by B. L. Hewlett. New York: Routledge.

Tereshima, H. 1987. "Why Efe Girls Marry Farmers: Socio-ecological Backgrounds of Inter-ethnic Marriage in the Ituri Forest. *African Study Monographs* 6:65–84.

Trivers, R. L. 1974. "Parent-offspring Conflict." *American Zoology* 14:249–64.

Tronick, Edward Z., Gilda Morelli, and Steve Winn. 1987. "Multiple Care-taking of Efe (Pygmy) Infants." *American Anthropologist* 89:96–106.

Tronick, Edward Z., Steve Winn, and Gilda A. Morelli. 1985. "Multiple Caretaking in the Context of Human Evolution: Why Don't the Efe Know the Western Prescription for Child Care?" In *The Psychobiology of Attachment and Separation,* edited by M. Reite and T. Field. New York: Academic Press, 293–322.

Turnbull, Colin M. 1957. "Initiation among the Bambuti Pygmies of the Central Ituri." *Journal of the Royal Anthropological Institute* 87:191–216.

———. 1960. "The Elima: A Premarital Festival among the Bambuti Pygmies." *Zaire* 14:75–192.

———. 1965. *Wayward Servants: The Two Worlds of the African Pygmies.* New York: The Natural History Press.

———. 1978. "The Politics of Non-aggression." In *Learning Non-aggression,* edited by A. Montague. New York: Oxford University Press.

Van De Koppel, Jan M. H. 1983. A *Developmental Study of the Biaka Pygmies and the Bangndu*. Lisse: Swets and Zeitlinger.
Vallois, Henri V., and Paulette Marquer. 1976. *Les Pygmées Baká du Cameroun: Anthropologie et Ethnographie avec Une Annexe Démographique*. Muséum National d'Histoire Naturelle, Paris (Mémoires, Série a Zoologie Tome C).
Vansina, Jan. 1990. *Paths in the Rainforests: Toward a History of Political Tradition in Equatorial Africa*. Madison: University of Wisconsin Press.
Weisner, T. S. and R. Gallimore. 1977. "My Brother's Keeper: Child and Sibling Caretaking." *Current Anthropology* 18:169–90.
Whiting, B. B., and C. P. Edwards. 1992. *Children of Different Worlds: The Formation of Social Behavior*. Cambridge, MA: Harvard University Press.
Wilkie, D. S., and G. A. Morelli. 1991. "Coming of Age in the Ituri." *Natural History* 100:55–62.
Winn, S., E. Z. Tronick, G. A. Morelli. 1989. "The Infant and the Group: A Look at Efe Caretaking Practices in Zaire." In *The Cultural Context of Infancy*, edited by J. K. Nugent, B. M. Lester, T. B. Brazelton. Norwood, NJ: Ablex.
Witkin, H. A., R. B. Dyk, H. F. Paterson, D. R. Goodenough, and S. Karp. 1962. *Psychological Differentiation*. New York: Wiley.
Yamauchi, T., H. Sato, and K. Kawamura. 2000. "Nutritional Status, Activity Pattern, and Dietary Intake among the Baka Hunter-gatherers in the Village Camps in Cameroon." *African Study Monographs* 21:67–82.

10

Multiangular Identities among Congo River Basin Forest Peoples

Stephanie Rupp

The present collection of essays rests on the assumption that the category "hunter-gatherer" is a useful conceptual net for gathering Congo River Basin forest communities together under a common rubric. The research reflected in this volume highlights commonalities among communities of "hunter-gatherers" in equatorial Africa that mark boundaries of inclusion within the category of "hunter-gatherer." In keeping with the decades of research that precede it, this volume takes the analytical category of "hunter-gatherer" as a social reality; "hunter-gatherer" emerges as not just an analytical tool, but as a designation of identity. Among the core issues inherent in the "hunter-gatherer" paradigm is that this model assumes that people's identities are oriented vertically: "hunter-gatherer" identities are presented as rooted in place (such as a forest) and determined by descent (such as ethnicity). Analyses that take "hunter-gatherer" and "pygmy" as categorical givens also project the assumption that the identities of forest dwellers are static. In contrast, this chapter argues that identities among forest communities of the Congo River Basin are multiangular and malleable: identities span multiple spaces, cut across vertical lines of descent through alliances and associations, and change over time depending on historical contexts and political interests. This research calls for analysis of identities that makes space for the multiple ways in which forest peoples perceive self and others, their multiple strategies for deploying particular identities and for navigating particular relationships, and the multiple contexts in which forest people experience and express their identities. Not only do people experience and express multiple identities, the identities themselves are multiangular—changing over time and with context. This essay seeks to draw more sustained, analytically rigorous attention to the multiple facets of identities held by forest peoples, as held by people throughout the world.

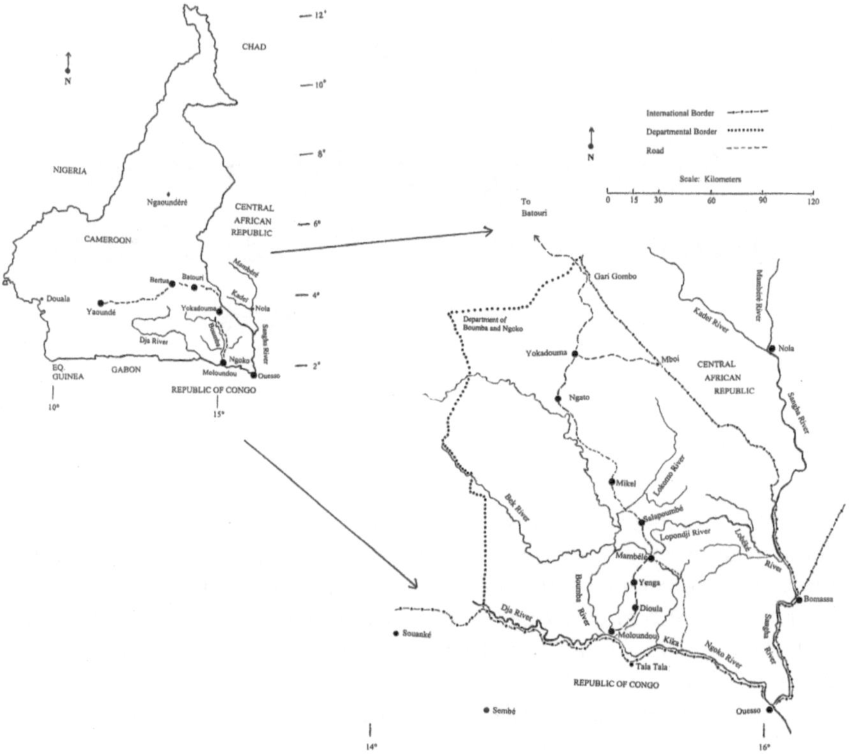

Figure 10.1. Lobéké forest region of southeastern Cameroon.

The Lobéké forest region of southeastern Cameroon is distinguished by "Lake" Lobéké, which seasonally dries into a swampy marsh (figure 10.1). The forest surrounding Lobéké is a pocket of semideciduous, tropical forest within a much larger cultural and ecological sphere of interaction on the western rim of the Congo River Basin. The Lobéké forest region, located near the triangulation of national borders of Cameroon, the Republic of Congo, and the Central African Republic, is home to four locally based ethnic groups: Bangando, Baka, Bakwélé, and Mbomam. Scholars, conservation and development workers, missionaries, and travelers consistently identify Bangando, Bakwélé, and Mbomam as "villagers" or "farmers" and their numerically dominant neighbors, the Baka, as "pygmies" or "hunter-gatherers." From oral historical accounts gathered among the four main communities in southeastern Cameroon today, consistent accounts emerge indicating that the Bangando, Baka, Bakwélé, and Mbomam have coexisted as neighbors in various relations of mutual support and cooperation as well as tension and competition for nearly two centuries.[1]

In the Lobéké forest region, people of the four primary ethnic communities cultivate varied senses of belonging that straddle vertical lines of ethnic descent. The first set of cases presented in this chapter demonstrate that, although Bangando, Baka, Bakwélé, and Mbomam individuals recognize ethnic sentiments of belonging within particular communities defined by those who speak a shared primary language, meaningful senses of belonging are not limited to vertical dimensions of belonging, such as language and ethnicity, that are derived from descent (see figure 10.2). Three interethnic associations—*Éésòngé*, *Mbékà*, and *Bálè*—gather together people of all four ethnic communities to support one another in times of particular stress or strain. The second set of cases highlights that multiangular identities are not

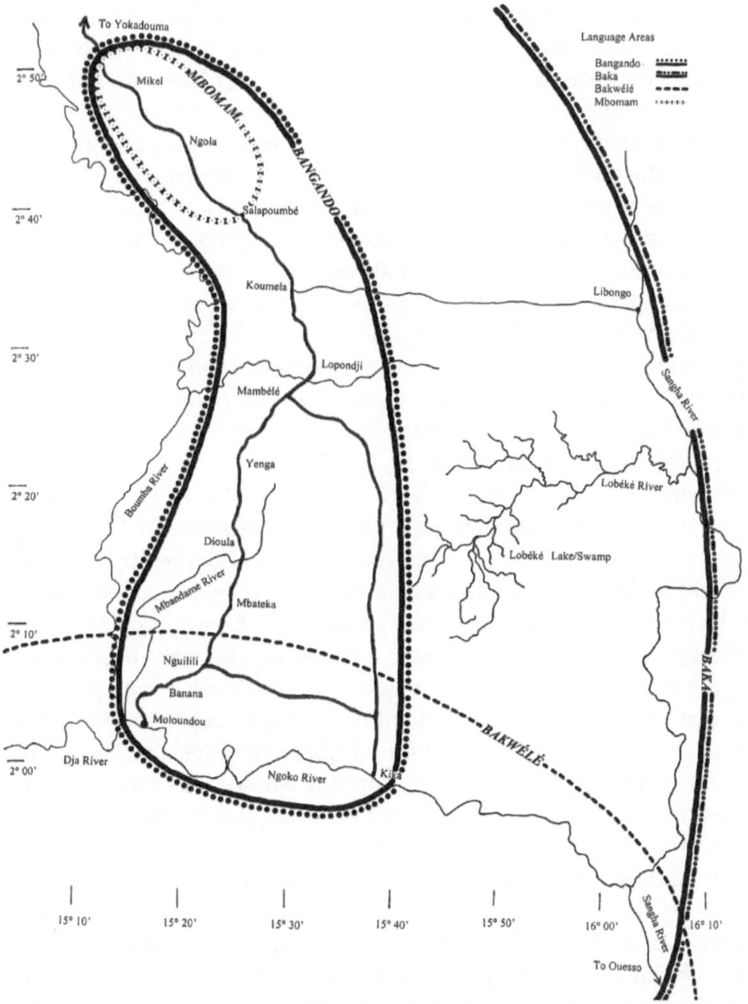

Figure 10.2. Language communities in the Lobéké region, southeastern Cameroon.

rooted to particular places, but span spaces within and beyond the forest. For example, within the forest region, ethnic solidarities are variable; in Ndongo, a region of the forest along the Ngoko River that is primarily dominated by Bakwélé, Baka residents view interethnic relationships with Bangando favorably, even going so far as to weave the Bangando ethnic affiliation into their own sense of identity. The third set of cases underscores that the sense of social belonging within the forest, expressed through the expression of identity *bò kóà* or "person of the forest," is variable in different ecological, social, and institutional spaces of the forest. And a final set of cases highlights the multiple ways and contexts in which the stereotype "pygmy" can be ascribed to people of the Lobéké forest region. Taken together, these diverse cases highlight the different registers of identity among forest peoples, including social identities, ethnic affiliations, and stereotypes. The cases underscore that any given person's identity is multiangular *and* that categories of identity are ascribed and described in multiple ways for multiple purposes in multiple contexts. Recognizing the multiangularity of forest people's identities necessitates sustained, analytical attention to the ways in which identities, relationships, and belonging go beyond identities that are rooted to place (such as "indigenous people") and undermine categories that rest on simple economic orientations (such as "hunter-gatherers"), and transcend expectations of narrow, vertical lines of descent-based ethnic affiliation.

In previous work, I examined the differences among stereotypes, ethnic affiliations, and social identities among forest peoples of southeastern Cameroon (Rupp 2001, 2011). The central argument in these works is that forest peoples' identities have different qualities depending on circumstances in which particular identities are deployed. *Social identities* are designations of identity that are malleable, fluid, and context specific, enabling people to move in and out of social categories, relationships, and positions with relative ease. Examples of social identities held by forest people include, but are not limited to, identities related to people's means of earning income (people of the timber camp), identities related to social groups (people of the ball, or members of a soccer team), and identities related to people's relative position within domestic or village contexts (people of the kitchen; people of the verandah). This work also demonstrates that *ethnic affiliations*—designations of identity that reference language and history of origins—tend to be more viscous than social identities, changing sluggishly over time (Rupp 2001, 2011). For example, the ethnic affiliation held by contemporary Bangando is not the same ethnic affiliation held by their forebears, Ngombe, who migrated into the forests of southeastern Cameroon during the early nineteenth century. *Stereotypes*, finally, are the most rigid form of identity, comprising reified perceptions of common characteristics that outside observers—missionaries, government officials, conservation and development workers, tourists and travelers, and even fellow citizens from other regions of the nation—presume are overarching characteristics of members of a particular ethnic group. Stereotypes are based on distillations of elements of who people are that outside observers have determined are relevant (Rupp 2001, 2011; Allport 1954). For example, social characteristics such as living in the forest, hunting and gathering, and filing one's teeth serve as simplified abstractions of identities of "pygmies," "hunter-gatherers," or "indigenous people" that have become crystallized into stereotypes, conceptual simplifications, and rigid expectations about who is who in equatorial forest contexts.[2]

In examining social identities, ethnic affiliations, and stereotypes among forest peoples of southeastern Cameroon, I based my ethnographic research on the concept of "multiangularity." By multiangularity, I mean that fine-grained research on dynamics of belonging and identification must not examine one community in isolation but rather must take the multiethnic community as the unit of analysis, looking at individuals and groups as they interrelate across ethnic boundaries (see figure 10.3). Multiangular analysis also looks at individuals, groups, and communities of the region as they relate to their shared natural and social environments and as they relate to institutions, policies, and ideologies that affect their lives. In the specific context of the Lobéké forest region of southeastern Cameroon, this approach required a multiethnic examination of points of articulation among Bangando, Baka, Bakwélé, and Mbomam communities and examination of the larger social, economic, and political contexts in which these communities are situated.[3]

In this chapter, I argue that the notion of multiangularity provides not only a helpful orientation that insists on multiethnic analysis of forest communities, thereby counterbalancing research that tends to prioritize lopsided accounts of social relationships within tropical forests; but more fundamentally, the idea of multiangularity offers an analytical framework that allows—indeed requires—research to extend beyond the vertical dimensions of spatially rooted and descent-based identities. Together with long-term ethnographic fieldwork, recognition that forest peoples' identities are multifaceted helps to drive research beyond considerations of subsistence, beyond binary relationships of presumed domination and subordination, and beyond oppositions between nature and culture. Multiangular perspectives on identity bring analysis of forest peoples into the flexible, globalized world of the twenty-first century, the context in which forest peoples actually live today.

A multiangular analytical perspective acknowledges that people's identities are shaped by the people around them, and as a result of their varied relationships with these other people; people's identities are shaped by social groups, institutions, and the ways that they fit into these groups; and people's identities are shaped by the different spatial and historical contexts in which they live and interact. What it means to be a particular kind of person is variable: being "Bangando" or "Baka" or a "pygmy" or "*bò kóà*" (person of the forest) does not necessarily mean the same thing, even for the same person, in different

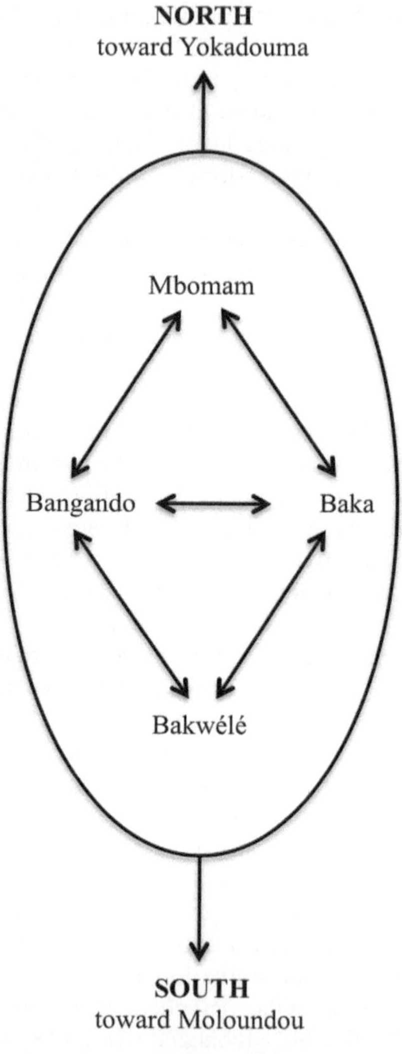

Figure 10.3. Multiangular relationships among communities of southeastern Cameroon.

contexts. In analyzing processes of identity formation and change among forest communities, it becomes clear that just as people hold multiple identities simultaneously and just as these identities cut across narrow expectations of place and descent, the identities themselves—the social identities, ethnic affiliations, and stereotypes—*also* consist of multiple facets, multiple angles.

Senses of Belonging, Dynamics of Identity in the Lobéké Forest Region

According to linear, generational models of "indigenous people" and ensuing policy prescriptions for conservation, development, and missionization in southeastern Cameroon,[4] it is commonly accepted that Baka "hunter-gatherers" are the singular original inhabitants of the Lobéké forest region. But contrary to this prevailing and stereotyped expectation, the forest communities of southeastern Cameroon are, and have been for centuries, diverse and dynamic with various distinct groups overlapping and merging, interdigitating and separating. These relations of interethnic cohesion stem from social relations sparked by fortuitous meetings, collaborative partnerships that culminate in shared ritual practices and ceremonial structures, intermarriages, and other intimate relations. Simultaneous relations of competition and conflict that cut across lines of amity produced countervailing tensions, giving social relations their dynamic qualities. A multiangular approach to analyzing social identities and relationships brings such horizontal relationships that cut across ethnic boundaries into view.

Multiethnic Associations: *Éésòngé*, *Mbékà*, and *Bálè*

The communities of southeastern Cameroon come together across ethnic boundaries in three associations; the core sentiments of belonging in these associations are gender, a sense of common purpose, and a shared sense of belonging to the forest of southeastern Cameroon. *Éésòngé* is an association of women "of the same heart" who live in the village of Dioula. Members of *Éésòngé* belong to each of the four primary ethnic communities; membership in the association is based on mutual support, development, and participation among women regardless of ethnic affiliation. Other villages in southeastern Cameroon have similar women's associations. Men throughout the region have the option of belonging to *Mbékà*, a society of men who step forward—enduring a difficult initiation process—to guide and protect the community as a whole. As with *Éésòngé*, participation in *Mbékà* is not ethnically bounded but determined by the choice of each individual man and by his willingness and ability to build social networks, amass economic resources, and undergo challenging ritual preparations for his role as a senior counselor for the community at large. Members of the association called *Bálè* include Bangando, Baka, Bakwélé, and Mbomam residents of Yaoundé, the capital of Cameroon. This association comprises individuals who migrated from the Lobéké forest region of southeastern Cameroon to Yaoundé for various professional reasons, but who find comfort in socializing and supporting people from their home region, which lies nearly 1,000 kilometers from Yaoundé. These associations highlight social, economic, and moral ties that connect individuals across the ethnic communities of the Lobéké forest region.

Éésòngé

Labor pains came earlier than she expected. Baleke, a first-time Mbomam mother, was on her way to her garden, but she turned back toward home, walking, shuffling, and lurching three kilometers back to her house along the path that meandered through the forest. Arriving home nearly an hour later, there was no fresh water to drink, no plantains to prepare, no meat in the rack above the kitchen fire. Because her stomach was not yet fully rounded, her mother had not yet come to assist her from the northern region of the Lobéké forest; that day, one Bangando sister-in-law was tending her dry-season garden (*sóbó*), while another sister-in-law had gone deep into the forest for dry-season barrage fishing. Baleke hunkered down to endure the contractions, and in between the pains, she positioned herself near the doorway of her kitchen house so that she could alert a passer-by along the road to inform Molina, a Bangando neighbor and the *présidente* of *Éésòngé*, that she had gone into labor.

Within an hour, Molina received the news and came quickly to Baleke's aid. Molina also informed the women of *Éésòngé* to prepare supplies and other donations for Baleke and her new baby. By this time, the sun was high and hot in the sky; women of all four ethnic affiliations began filtering back to their kitchen houses from their morning activities in the forest. News of Baleke's sudden and premature labor spread quickly, from kitchen house to kitchen house. Sharing out portions of the plantains, palm nuts, fish, and *kúmbí* (*Gnetum africanum*, a forest leaf commonly cooked into a hearty stew) that they had gathered for their own families, women set aside food for Baleke for the day as well as provisions for days to come. An elderly Bakwélé woman who had not gone into the forest at all that morning filled an enamel basin with clear water for drinking and called to her grandchildren to fetch buckets of water for bathing. A young Baka woman wrapped a block of partially used grease soap in a broad, round *sàà té* (*Marantaceae*) leaf as her contribution; every new mother needs generous supplies of soap for washing baby's diapers. A middle-aged Bangando woman had kerosene for her hurricane lamp and poured the oil out of her own lamp and into a plastic bottle, ensuring that Baleke had fuel to keep the surroundings of her new baby lit, even throughout the night. A Bangando woman renowned throughout the region for her knowledge of medicinal plants arrived from her household on the northern edge of Dioula, bringing her supplies to tend to the new mother and her baby. Throughout the long, narrow village stretched along the road, the women of *Éésòngé* gathered supplies, washed their faces and legs from the morning's work, wrapped a clean *sándá* cloth around their waists, and headed out into the early afternoon heat to visit Baleke and bring her their bundles of resources wrapped securely in leaves, pots balanced expertly on their heads. Even without her own family present, Baleke and her new baby were surrounded and supported by the powerful network of multiethnic women of the village, who provided emotional support, material resources, and depth of experience.

Éésòngé is a voluntary association of "women of the same heart." Women of the village's four ethnic groups—Bangando, Baka, Bakwélé, and Mbomam—participate in the women's group, knowing that their active support of women in the village provides individuals with crucial support during times of crisis. When circumstances arise that a woman cannot bear alone—the death of a relative and requirements of preparing the funeral and hosting dozens and dozens of mourners, the birth of a baby, an unusually large ceremony or rite of passage—women of the same heart come together to support each other. Women are very clear that they support only women in this association, but they are also equally adamant that they support *all* women of the village, regardless of ethnicity. Through *Éésòngé,* women traverse and transcend the vertical lines of ethnic affiliation that distinguish speakers of different languages in southeastern Cameroon. The sense of belonging that unites women is a shared sense of loyalty, shared commitment, and common identity as women who give, provide, and nurture each other.

The men of the village also form a multiethnic association that extends support to all. Bangando, Baka, Bakwélé, and Mbomam men who are between the approximate ages of twenty and thirty-five may choose to be initiated into *Mbékà*, an association of men that cares for the long-term concerns of the village.[5] The initiation process requires each new inductee to amass resources—meat, oil, salt, plantains, alcohol—provisions for the ceremony and symbols of the man's ability to support the community. While all men are expected to contribute staple resources for the ceremony and celebration, depending on their access to certain goods, men may provide resources in various proportions. In Dioula, for example, the most prolific grower of plantains is a Baka man, while the most successful hunter is Bangando. Each man contributes to *Mbékà* according to his ability. Men form alliances with friends, relatives, and neighbors to gather the resources for their induction into *Mbékà*; the wealth in goods is based on wealth in people (Guyer 1995), with the underlying strength of individual men, and of the community overall, signaled by a man's ability to find support in other men. Relationships of financial and emotional support during *Mbékà* initiation are interethnic; often Bangando and Baka men who have close friendships and alliances will support each other as they amass the necessary resources and as they endure the physical challenges of the initiation process. The *Mbékà* society itself is also explicitly multiethnic. Overall, *Mbékà* society mirrors the structure of an ideal society, with men working together to ensure that clear and sensible decisions are made for the multiethnic community as a whole, in particular during times of crisis. Through their participation in *Mbékà*, Bangando, Baka, Bakwélé, and Mbomam men build relationships of mutual trust, confidence, and engagement.

While both *Éésòngé* and *Mbékà* associations take gender as the common core of shared identity among members, another multiethnic association among Bangando, Baka, Bakwélé, and Mbomam individuals has members' sense of belonging to the forest itself as its core, even as members live nearly 1,000 kilometers from the forest, in Yaoundé, the capital city of Cameroon. As is evident from relationships among members of this association, called *Bálè*, interethnic relationships differ more in intensity, organization, and dynamics in the context of the city than in the context of the forest-village nexus.

A cohesive and vibrant community of people from southeastern Cameroon lives and works in Yaoundé and Douala, pursuing professions that they hope ultimately will contribute to the development of their home villages in southeastern Cameroon. A dentist and a doctor, a policewoman and a government bureaucrat, three army sergeants and an army recruiter, as well as several teachers and students form the core of an organized association of "*ressortisants*" or literally "nationals" of the Ngoko River region. The name of this association—*Bálè*—is the Bangando term for the Ngoko River, providing a unifying symbol of belonging from southeastern Cameroon with which its thirty-odd members, including Baka, Bakwélé, Mbomam, and Bangando, identify. These southeastern Cameroonians have come to Yaoundé to establish or improve their lives through jobs, education, and marriage. Living far from the Lobéké forest region of southeastern Cameroon, many members of *Bálè* maintain deeply emotional—and ambivalent—sentiments about the Ngoko River region and the villages in the forest that they left behind.

> When you are in the village, you dream of the city as if it's *El Dorado*. When you arrive in the city, you are surprised about the many differences between the city and the village. In the village, nature is everywhere and keeps everything—the water, the swamps—unpolluted. You can find a papaya someplace and eat it without someone coming to threaten you. In the city you need money to rent a room, for water, electricity, to make phone calls, because when you live in the city, you need to be cultivated and educated. . . . Of course there are people in the village who think that life is rosy in the city. Don't be surprised if someone in the village asks you to buy them a tape recorder like that [indicating the one recording his voice], or a shirt, even if you don't have the money to buy gifts. The people in the village believe that the city is a paradise where we harvest things from the side of the road, that we just collect what we need and find things to take back to the village. "*Où sont les provisions?*" ["Where are the gifts?"] is the question that everyone asks you when you come back home from the city.[6]

City-based members of *Bálè* consistently cite economic hardship as one of the highest hurdles to overcome in the city. Before they came to Yaoundé they recall thinking of life in the big city as brimming with opportunities to develop oneself, to make money, and to prosper. But in their experiences living in Yaoundé, the reality of urban life is starkly different. Two young Bangando men, Gnangue and Kouamanda, are working toward their master's degrees at the École Normale Superièure in Yaoundé and share a dim, dank room with a third student not far from the university campus. The three students cook meager meals over a simple gas ring, share a single bed, and wash their faces in a communal bucket of water each morning and wash their feet from a communal bucket of water each evening. More than once, the students have been evicted from their rented room because, among three of them, they could not afford the rent of US $12 per month. Ironically, learning to live on the margins and to scrape by in the city is an intimate part of what Gnangue and his friends experience as they become "cultivated and educated" living in the city.

When faced with the realities of student life in Yaoundé, it is life in the village that seems idyllic:

> When I am here in the city, I miss many things: the village, the forest. I miss the gardens and the calm—there is no noise of cars and motorcycles. And I see the warmth of relations with my parents, my brothers. I play soccer with my brothers and friends. I also go to fish because I love to fish, and I go to hunt and I work in the garden too. I sit with the elders, my grandparents and talk with them, and they recount the history and marvels of the past.[7]

While from a distance the city seems cosmopolitan, exciting, and holding limitless possibilities for a future of wealth and influence, from the vantage point of the gutter along a rutted road in Yaoundé where the landlord has just turned the students out, the village is a rural, secure paradise where individuals live in a wealth of social resources through their families and in a plethora of basic resources provided by the forest and gardens.

These stories of the difficult transition from rural to urban life, the hankering of city-dwelling people for the bucolic countryside, and the ensuing conceptual tensions between city and forest are certainly not unique to people arriving in Yaoundé from southeastern Cameroon (see Williams 1973). More interesting is the way that these urban hardships and sense of dislocation have brought Bangando, Baka, Bakwélé, and Mbomam in Yaoundé together to form a multiethnic association of people from the Ngoko River region—*Bálè*. This group meets each month to discuss their collective and individual problems, drawing on the resources that each member has to offer. Members pay a monthly contribution to the association's kitty, which is reserved for emergencies among the Yaoundé-based community or for transportation back to the village should someone need to return home urgently. *Bálè* members celebrate major holidays together, much as their family and friends do in the village: at the end of the calendar year, their annual bonuses parallel the spurt of income that village-based families experience when they harvest and sell their cocoa, making the Christmas and New Year seasons particularly festive for all people whose roots lie near the Ngoko River, whether they live in cities or villages. At these celebrations, the *Bálè* members dance Bangando and Bakwélé dances and recount legends and histories; were there enough Baka women in Yaoundé to sing, *Bálè* members unanimously claim they would also include Baka dances in their parties, to include the traditions of their Baka members.

The differences in celebrations that simultaneously take place in Yaoundé and in the villages at the end of each calendar year reflect different degrees and dimensions of overlap among the interethnic communities. While in the forest, villagers of all ethnic backgrounds have friends and neighbors who belong to other groups, their holiday celebrations will primarily revolve around their own extended families and especially close friends, who are often, though certainly not exclusively, of the same ethnic community. In Yaoundé, on the other hand, families are fragmented and do not have extended networks of relatives with whom to celebrate; thus the members of *Bálè* celebrate the holidays in particular but support each other emotionally in many other general contexts, without regard to language and ethnicity. The common denominator of having origins in the forest near the Ngoko River unifies the urban-based Bangando, Baka, Bakwélé, and Mbomam, bringing their solidarity to a level unique to the urban setting. As expressed by Gnangue, a Bangando,

> There is a big difference in our relationships in the village and in the city. In the city there is no difference between Bangando, Bakwélé and the others; there are no barriers between us, because we all come from the same place. I feel like I'm at home at Naomi's house (Mbomam) just as I feel at home at Mota's house (Bakwélé). In the village, there are distinctions between families. In Yaoundé, these differences don't appear.[8]

The comments of a Baka army sergeant who lives in Yaoundé echoes the sense of connectedness that he feels to others, mostly non-Baka, from the Ngoko region and hints at his wonder at the utter sense of acceptance that he feels among the *Bálè* association members:

> Here in the city I call Bangando and Baka and Bakwélé my brothers. If I introduce a friend to my military colleagues, I introduce him as my brother. Because all of the people from the Ngoko call me "brother." Maybe our relations are different because we're in the city. In the city it's different. In the village, I never belong to Bangando associations. I don't know why it's different here in the city. The Bangando living in Yaoundé say it's only the problem of a name, that we should all be unified.[9]

When *Bálè* members leave the context of the city and return to their home villages in the forest region of Lobéké, members say that they still feel the intimate connections of their shared experiences in the city. *Bálè* members will visit each other's homes when they return to the village, evidently seeking out companionship and nurturing alliances with people who have common interests and perspectives. Urban Bangando, Baka, Bakwélé, and Mbomam cherish the close interethnic relations that they have fostered in the context of Yaoundé and make an effort to maintain their shared sense of partnership and affiliation, the angles of identification and belonging that are inflected by their shared city experiences.

In addition to illustrating the multiethnic ties that create a tight, horizontal lattice of relations and affections across the ethnic groups, the urban association of Lobéké forest peoples highlights the multiangularity of spatial senses of belonging. For members of *Bálè*, the spatial nexus that is relevant for their sense of communal belonging positions the forest region overall, including—not separating—the forest and its villages over and against the city, Yaoundé. Finally, urban-based Bangando, Baka, Bakwélé, and Mbomam clearly indicate that their sense of belonging together as a community straddles space, rather than being vertically rooted to it.

Multiangular Ethnic Affiliations

The examples of multiethnic associations—*Éésònge*, *Mbékà*, and *Bálè*—demonstrate that a shared core of belonging unites people of different ethnic backgrounds in cohesive relationships based on common gender, collective visions of a mutually supportive community, and shared experiences of urban living. Just as sentiments of ethnic affiliation can be transcended by such common experiences and expectations among groups of individuals, so too can sentiments of ethnic affiliation vary in the experiences of individuals. What it means to *be* Bangando, Baka, Bakwélé, or Mbomam—even for a given individual—may vary across spatial and social contexts.

The village of Ndongo lies far upriver along the Ngoko River, requiring a six-hour journey in a canoe with an outboard motor from Moloundou, the last town on the southern Cameroonian border before crossing the Ngoko River into the Republic of Congo. Because of the roadwork provided by a timber company, the village and its satellite settlements were once connected to the main Moloundou-Yokadouma road; when the timber company abandoned its concession in the early 1990s, the road fell into disrepair and became impassable once its bridges collapsed. The people of Ndongo are primarily Bakwélé who make their living from catching fish on the river, growing subsistence crops in their gardens, and cultivating small plantations of cocoa. For others living at Ndongo—Baka and a few Bangando who stayed behind after the timber company closed—Ndongo feels very remote.

Visiting Ndongo, I was surprised by the very different quality of relationships among the communities—Bangando, Baka, Bakwélé—that I had come to know in villages along the Moloundou road. Where I had become accustomed to interethnic relationships that were affable and integrated, in Ndongo, the social context felt tense, even toxic. Bakwélé and Baka had lived for numerous generations in Ndongo in roughly equal numbers, with just a half-dozen or so Bangando families living in the region. However, social relations soured in the context of the timber and cocoa boom during the 1970s and 1980s, and Baka living in Ndongo moved away from their Bakwélé neighbors in the main village to settle their own villages two kilometers away. Finding it difficult to gather detailed information about these complex social relationships because of my own linguistic limitations (I speak Bangando well, but neither Baka nor Bakwélé), I was relieved to learn that a "young Bangando woman" lived in one of the Baka villages. My sense of puzzlement on finding a Bangando woman living in a Baka village in the Bakwélé region of Ndongo was both eased and amplified when I met Nana, who in all respects—lilting Baka accent when she spoke, body decorations—appeared to be a Baka woman. But she spoke fluent Bangando, and embraced me literally (a left- then right-leaning hug of each shoulder, a typical Bangando greeting) and also embraced me in her willingness to help me learn. Yet why would a Baka woman identify herself—adamantly—as Bangando?

Nana had come to Ndongo when she married her husband, a Baka man whose family and social networks are situated along the Boumba River. After the early years of marriage, when she and her husband lived among her relatives in the village of Ngombe along the Moloundou road, Nana moved with her husband back to his village and family in Ndongo. But even after living for many years in Ndongo, when Nana introduced herself to me, she said, "My name is Nana. I am a Bangando girl!" with a smile radiating welcome and enthusiasm.[10] Living in Ndongo among Bakwélé neighbors had been difficult for Nana, who has friends and neighbors who are both Baka as well as Bangando in her village of Ngombe. Rather than amicable, interethnic relations that are based on generalized reciprocity as in Ngombe, relations between Baka and Bakwélé at Ndongo were tense and sometimes outright demeaning for the Baka.[11] Nana was very enthusiastic about identifying herself as Bangando, explaining "*mi tòlì ná Bángàndò*," (or, "I come from Bangando"; here "Bangando" was used as a place marker.) In the taut social environment of Ndongo, where interethnic relations between Baka and Bakwélé were characterized by mistrust and animosity, Nana was eager to emphasize the angle of her own identity that affiliated her geographically, socially, and morally with the multiethnic communities of the region near her home village in the "Bangando" region of the forest.

As I spent more time in Ndongo, I met several other Baka individuals who identified themselves enthusiastically as Bangando. In each case, they were

Baka who had moved from villages along the Moloundou road to the Ndongo region following their spouses. In some cases, Baka individuals did, in fact, have biological relatives who were Bangando; but after very carefully tracing these kinship relations, it was clear that these Bangando relatives were affines or maternal relatives whose structural kinship positions would not affect the formal, partilineal designations of these individuals as Baka. Finally, one Baka man from Salapoumbé, the northern-most Bangando village along the main road to Yokadouma, offered an explanation for the ethnic intimacy felt between Baka and Bangando: "We are one thing, one concept, with Bangando. We come from the same soil. We come from the same stomach."[12] Bangando living in Ndongo also praised the interethnic intimacy that they share with Baka: "For Bangando and Baka, our villages and even our houses are mixed. We eat together. You rarely see Bakwélé and Baka sitting together to eat. We (Bangando) speak Baka, so we *are* also Baka."[13] These examples of multiangular ethnic affiliations signal that ethnicity is more than something that individuals are born into in a narrow, descent-derived, primordial sense. Senses of belonging among friends and neighbors who share common experiences, networks, attitudes, and interests can foster ethnic affiliations that have multiple dimensions and multiple angles, enabling these individuals to embrace more than one ethnic affiliation as their own. What it means to belong to any particular ethnic group may change depending on social and spatial circumstances, lending a surprisingly multiangular quality to ethnic affiliation.

Multiangular Identity of *Bò Kóà*, a Person of the Forest

Belonging to the forest is a quality of identity that prevails among individuals of all four communities of the Lobéké forest region. While there is no equivalent for "hunter-gatherer" or "pygmy" in the local languages Bangando, Baka, Bakwélé, or Mbomam, the notion of "people of the forest"—or *bò kóà* in Bangando—serves as a gloss for those who derive their livelihoods and their economic, symbolic, social sustenance from the forest, or who spend a significant portion of their time within the forest. In contrast to the tight link drawn by external researchers between the categories of "hunter-gatherer" or "pygmy" and particular ethnic groups in the Congo River Basin, such as the Baka of southeastern Cameroon, the idea of *bò kóà*—people of the forest—can apply to individuals from each of the four ethnic groups in southeastern Cameroon. "People of the forest" is a multiethnic category. Furthermore, *bò kóà* is a multiangular identity because the meanings and attributes of this identity vary in different historical eras and in different ecological, social, and political spaces of the forest.

The social identity *bò kóà* has been used by particular groups of people to describe themselves and by some groups to ascribe identities of others. Over all, the notion of *bò kóà* or "person of the forest" reflects a deep sense of connection to and involvement in the forest. What varies is who is connected to and involved in the forest at a particular point of time in a particular context, and why.

Historically, the category of *bò kóà* was used by Bangando to refer to Baka partners and neighbors who lived in small settlements that were set apart from their more centralized villages that were located along the Boumba River. During the precolonial era the architecture and layout of Bangando villages—rectangular houses of pounded bark, built in rows on opposite sides of a central pathway—contrasted from the more dispersed, more mobile villages of Baka, who built their rounded houses in small clusters to the sides of the larger Bangando villages. During this time, *bò kóà* referred to people's place of primary habitation (the forest), not a particular or specialized set of techniques for subsisting in the forest, for Bangando were also accomplished hunters. When French colonial officials insisted on the migration of villages from the main path along the Boumba River to the side of a new road (which Bangando, Bakwélé, and Mbomam "villagers" were forced to build by hand), the identity *bò kóà* was further accentuated; after roadside resettlement, Baka houses remained in the forest behind Bangando (and Bakwélé, Mbomam) villages. When Baka were finally forcibly settled in villages along the road during the 1970s, the notion of *bò kóà* ceased to link a particular ethnic community (Baka) with a particular space (the forest). By this time, all people were *le*-based, based in the village, and were collectively referred to as *bò le*, people of the village. *Bò kóà* shifted in meaning to refer to people (of any ethnic community) who went into the forest for an extended period of time, for example, for the harvesting of wild mangoes during the short dry season.

During the years just after independence, the identity "people of the forest" took on new meaning that was unrelated to local identities. The Marxist rebel faction led by Ossende Afana infiltrated Cameroon through its southeastern border with the Republic of Congo (Brazzaville), living in the forest while percolating their plan to topple Prime Minister Ahmadou Ahidjo from power. As the Cameroonian military moved into the Lobéké forest region to confront the threat, local people took steps to disassociate themselves from the Marxist rebels, even as they assisted the rebels by providing supplies and guidance through the forest. Bangando, Baka, Bakwélé, and Mbomam villagers referred to the rebels as *bò kóà*—"people of the forest," referencing their use of the forest as cover for their activities—and contrasting them with local people of all communities, *bò le*, who remained in the safety of their villages.

Today, people of southeastern Cameroon continue to employ the notion of "people of the forest" in multiangular ways; *bò kóà* is an identity that reflects variable meanings in different contexts of the forest. For example, Bangando who spend the majority of their time living in their forest gardens describe themselves as *bò kóà*; Bangando and Baka men who base their livelihood on a shared pursuit of hunting describe themselves as *bò kóà*; and local people of all four communities who work for international conservation organizations or for safari hunting companies describe themselves as *bò kóà*.

A Bangando couple named Zoduma and Besombo have departed from the village-based lifestyle that now predominates among people of all four communities in southeastern Cameroon. Zoduma and Besombo focus their lives and work on their forest garden, *mbándá*. Their garden is located fifteen kilometers east of the village of Dioula, where the rest of their family members live. Zoduma and Besombo spend the vast majority of their time in the forest, coming back to Dioula only for holiday celebrations or in case of illness or emergency. In the forest, Zoduma and Besombo grow food for both subsistence and for sale on the regional market. They carry heavy loads of plantains to the roadside at Dioula or arrange from transportation of their crops to the larger market in Moloundou. In addition to cultivating their garden, Zoduma is an active hunter and fisher; the couple eats well in the forest and still has meat to dry and sell. Zoduma describes himself as *bò kóà*, a person of the forest. But rather than emphasizing the obvious contrast in their choice to live in the forest rather than in the village, Zoduma explains that being "of the forest" signals a disciplined approach to work, an openness about working cooperatively and sharing with each other.

Another set of people describes themselves today as people of the forest, or *bò kóà*: contemporary hunters. A small subset of men—Bangando and Baka—in southeastern Cameroon continue to live primarily by hunting, relying on hunting both to feed their families and to provide income to support their other needs. Hunting teams combine men who have complementary skills and resources; sometimes this complementarity forges lasting partnerships between men of different ethnic affiliations. For example, a hunting team between an elderly Baka man and a younger Bangando man match the Baka partner's deep knowledge of paths within the forest and behavior of forest animals with the Bangando partner's physical stamina, knowledge of logging tracks in the forest, as well as his access to resources such as a gun for hunting, suppliers of bullets, and networks among "patrons" who provide markets for meat and animal products and some degree of protection from accusations of poaching. Both partners are deeply experienced in hunting in the forest, spend a large portion of their time in the forest, and speak with clear fondness for each other and respect for the other's contribution to the partnership.

Another group of contemporary *bò kóà* includes local people who work for international conservation organizations, which have been active in the Lobéké forest region since the early 1990s. Lopondji, a subhamlet of the village of Mambélé, has served as the geographical focus of conservation efforts undertaken by both the Wildlife Conservation Society (from the early 1990s until 1996), and the World Wide Fund for Nature (from 1996 until the present). The conservation organizations maintain a tented camp at a clearing in the forest roughly two kilometers from Lopondji and employ men from this small neighborhood as camp assistants, guards, and field research assistants. As the conservation efforts expanded throughout the region—village conservation committees were formed in each village in the Lobéké forest region, for example, in preparation for the legalization of the Lobéké National Park in 2001—the conservation and development organizations began to engage men from various locations more actively in their work. However, the focal team of local conservation employees still comprises primarily men of Lopondji and Mambélé, where the conservation work first began. In Lopondji, a Bangando man named Mokogwea describes himself as a forest expert, *bò kóà*, saying: "Because WWF has seen that we (the men of Lopondji) know the forest, they come each time to look for us. Especially me. Because in the forest, it is I who shows them where to go. . . . I know this entire forest. So the work that I do for WWF is the work of the forest."[14] Conservation efforts have provided new opportunities for local people to deepen their knowledge of the forest and new ways for local people to become forest experts, cutting new facets on the already multiangular identity "people of the forest."

Multiangularity of "Pygmy" Stereotype

Of the three modes of identification—viscous ethnic affiliations (such as Bangando, Baka, Bakwélé, and Mbomam), fluid social identities (such as *bò kóà*, person of the forest), and rigid stereotype (such as "hunter-gatherer" or "pygmy")—even the most rigid form of identity, stereotype, is multiangular in its meanings and uses. The stereotype "pygmy" may convey multiple, even contradictory values when it is used in varying contexts to describe different kinds of individuals.

In the local setting of southeastern Cameroon, there is no equivalent term in the local languages to convey the notion of "pygmy." Only once did I hear the idea of "pygmy"—expressed in French—used by a local resident to identify another person in a context that had no reference to external factors such as conservation, development, or missionization, settings in which stereotypes such as "pygmies" and "hunter-gatherers" are employed frequently. In this case, in an inebriated rage, a Baka man derided his (similarly inebriated) friend shouting, "*Tu es pygmée*!" In the context of their argument, this insult carried the connotation that his friend embodied the worst of the characteristics attributed to "pygmies" by outsiders: depravity, savagery, and lack of control. The Baka man hurled an insult that conveyed his comprehension of the negative associations with the stereotyped identity of "pygmy;" he utilized the rigid, negative attributes in of this essentialized identity a strategic way to position his drunk friend as the lowest of the low, a "pygmy" (Spivak 1988).

From a national perspective in Cameroon, the stereotype "pygmy" has another, related facet and is applied to all people of the Lobéké forest region, regardless of their ethnic affiliation. A young woman of mixed Bangando (maternal) and Bamileke (paternal) parentage recalled the bitter arguments that her parents waged over where she should be raised as a child: among her maternal relatives in the Lobéké region or among her paternal relatives in the grasslands of western Cameroon. Her maternal grandmother was determined that the girl not be raised among her Bamileke relatives, fearing that the girl would be considered a "pygmy" and forced to eat unclean and poorly prepared food. The stereotype "pygmy" is often attributed by Cameroonians from other regions to all residents of the southeastern corner of Cameroon, stigmatizing southeastern Cameroonians as dirty, underdeveloped, and backward (Rupp 2001, 2011).

Simultaneously, however, positive stereotypes of "pygmies" proliferate in ideologies and rhetoric of conservation and development organizations. During the late 1990s conservation and development policies in southeastern Cameroon reflected the widespread and largely unchallenged popular perceptions of "indigenous communities" and "hunter-gatherers" as monolithic and marginalized communities to be protected alongside of declining natural species. Inherent in these descriptions are value judgments of both people and other animals. Just as elephants, gorillas, chimpanzees, and leopards are innocent creatures facing imminent extinction at the hands of poachers and human consumers of animal products, the indigenous people of the forest are innocent victims of overhunting, ecosystem destruction, and commercialization of forest products. In this view, "primitive" people exist side by side with "pristine" nature; both deserve particular attention, dedication, and protection. As described by the World Wide Fund for Nature, the Lobéké forest region "is the traditional home of the Baka pygmies, and it is hoped that by preserving their forest, it will help to preserve their threatened culture and way of life. . . . The Baka are the indigenous people of this forest; they depend on it and their way of life is deeply intertwined and enmeshed with the forest ecosystem."[15] From this stereotyped perspective, Baka "pygmies" appear to be relics of a noble and primeval past, another forest species to be protected and preserved.

Setting these three examples (of many) of stereotypes of "pygmies" side by side, it is clear that even rigid markers of identity such as stereotypes take on different meanings depending on the angles of relationships and positions of the people involved in ascribing, or being described by, the identity. Taking a rigid category of identity such as "pygmy" (or "hunter-gatherer") as a given reality overlooks the multiangular facets of the stereotype itself, and ignores the complex, inherently political ways in which stereotypes are utilized to ascribe identities of multiple groups of people in varying contexts.

Conclusion

The reality of people's relationships with the places they inhabit, as with their relationships with the people among whom they live, is that people move and interact simultaneously in various and multiple dimensions. While stereotyped models *for* (in Geertz's 1973 sense) "hunter-gatherers"—as with the globalized model *for* indigeneity—are premised on expectations of temporal and spatial rootedness in the forest and ancestry in aboriginal, ethnic groups, the realities of human relationships with places and people tend to involve movement across and through space, and relationships, networks, and alliances across lines of descent. The lived experiences of Bangando, Baka, Bakwélé, and Mbomam in southeastern Cameroon tend to reflect movement, fluidity, and flux across boundaries and borders, both spatial and social. This chapter has argued that it is analytically helpful to disaggregate processes of identity into distinct concepts such as social identities, which are fluid and malleable; ethnic affiliations, which are viscous and thick but do change over time; and stereotypes, which are simplifications of reality that crystallize into rigid expectations. By looking at identities as processes rather than categories, we are able to analyze how forest peoples of the Congo River Basin create social relationships and assume new and changing social identities depending on social, economic, political, and geographic contexts. We are able to appreciate that ethnic relations are not always and everywhere the same, but change in historical and social contexts. And we are able to see that even stereotypes can be projected and assumed by forest peoples themselves in variable ways. A multiangular perspective invites us—requires us—to look at human relations in the diversity of their actual contexts, enabling us to see the many ways that people form relationships and shape their senses of self and other. This perspective also helps us to appreciate that forest peoples experience and express social identities, ethnic affiliations, and even stereotypes in malleable, diverse ways.

Notes

1. See Rupp 2001, 2011 (chapter two) for detailed discussion of historical confluences among the communities of southeastern Cameroon.
2. I apologize for the unwieldy (and annoying) use of quotation marks around terms that reference stereotyped identities, such as "pygmies," "hunter-gatherers," and "indigenous people." I specifically use quotation marks to signal that these terms are identities constructed by outsiders and projected onto forest peoples. In some contexts, people may utilize such stereotyped identities strategically to align themselves with particular (usually external) categories (Spivak 1988).
3. Out of linguistic necessity, my fieldwork focused on one of these four communities: Bangando. The Bangando language has not been studied before, and I invested my linguistic efforts in achieving proficiency in this one language. As a result, my fieldwork among Baka, Bakwélé, and Mbomam individuals used a combination of Bangando, French, and local multilingual friends and translators. In southeastern Cameroon, most individuals are proficient in the languages of their neighbors, allowing me to conduct interviews and gather research among all four communities, though not to the same depth. While out of methodological necessity my research emphasized

perspectives from the Bangando community, the premise at the outset of my research was that I must also actively analyze the intersections across the communities. As a result, I learned greetings and basic vocabulary in all four languages and cultivated research and social relationships with many individuals in all four communities (see figure 10.2).

4. See Rupp 2001, 2011 for detailed discussion of categories such as "pygmies," "hunter-gatherers," and "indigenous people" in conservation, development, and missionary work in southeastern Cameroon.
5. Because Mbékà is a men's society and I am a woman, I was unable to gather detailed information about the Mbékà initiation process and secret society. Ironically, as I became proficient and then quite fluent in Bangando, men were increasingly hesitant to tell me information about Mbékà, saying that they could not reveal this information to their "daughter." Some details men refused to discuss at all; some details men shared with me, but requested that I not publish in detail.
6. Discussion with Kouamanda, recorded on Field Cassette #43. Yaoundé, September 1998.
7. Discussion with Gnangue, recorded on Field Cassette #43. Yaoundé, September 1998.
8. Ibid.
9. Discussion with Limé, recorded on Field Cassette #45. Yaoundé, September 1998.
10. "*Lìmú à Nana. Mì bò bèmbèkò Bángàndò*!" Discussion with Nana, recorded in Field Notebook #5, p. 106. Bàkà, village near Ndongo, March 14, 1999. See also extended discussion in Rupp 2011.
11. For detailed discussion of variables of interethnic relationships, including tension between Bakwélé and Baka in the region of Ndongo, please see Rupp 2001, 2011.
12. Discussion with Lapo, recorded in Field Notebook #5, p. 142. Ndongo, March 21, 1999. See also extended discussion in Rupp 2011.
13. Discussion with Ngongo, Sakombo, and Albert, recorded in Field Notebook #5, p. 146–9. Ndongo, March 21, 1999.
14. Discussion with Mokogwea, recorded on Field Cassette #17. Lopondji, May 4, 1999.
15. http://ld.panda.org/jengi_news.html. Accessed July 15, 2005.

References

Bringa, T. 1993. "Nationality Categories, National Identification and Identity Formation in 'Multinational' Bosnia." *The Anthropology of East Europe Review*, Special Issue "War among the Yugoslavs," 11(1–2).

Carsten, J. 1995. The Substance of Kinship and the Heat of the Hearth: Feeding, Personhood, and Relatedness among Malays in Pulau Langkawi. *American Ethnologist* 22(2):223–41.

——, ed. 2000. *Cultures of Relatedness: New Approaches to the Study of Kinship*. Cambridge: Cambridge University Press.

Guyer, J. 1995. "Wealth in People as Wealth in Knowledge: Accumulation and Composition in Equatorial Africa." *The Journal of African History*, 1995(36):91–120.

Ingold, T. 2000. "Ancestry, Generation, Substance, Memory, Land." *Perceptions of the Environment: Essays in Livelihood, Dwelling, and Skill.* London: Routledge.

Mbembe, A. 2000. "A propos des ecritures africaines des soi." *Politique Africaine* 77:16–43.
Rupp, S. 2001. *I, You, We, They: Forests of Identity in Southeastern Cameroon.* PhD diss., Yale University.
———. 2011. *Forests of Belonging: Social Identities, Ethnic Affiliations, and Stereotypes in the Congo River Basin.* Seattle, WA: University of Washington Press.
Sharp, B. 1998. First the Forest: Conservation, 'Community,' and 'Participation' in South-west "Cameroon." *Africa* 68:25–45.
Spivak, G. 1988. *In Other Worlds: Essays on Cultural Politics.* New York: Routledge.
Williams, R. 1973. *The Country and the City.* Chatto and Windus.

11

Interethnic Relationships between Pygmies and Farmers

Kiyoshi Takeuchi

Introduction

There are at least fifteen Pygmy ethnic groups living in different areas throughout the tropical rainforests of the Congo Basin (see Bahuchet, chapter 1, figure 1.1). Some groups have thus far been studied very little or not at all, due in most cases to either their small population size or to recent and significant changes in their nomadic life styles.[1] It has been determined, however, that all of the Pygmy ethnic groups, without exception, maintain relations with neighboring farmer groups (table 11.1). Though, as has been noted by Hewlett (1996), these relations are not uniform, exhibiting extensive intra- and interethnic group variation, and have been subject to historical changes (Bahuchet and Guillaume 1982; Klieman 2003; Wilkie and Curran 1993).

The Pygmy population first separated genetically from non-Pygmy populations about 70,000 years ago, and subsequently divided into ancestral Western and Eastern populations around 27,000 years ago (Batini et al. 2011). Furthermore, a third split into smaller groups occurred, probably because of the expansion of Bantu farming populations into the Congo Basin, roughly 3,000 years ago (Verdu et al. 2009; see also chapter 2 in this book). Current Western Pygmy groups are genetically much closer to their farming neighbors than to the Eastern Pygmies (Patin et al. 2009). These findings suggest that Pygmies have had contact with farmers for several thousand years. These relationships are described from archaeological, historical, and historical ecological perspectives in other chapters (3, 4, and 6, respectively). In this chapter, three facets of contemporary relations between Pygmy and farmers are described and analyzed: cultural and ideological relationships; economic relationships that focus on subsistence activities;

Table 11.1. Pygmy-farmer relationships.

Ethnic Group[a]	Country[b]	Associated Farmer Group[c]	References[d]
Mbuti	DRC	Bila, Pakombe, Nande	Hart and Hart 1986; Ichikawa 1986; Turnbull 1961
Efe	DRC	Lese	Bailey and Peacock 1988; Grinker 1994; Terashima 1987; Wæhle 1986
Cwa	DRC	Hemba, Luba	Kazadi 1981
Twa	DRC	Oto	Pagezy 1988; Schultz 1986
Aka (Mbendjele)	RC; CAR	Ngando, Mbati, Kako, MimpoutouBongili, Kabunga, Sangha-Sangha,	Bahuchet 1985; Delobeau 1989; Hewlett 1991; Lewis 2002; Takeuchi 2005
Baka	Cameroon	Bangando, Fang, Nzime, Bakwele, Konabembe	Dodd, 1986; Hattori 2012; Joiris 1998; Rupp 2011
Kola (BaGyeli)	Cameroon	Fang	Biesbrouck 1999; Van de Sandt 1999
Koya	Gabon	Kwele, Makota,Mwesa, Ngom, Mahongwe	Soengas 2009
Bongo	Gabon	Masango, Mitsogo	Matsuura 2006

Note: (a) Pygmy ethnic groups referred to in this chapter, names in parentheses are other ethnonyms by which the group is known; (b) DRC: Democratic Republic of Congo, RC: Republic of the Congo, CAR: Central African Republic; (c) Mainly groups referred in this chapter; (d) Main references in this chapter.

and social relationships. Socioeconomic factors leading to diversity and change are also discussed.

I have used the term "Pygmy" in this chapter to address forest foragers, or post-forest foragers. The term "Pygmy" has a somewhat contemptuous implication referring to their short stature. This tends to lead the reader to a mistaken biological reductionism, in which various Pygmy groups are supposed to share a single culture, with the shortest of these having the purest form of such culture. However, I employ the term "Pygmy" not only because an increasing number of Pygmy groups have settled down and are engaged mainly in farming rather than foraging, but because Pygmies have been categorized by their neighbors and themselves as being culturally, economically, and socially different from their neighbors everywhere in the Congo Basin.[2] This distinction, which is relevant to relations with farmers, will be clarified below.

Cultural Relations

The relations between Pygmies and farmers in the Congo Basin have been among the most studied subjects since the early days of foraging population research. Schebesta (1936), who investigated the Ituri region in the eastern Congo Basin during the 1920s, described the foraging populations in this region as subservient to, and economically and culturally dependent on, their farming neighbors.

In contrast, Turnbull, who studied the same region during the 1950s and 60s, found not only that Mbuti Pygmies were autonomous from neighboring Bila farmers, but that their relationship was fundamentally dichotomous: "Pygmies are the forest world" and "farmers are the village world" (Turnbull 1961, 1965). According to his descriptions, for farmers, "the forest was hostile, something to be feared and fought" and for Mbuti, the village was "an isolated and foreign territory; hostile but ultimately impotent, to be regarded with indifference, amusement, and occasionally with annoyance" (1961).

Turnbull's opposing dichotomy between Pygmies and farmers has been employed by many subsequent ethnographers working in Congo Basin forests as one of the most essential features of ethnic difference. However, even Turnbull himself indicated that there have been historical changes in the cultural dichotomy between the "forest" and "village" worlds, an assertion more recently supported by Kenrick. In the 1950s, when Turnbull studied the Ituri region, the Belgian colonial administration forced Bila farmers to migrate to roadside areas and cultivate cotton, which was then used to pay taxes. The Bila farmers needed Mbuti labor to tend their crops. However, the Mbuti, who unlike the Bila were not interfered with by the administration, refused the farmers' demands for their labor. Thus, the opposition between the two worlds was brought to the fore in the colonial context (Turnbull 1983). However, after Zaire achieved independence, and as the meat trade by farmers increased, cultural antagonism between the two seems to have weakened (Kenrick 2005).

The notion that all farmers belong to the "village world" also seems to be an excessive generalization of Turnbull's schema. In contrast to the Bila that have immigrated relatively recently into forest areas from savannas (Sporcq 1975), most of the farmer groups in the northeastern part of the Republic of Congo (RC) have deep knowledge of forests and are highly adapted to the forest environment. Similarly, according to Dodd (1986), the Nzime farmers in the eastern part of Cameroon are "forest people" from the perspective of socio-economic adaptation, like the Baka Pygmies with whom they have relations. Thus, it is important to recognize that the "forest world" and "village world" are historical constructs and this dichotomy is a conceptual framework applicable in limited cases. Nevertheless, it is a fact that ideological oppositions are widely observed between foraging populations and farmer groups.

The cultural antagonism between foraging populations and farmer groups is seemingly reflected explicitly in the discourse of farmers, many of whom regard Pygmies as quasi-human. For example, the Efe Pygmies in the Ituri region and the Cwa Pygmies in the central part of the Democratic Republic of Congo (DRC) are considered by their respective farming neighbors to be "savage," or as a sort of intermediate species, somewhere between human and wild animal (Grinker 1990; Kazadi 1981; Wæhle 1986). On one hand, the Aka Pygmies in the northeastern part of the RC, which I have been studying, and their neighboring farmers regard and insult each other in similar terms. For the Aka, farmers are lazy and wild subhumans, males having the nature of gorillas, females that of chimpanzees, and children that of bush babies. The Aka believe that the farmers, after they die, reincarnate in accordance with their true inner nature. On the other hand, farmers say that the Aka do not keep promises, are not civilized, and are impure and fertile like the wild animals in the forests. Similar ideologies and patterns of reciprocal insult were observed among Baka/farmer dyads in southeastern Cameroon and northwestern RC by Oishi (2012) and Koehler (2005), respectively.

While the insults traded between Pygmies and farmers are certainly demonstrative of their mutual antagonism, they also convey implicitly their awe of each other. Farmers believe Pygmies are in control of supernatural powers derived from the forest in which they live (Bahuchet and Guillaume 1982; Wæhle 1986), whereas Pygmies credit farmers as powerful sorcerers (Lewis 2002; Takeuchi 2005).

It is also noteworthy that new metaphorical expressions have emerged in some Pygmy groups, reflecting the entrance of newcomers into the local communities. According to Giles-Vernick and Rupp (2006), there are some among the Baka who believe farmers reincarnate as gorillas after they die, whereas the Baka believe themselves to come back to the village as "white" after their death. The authors attribute this change to the presence of forest conservation NGOs and the Christian Church in the area. Lewis (2002) reported that the Mbendjele Pygmies in the northwestern part of the RC have adopted the metaphor of red river hogs (which carry massive fat stores and travel in herds) to refer to logging company employees, forest conservation NGOs and researchers, who come in groups and who, to the Mbendjele, appear incredibly rich.

Language and Historical Contacts

Another issue that must be discussed here regarding the cultural relationships between Pygmies and farmers is that no Pygmy ethnic group has its own language; rather they employ the one that is used by the farmers with which the group has or has had relationships. For example, the Mbuti and the Efe Pygmies have relations with several farmer groups, which between them speak a total of eight or nine languages. From among these languages, the two foraging groups have adopted the languages spoken by the Bila and the Lese farmers, with whom they seem to have had relatively close relationships for a long time.

The Aka and the Baka Pygmies do not speak the same languages as any of the neighboring farmer groups with whom they have relations. The Aka's language, part of the Bantu C10 language group, is most closely related to that spoken by the Ngando farmers in the southern Central African Republic (CAR). The Baka's language, in the Ubangian Gbanzili group, closely resembles the language of the Ngbaka farmers, also in the southern CAR. The area where the Baka live is several hundred kilometers away from the southern CAR, however, according to ethnolinguistic studies by Bahuchet, the Ngando and Baka vocabularies share commonalities related to both the forest and the crops that were introduced from America and by the ivory trade. From these data, Bahuchet supposed that at some time prior to the fifteenth century, a single ancestral group lived in the southern CAR, where they had contact with the Ngando and the Ngbaka and borrowed their languages. This single group then divided into two separate linguistic groups—what are today the Aka and Baka. Later, Baka migrated to the southwest, finally coming into contact with the farmer groups with which they currently have relations (1993).

In addition, Bahuchet supposes from his analyses of the oral histories of farmer groups that the Kola Pygmies of coastal Cameroon immigrated there with other farmer groups before the 1880s from southwestern Cameroon and then established new relationships with other farmer groups in that area (2012).

During my surveys in the Motaba River region in northeastern RC, I observed Aka groups in the upper reaches of the river. They had relations with a Bantu A–speaking Bakwele farmer group that migrated from northwestern Congo during World War I. Some of the Aka clan names there are the same or similar to those of a Bantu C–speaking farming group living along the river's middle reaches (figure 11.1.). From this fact it is inferred that the Aka in the upper reaches of the Motaba River for some reason migrated from the middle reaches after WWI.

The question as to why Pygmies have lost their own languages and are employing languages of farmers is beyond the scope of this chapter. Nevertheless, it is noteworthy that Pygmies and farmers in the Congo Basin have a history of relationships going back at least several centuries, and it is possible that both Pygmy groups and farmer groups have had relations with partners different from the present ones.

Many Pygmy groups have different languages than the ones spoken by the farming neighbor groups with whom they have relations, and in such cases the Pygmies generally use the farmers' language for interethnic communication.

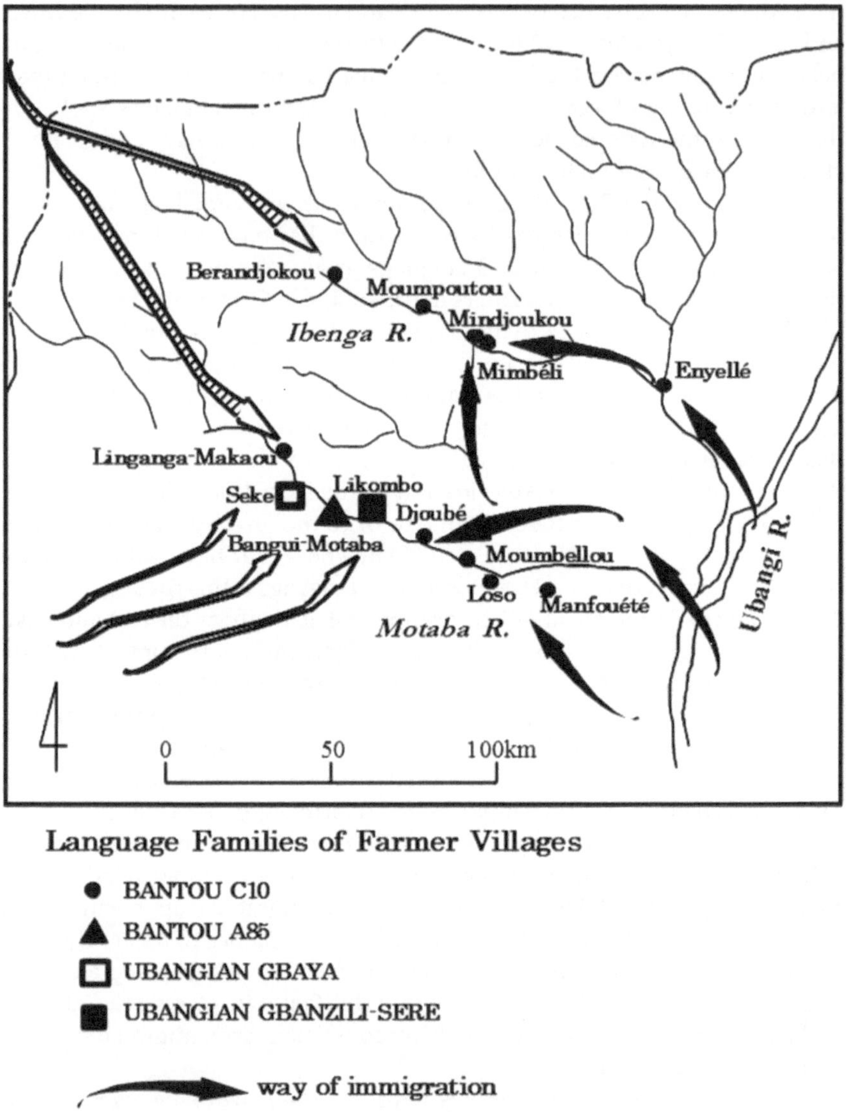

Figure 11.1. Distribution of horticulturalist villages in northeastern Republic of Congo (Takeuchi 2005).

Economic Relations

Economic role sharing and complementarity, centering on exchanges of goods of services, are the basic principles underlying the relations between Pygmies and farmers. During the colonial period, trade between the two groups was based primarily on the exchange of ironware, such as spearheads and axe blades provided by farmers for ivories and then later for duiker skins provided by Pygmies (Bahuchet 1985; Bahuchet and Guillaume 1982; also see chapter 4).

Today, in general, forest Pygmies provide farmers with forest products, such as meat, or labor for cultivation to the farmers in exchange for crops or industrial products such as pots, and more recently cash to Pygmies. According to Ichikawa, the Mbuti in the Ituri region trade nearly half of the meat that they get by net hunting with farmers or bushmeat traders, and thereby receive 71.7 percent of the total calories consumed at their hunting camps from the farmers' crops. Farmers do not often hunt, and the meat provided by the Mbuti constitutes a major source of protein. The nomadic lifestyle of Pygmies such as the Mbuti is advantageous for capturing wild animals in the forest, while the sedentary lifestyle of farmers is suited to the practice of cultivation. In other words, the Mbuti and neighboring farmers have established mutually beneficial interdependent relationships in which the Pygmies supplement the farmers' diets with protein in exchange for the calories provided by the farmers' crops (Ichikawa 1986).

In the area where I have conducted my studies, the Aka, similarly to the Mbuti, practice net hunting. Of some 980 kilograms of meat procured through net hunting over a forty-two day period, only a few kilograms reached the hands of farmers; the rest was consumed by the Aka (Takeuchi 1995). However, any game hunted with shotguns, which are provided to the Aka by farmers, is considered the property of the gun owner. According to the customary rules of game distribution, only the heads and innards of the animals hunted with guns are granted to the actual hunters. In terms of maintaining subsistence, it can be said that the Aka and villagers procure wild meat independently for themselves.

In addition to gun hunting, the Aka provide the villagers with labor in a variety of other forms, including field clearing, weeding, harvesting and transporting produce, harvesting palm nuts and making palm oil, assisting in canoe construction, and daily chores like gathering firewood. In return for either immediate labor or the promise of labor at some time in the future, the farmers give the Aka foodstuffs or are treated to simple cooking. When the Aka help themselves to young cassava leaves or yams directly from the farmers' fields, the farmers will turn a blind eye so long as the amount is not too great. Simply put, in return for their labor services, the Aka are recompensed with daily food supplies. My studies have shown that cultivated crops supplied by farmers account for 53 percent of the foods used at dinnertime. From the viewpoint of the Aka, who have no stable source of calorie-rich foodstuffs of their own, the village is like a great storehouse, which can be relied upon in times of need. The villagers, on the other hand, by sharing excess agricultural products such as cassava and its leaves, can avail themselves of the Aka's excellent hunting skills and thereby obtain a source of animal protein, while at the same time gaining manpower for farming and other activities.

It has also been reported that 43.5 percent of the total calories consumed by the Efe were supplied by agricultural foods that they received from neighboring farmers in return for labor (Bailey and Peacock 1988). In this way, the Aka and the Efe have economic relations with farmers centering on the trade of labor for crops, and they are like the Mbuti in that their relations are based on role sharing and interdependence.

However, the Mbuti, who are influenced by the meat trade, exchange meat and crops at fixed rates, whereas in the exchange of labor for crops between the Aka and neighboring farmers there are no fixed rates for how much produce is to be supplied for a given type or amount of labor. Moreover, the exchange of labor for produce is not always synchronous. Similar reciprocal gift giving is observed in the relationship between the Efe and their neighboring farmers (Terashima 1986).

Quarrels commonly arise between the Aka and farmers because of differences in how each group perceives their trade relations. While for the Aka food exchanges are not based on dyadic relationships between individuals, farmers expect to be personally recompensed for crops given to the Aka. If the Aka receive crops but do not reciprocate with labor farmers consider to be equivalent in value, the farmers abuse the Aka as thieves.

Under the government's sedentarization and agricultural policies, which began to be instituted in the 1960s, many among the Baka in Cameroon have settled along main roads. The Baka, like the Aka and the Efe, have been engaged in trading labor for crops with neighboring farmer groups, but as they become increasingly sedentary, they tend to add agriculture to their subsistence activities, though unsystematically and with unplanned labor input (Joiris 1998; Kitanishi 2003).With the additional impact of the commodity-based economy in recent years, the interdependent economic relationships between Pygmies and farmers described above are changing. For example, the Baka in southeastern Cameroon and northwestern RC and Aka in southwestern CAR are earning cash through farm work, selling bushmeat, and cacao cultivation (Kitanishi 2006; Kohler 2005; Noss 1997; Oishi 2012; Yasuoka 2006).

While the changing economy and sedentarization have led some groups such as these to a greater degree of autonomy from neighboring farmers, in regions not integrated into the cash crop distribution network some Baka groups have instead become more dependent on farmers for cash and industrial products. For example, the Baka that have relations with the Konabembe farmer group are now sedentary, but rarely work the land and instead trade meat, fruits, nuts, tools, and labor for cash and crops with the Konabembe and other traders. They obtain 94 percent of their principal food and 62 percent of their cash income from the Konabembe, showing a marked increase in economic dependency on farmers from groups that remain nomadic (Hattori 2012).

The socioeconomic impact of the outside world varies depending on the demographic circumstances and the volume of local traffic; Baka in different regions or groups are influenced in varying ways and degrees.

The variance in effect is demonstrated by the contrast between the Bongo Pygmies in eastern Gabon and Kola Pygmies in Cameroon. Matsuura reported the very unique case of the Bongo, who began transitioning toward sedentarism during the 1960s and now live in forest camps for only a short period and dedicate almost half their labor to farming. The Bongo in this region live either near or in Masango farming villages, but the two groups do not have economically interdependent relations (2006). Meanwhile, the Kola, who also live in farming villages and subsist through agriculture, are often involved in property disputes with the farmers because of high population density and the limited availability of arable land (Biesbrouck 1999; Van de Sandt 1999). However, it is notable that both Bongo and Kola earn cash through traditional medicine, the demand for which is substantial, particularly for the Bongo.

Social Relationships

It has been widely observed that personal, dyadic relationships often exist between individual Pygmies and farmers as well as between families. Though the nature of these relationships is diverse according to ethnic group and region, they are generally unequal, with Pygmies in the lower stratum of farming society, subservient to farmers. In some cases, however, Pygmies and farmers form mixed communities on relatively equal terms. Below, Pygmy-farmer relations recorded in ethnographic accounts are classified into four types based on the relative subordinance or autonomy of Pygmies.

Subordinate and Heteronomous Relationships

The Twa Pygmies along the central basin of the Congo River in the DRC live in subservient status to Oto fisher-farmers, maintained through fictive kinship. For example, when an Oto passes, the Twa must get out the way for him or her and drop their eyes, whereas the Oto need not do so for the Twa in the reverse situation. According to Pagezy (1988) and Schultz (1986), the Twa had already begun to cultivate at the time they conducted their research in the 1970s and the early 1980s, but the Twa could not produce sufficient crops to support themselves and were able to obtain only a little cash, thus depending heavily upon wage labor and crops offered by the Oto. Pagezy indicated that the Twa constitute a sort of a lower caste within a single society. Similarly, Delobeau described social relations between the Aka and Monzombo fisher-farmers along the upper basin of the Ubangi River in southeastern CAR as a caste system that emerged due to the sedentarization of the Aka and the increase in demand for labor for coffee cultivation around the time of independence (Delobeau 1989). However, the Twa began to earn cash as laborers in large plantations and have become somewhat autonomous from the farmers and fictive kinship bonds by virtue of widespread penetration of the monetary economy in the late 1980s.

In the instance observed by Hattori cited above in which Baka have become more and more economically dependent on farmers, she also describes that in the region of her research, farmers were exceedingly dominant over Baka both socially and politically (Hattori 2012).

Subordinate and Autonomous Relationships

Relationships between the Aka and their neighboring farmers in the northeastern part of the RC, which I have researched, were as follows: Pygmy groups were socially differentiated from farmer groups, but farmers were socially predominant over Pygmies. As described above, Aka and farmers are mutually exclusive though economically interdependent, exchanging crops and labor. However, their relationships and interactions in daily life are asymmetrical, like those of a patron and client (Bahuchet 1999), very similar to what has been observed with the Twa. In addition, while farmers will help themselves to the Aka's palm wine, if the Aka partake of the villagers' palm wine it would be viewed as an act of theft. Underlying such relationships is a fictive kinship in which farmers are pseudo-parents and Aka their pseudo-children; in other words, patriarchal farmers order their Aka "children" to do labor, offering food and protection in return (Takeuchi 2005). This framework characterizes the relationships between Aka and farmer family groups and is transmitted from generation to generation.

Similar cases have been observed between Efe and Lese farmers, as reported by Wæhle (1986) and Grinker (1994). According to Grinker, Lese incorporate Efe into each of their phratry and confer them with a lower status, such as juniors or females. Grinker called these socioeconomic units "houses."

Coordinative and Autonomous Relationships

Contrary to Wæhle and Grinker, Terashima (1986), who conducted research at a village some tens of kilometers away from their research sites at around the same time, emphasized equality and solidarity in the relationships between Efe and Lese. Terashima also described patron-client relationships transmitted patrilineally in both Pygmy and farmer families; however, he claimed that these relationships were important in regulating the reciprocal exchanges of goods and services and noted the sense of mutual propriety inherent in their fictive kinship terms for one another: *muto maia* (my farmer) and *efe maia* (my Efe). The relations between Efe and Lese appear more equal and friendly than the relations depicted by Wæhle and Grinker. This difference may be explained by the fact that research sites of Wæhle and Grinker were both roadside villages where farmers could retain more dominance and control over the commercial economy and official services than was the case in Terashima's site, which was situated in the deep forest.

The coordinate relations between the Efe and the Lese are also different from the relations between Aka and neighboring farmers, though they are similar in that they have been constructed through fictive kinships. According to Joiris (2003), there are also fictive kinships between Baka and Bakwele farmers, similar to the *muto maia-efe maia* relations of the Lese and Efe, that is, the relationship between Baka pseudo-brothers and Bakwele pseudo-sisters. Additionally, Dodd (1986) suggested that, in the 1980s, fictive kinships also existed between Baka and Nzime farmers. However, neither fictive kinships between Baka and farmers nor patron-client relationships have been observed in more recent studies (Hattori 2012; Kimura 2001).

Regarding the relationship between Baka and Bakwele farmers, Joiris emphasized institutionalized friendships and solidarity among people that participate together in secret society initiation rituals. Institutionalized friendship is a bond between two persons of the same sex who mutually choose one another as partners during childhood or adolescence and privilege each other in the exchange of goods and services. Rupp, who researched the relationships between Baka and Bangando farmers, found the exact same relationship pattern. She claimed that though there were hierarchical relationships between the Baka and Bangando, generally speaking, friendships based on personal affection and trust and the bond generated through rituals are principal factors in interethnic relationships (2003, 2011)[3]. In southern Cameroon, the Baka harvested the Fang farmers' cacao plantations and obtained cash, crops, liquor, and cigarettes as payment. Many of the Baka employed were friends of the owners of the plantations on which they worked (Sakanashi 2010). Patron-client relationships still exist among the Koya in Gabon as well, though personal friendships and interethnic marriages with farmers are also common (Soengas 2009).

Since the late 1950s, the Mbuti of the Ituri in the eastern DRC have been affected by the commercial meat trade and have exchanged meat, crops and clothes more often with traders than with farmers (Hart 1978; Ichikawa 1991). Though in the past, Mbuti and Bila farmers had relatively strong and egalitarian ties based on fictive kinship, with the advance of the commercial economy these relationships had dwindled severely by the late 1970s (Ichikawa 1978). The political instabilities and disputes in the Democratic Republic of Congo have made it difficult to conduct research there since the 1990s. Both Pygmies and farmers in the Ituri region were involved in the violent disturbances during the second Congo War from 1998 to 2003. The Pygmies, especially, suffered oppressive persecution by armed groups. The ethnic relations that were prevalent up to that time are said to have radically changed (Pottier 2007; 2008), but the present situations on Pygmy-farmer relations remains unclear.

Coordinative and Assimilative Relationships

The Bongo in Gabon live in mixed communities with farmers, as do the Koya. Until the middle of the twentieth century, Bongo and neighboring farmers seemed to have interdependent relationships that were characterized, similarly to many others, by the exchange of forest products or labor for crops (Andersson 1983). Presently, their main subsistence activity is farming, and they do not maintain fictive kinship or patron-client relationships with farmers. According to Matsuura (2006), the Bongo are beginning to assimilate the social culture of Masango farmers, for example, their clan structures, rituals, and names, among other aspects. In daily life, they usually use the Masango's language. Sometimes, the Bongo host rituals that originated with the Masango and solidarity is generated through cooperative and equal participation in these rituals by both groups, though relationships between Bongo and Mitsogo farmers seem rather unequal (Bonhomme, De Ruyter and Moussavou 2012). Moreover, as will be discussed later, intermarriage between the Bongo and the Masango is increasingly common and ethnic boundaries are dissipating.

Interethnic Marriage

Table 11.2 is an index of intermarriages between Pygmies and farmers. With regard to the Aka and Mbuti, intermarriage with neighboring farmer groups is occasional. There are some cases of intermarriage between Baka women and farmer men. Among the Efe, 17–28 percent of the women have married farmer men, whereas no Efe men have married farmer women. These intermarriages between Pygmy women and farmer men are thought to be hypergamous, as no intermarriage between Pygmy men and farmer women have been reported among these Pygmy groups. However, among the Bongo, over 40 percent of women have married farmer men, and some Bongo men have married farmer women. These statistics indicate a positive correlation between the level of assimilation of Pygmy groups into farmer groups and the frequency of interethnic marriage. It is also noteworthy that the tendency for interethnic marriage among western Pygmy groups (Aka, Baka, and Bongo) is clearly correlated with the pattern of genetic diversity among Pygmy populations and farming populations, showing that the Bongo are genetically closer to neighboring farmer groups than other Pygmy populations (Quintana-Murci et al. 2008; also see chapter 2, figure 2.2).

The frequency of interethnic marriage is also related to sedentarization and subsistence patterns. Among sedentary groups that subsist through farming, marriage with farmers is common. One ethnographic example is provided by the Efe Pygmies and Lese farmers. As an aspect of the *muto maia-efe maia* relationship described above, an Efe daughter whose father was aided his bride price by a Lese farmer becomes an adopted daughter of the Lese. Such Efe females that have been raised in a farming village and have learned daily habits

Table 11.2. Interethnic marriage among Pygmy ethnic groups.

Ethnic Group	Counter Farmer Group	Area	Intermarriage Index		References
			Female	Male	
Aka	Mimpoutou	Ibenga	0(/172)	0(/151)	Takeuchi (this study)
Aka	Ngandu	Bokoka	rare	0	Hewlett 1991
Mbuti	Kaheku, Nande, Pakombe	Southern Ituri	not frequent	0	Hart 1979
Baka	Bangando	Koumela	0.02 (3/140)	0(/120)	Matsuura 2012
Baka	–	Lomie	0.045	–	Hewlett 1996
Baka	–	Abong Mbang	0.086	–	Hewlett 1996
Efe	Lese	Malembi	0.17(21/123)	0(/97)	calculated from Bailey 1985
Efe	Lese	Mangala	0.25(14/55)	0	Terashima 1987
Efe	Lese	Andiri	0.28(17/60)	0	Terashima 1987
Bongo	Masango	Ogoulou	0.42(18/43)	0.11(3/28)	Matsuura 2012

Intermarriage index: ratio of persons married to farmers in the total number of married persons

of farmers often marry Lese men (Terashima 1987). Similarly, Rupp (2011) has pointed out that Bangando farmers are not reluctant to marry Baka females if the Baka group from which they come is as developed and modern as the Bangando themselves. As indicated by Hewlett (1996), when Pygmies become sedentary and adopt farming, that is, when their lifestyle begins to resemble that of farmers, then farmers' resistance to interethnic marriage decreases.

Pygmy-farmer relations are very complex. Turnbull has described the relationship between the Mbuti and farmers as "a strange one, full of ambivalence and uncertainty" (1961). Dodd expressed the relationships between the Baka and Nzime farmers as hierarchical and hostile, though he also wrote: "the women of both groups enjoy a cheerful friendship and when a Baka died, the farmers show sympathy" (1986). Another example from my own work occurred when I was accompanied by two Aka from my field site on a visit to a farmer village that they had never been to before. Throughout the journey, the Aka poured forth an endless stream of verbal abuse about the farmers. Yet, upon reaching our destination, their first action was to inquire among the Aka there as to whether there were any relatives of their patron farmers in the village who can guarantee their safety. These instances demonstrate the resistance of Pygmy-farmer relations to straightforward classification, and it should be noted that the classifications I have described in previous sections are based on predominant characteristics among various factors involved in these complicated and multifaceted relationships.

Dialectic Symbiosis in Local Context

As Köhler and Lewis pointed out, most anthropological research on Pygmy-farmer relations had adopted the view that these relations are in the nature of preestablished harmony that is grounded on complementary and specialized subsistence practices. The cultural and ideological dimensions of these relations have hardly been considered, although there are a few exceptions, such as Grinker (1990) and Joiris (1998) and their ethno-anthropological analysis of mutual perspectives in the relations between the Mbendjele and neighboring farmers (Köhler and Lewis 2002).[4] I propose a dialectic model to Pygmy-farmer relations in the local context during the period between when the ethnographic survey on Pygmies had started in the 1950s and when socioeconomic changes occurred in recent times, where economic interdependence and cultural antagonism have been integrated through social relations to create a compound regional community.

Economic relationships (i.e., exchange of goods and services through the division of roles in resource use) and social relationships (i.e., fictive kinship bonds) form two sides of the same coin and each reinforces the other. Subsistence differentiation and interdependence between Pygmy and farmer groups are necessary in order for them to coexist, whether Pygmies are trading bushmeat or labor for farmers' crops. Social relationships, such as fictive kinship bonds, facilitate the continuance of subsistence role differentiation as well as interdependence. The shape these social relations take is largely determined by ecological and demographic contexts and conditions. For example, farmers in the northeastern RC are generalists who employ multiple modes of subsistence and are well adapted to the forest environment. These farmers consider themselves dominant over the Aka, and they consider the Aka their pseudo-children. Farmers insist along their patriarchal views that the Aka serve them, thus providing the social basis for the exchange of meat and labor for crops. Figure 11.2 shows fictive kinship relations between 200 adult Aka and lineages of Mimpoutou farmers in 1989 and indicate fictive kinships organize reciprocal economic relationships among Aka and farmers. On the other hand, the Lese farmers that have migrated from the savanna to the forest have more equal personal relationships with the Efe (Terashima's case). In this case, ethnic boundaries with farmers are more flexible and interethnic marriage is frequent.

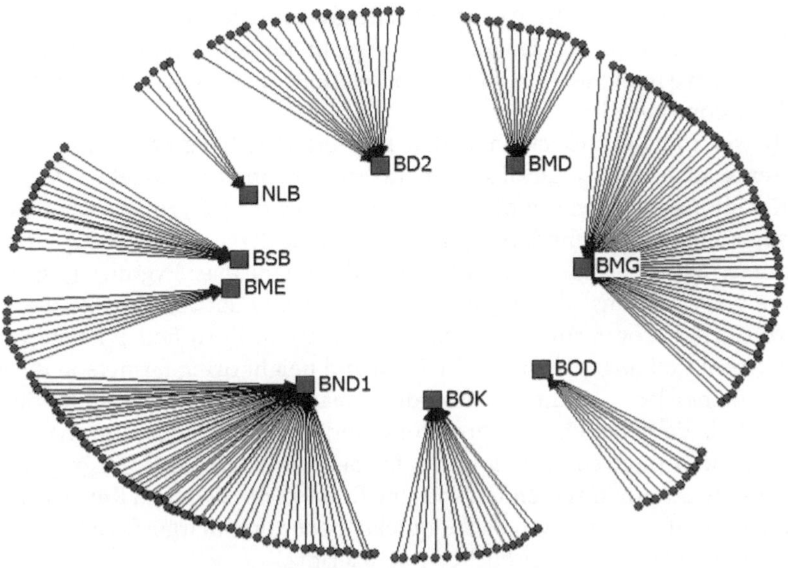

Figure 11.2. Fictive kinship relations between Aka and farmers in 1989.
Note: Circles represent Aka adults and squares represent lineages of farmers

Role differentiation in subsistence activities reinforces Pygmies' connection with the forest, as mentioned in the section on cultural relations, which results in clear cultural divisions between Pygmies and farmers. On the one hand, this differentiation creates cultural antagonism between Pygmies and farmers—evident in their verbal abuse of one another—while on the other, it helps to strengthen each group's ethnic identity. Cultural antagonism is particularly important for Pygmies, who have highly fluid group membership and no centralized power, in maintaining a strong sense of ethnic identity. Through the dialectic process described above, a regional symbiotic system has been constructed between Pygmies and farmers, characterized simultaneously by cultural and ideological antagonism and economic interdependence. Social logics of farmers such as patriarchy are liable to be applied to form the basis of the interethnic system. This is because Pygmy societies characterized by the homogenization and equalization of differences between members lack formalized institutions necessary for drawing distinct boundaries with farmers.

Changes in Global Context

The regional symbiotic systems formed by Pygmies and farmers has been changing dramatically in recent times because of the impact of the market economy, commercial lumbering, forest conservation activities, and government policies promoting a sedentary lifestyle and other influences of the outside world. Under these situations, neighboring farmers are no longer the sole actors for Pygmies. Emergence of new metaphors among Pygmies mentioned above in the section on cultural relations reflects these recent changes in local communities.

Lewis and Kenrick claimed that farmers that have embraced modern systems, such as the judicial system, tax system, and the monetary economy, generally have become increasingly dominant over Pygmies, who cannot easily adjust to the modern systems (Kenrick 2005; Lewis 2005). Though farmers might become more and more dominant over Pygmies in general, there are variations to this pattern. As described above, some of the Baka have started crop farming and cacao cultivation. Both fictive kinships and reciprocal exchanges between the Baka and neighboring farmers in the area near the border with the RC have disappeared, the regional symbiotic system has become an empty framework, and economic disparities have been generated through competition with farmers and merchants (Oishi, personal communication). In agreement with the findings of Lewis and Kenrick, under these circumstances Pygmies are quickly becoming marginalized under the nation-state (Sinang, unpublished manuscript).

Severe conflicts over land and resources have been observed between the Kola Pygmies in Gabon who have settled and are farming and neighboring farmers. According to Van de Sandt (1999), conflicts did not occur in villages where fictive kinships were still maintained. In villages with a large population and where fictive kinship relationships have disappeared, the Kola have left the villages to avoid conflicts with farmers. This case indicates that conflicts may be avoided when formalized bonds remain between Pygmies and farmers.

On the other hand, as Hattori has reported, there are some cases where farmers dominate over Baka that rarely do farming. Such cases can be explained by the disappearance of formalized interethnic bonds resulting from population fluctuations, as well as by the advance of the monetary economy and forest conservation activities into the region. The farmers, who have opportunities for getting cash and negotiating with conservation agents about forest resource management, have become dominant over the Baka economically and politically.

Another case, reported by Rupp (2003) and Joiris (1998), in which Baka that had rather harmonious relationships with Bangando and Bakwele farmers through institutionalized friendship, also deserves examination. This friendly relationship might be explained in ecological terms, considering that Bangando farmers, like Lese farmers associated with Efe, have migrated from the savanna to the forest relatively recently and have needed the Bakas' support. However, in the modern context of the Bakas' transition to agriculturalism and the accompanying dissolution of cultural barriers, it seems more likely that the friendly relationships between the two groups are possible because the economic competition between them has not been as severe as it in the region near the border with RC.

While the Bongo in Gabon, described by Matsuura (2006), are gradually becoming horticulturalists and the ethnic boundaries with farmers have been dissolving, the trade of bushmeat and other forest products has not penetrated into their region, there are no forest conservation activities, and the population density is low. Therefore, competition between the Bongo and farmers over land, cash, and forest resources is not severe, and socially equal relations of the Bongo with farmers has been promoted.

Conclusion

Many groups of Pygmies in the Congo Basin have recently lost regional symbiosis with neighboring farmers, which had previously been supported by reciprocal and interdependent economic relationships and social bonds, through various impacts from the outside world. Pygmies have nonetheless maintained their cultural connection with the forest, though it is no longer reinforced through role differentiation. For example, the Bongo people, who are now mostly agriculturalists, are considered to have supernatural powers that originate from the forest, unlike their farming neighbors. As long as Pygmies maintain such cultural bonds with the forest, even when farming has become the primary mode of subsistence, assimilation into farming villages is not entirely welcome or promoted.

The Pygmies' deep connection with the forest, however, may be an unfortunate liability with concern to preserving their own interests. Their voices are not often heard, and their interests not often understood or considered in the discourse concerning forest conservation and other issues that have significant impacts on their lives and livelihoods. Human rights movements advocating for the rights of Pygmies have been active since the 2000s. However, these movements may inadvertently damage the situation by aggravating farmers whose interests may suffer as the rights of Pygmies as indigenous forest inhabitants are augmented. In this context, farmers' "traditional" stereotyped notion of Pygmies as quasi-human and uncivilized may be reinforced. This may result in further marginalization of Pygmies by both farmers and the outside world generally. It is a fact that new stakeholders in the forest such as merchants, timber companies, forest conservation NGOs, human rights NGOs, and town residents have become important agents in determining the survival of Pygmy societies and cultures; however, the relationship with neighboring farmers is still crucial for Pygmies.

As debate continues over how Pygmies may best adapt themselves to the modern political and economic environment while maintaining their forest ties, the importance of these compound regional communities—in which both Pygmies and farmers have constructed their cultural identities as well as their livelihoods—should not be overlooked.

Notes

1. For example, Baluma and Mikaya in the northwestern part of Republic of the Congo, and Tembo in the eastern part of the Democratic Republic Congo have not been studied.
2. Vansina (1990) and Schadeberg (1999) suggested the term "twa" is widely used among Sub-Saharan languages, to designate Pygmy populations. This indicates that the category denoted by the word "Pygmy" is not a product of the Europeans or of anthropologists but is instead a folk category in Congo Basin history (Köhler and Lewis 2002).
3. Robillard reported the recent tendency for Baka to be reluctant to engage in the personal alliance with Bangando farmers (2010).
4. Recently some anthropologists have advanced research on Pygmy-farmer relations in rituals (Bonhomme, De Ruyter, and Moussavou 2012; Rupp 2011).

References

Andersson, E. 1983. *Les Bongo-Rimba*. Vol. 4, *Occasional Papers*. Uppsala: Uppsala University

Bahuchet, S. 1985. *Les Pygmées aka et la forêt centrafricaine*. Paris: Peeters.

———. 1993. *La rencontre des agriculteurs. Les Pygmées parmi les peuples d'Afrique Centrale*. Paris: SELAF.

———. 1999. "Aka Pygmies." In *The Cambridge Encyclopedia of Hunters and Gatherers*, edited by R. B. Lee and R. Daly. Cambridge: Cambridge University Press, 190–4.

———. 2012. "Changing Language, Remaining Pygmy." *Hum Biol* 84(1):11–43.

Bahuchet, S., and H. Guillaume. 1982. "Aka-farmer Relations in the Northwest Congo Basin." In *Politics and History in Band Societies*, edited by E. Leacock and R. Lee. Cambridge: Cambridge University Press, 189–211.

Bailey, R. C. 1985. *The Socioecology of Efe Pygmy Men in the Ituri Forest, Zaire*. PhD Thesis, Harvard University.

Bailey, R. C., and N. R. Peacock. 1988. "Efe Pygmies of Northeast Zaire: Subsistence Strategies in the Ituri Forest." In *Coping with Uncertainty in Food Supply*, edited by I. de Garine and G. A. Harrison. Oxford: Clarendon Press, 88–117.

Batini, C., J. Lopes, D. M. Behar, F. Calafell, L. B. Jorde, et al. 2011. "Insights into the Demographic History of African Pygmies from Complete Mitochondrial Genomes." *Molecular Biology and Evolution* 28 (2):1099–110.

Biesbrouck, K. 1999. "Agriculture among Equatorial African Hunter-Gatherers and the Process of Sedentarization: The Case of the Bagyeli in Cameroon. In *Central African Hunter-gatherers in a Multidisciplinary Perspective: Challenging Elusiveness*, edited by K. Biesbrouck, S. Elders and G. Rossel. Leiden: Research School for Asian, African, and Amerindian Studies (CNWS), Universiteit Leiden, 189–206

Bonhomme, J., M. De Ruyter and G.-M. Moussavou. 2012. "Blurring the Lines: Ritual and Relationships between Babongo Pygmies and Their Neighbours (Gabon)." *Anthropos* 107 (2):387–405.

Delobeau, J. M. 1989. *Yamonzombo et Yandenga: Les relations entre les villages monzombo et les campements pygmées aka dans la sous-préfecture de Mongoumba, Centrafrique*. Paris: SELAF.

Dodd, R. 1986. "The Politics of Neghbourliness." *Paper Presented at the 4th International Conference on Hunting and Gathering Societies, London*. London School of Economics, UK.

Giles-Vernick, T., and S. Rupp. 2006. "Visions of Apes, Reflections on Change: Telling Tales of Great Apes in Equatorial Africa." *African Studies Review* 49(1):51–73.

Grinker, R. R. 1990. "Images of Denigration: Structuring Inequality between Foragers and Farmers in the Ituri Forest, Zaire." *American Ethnologist* 17(1):111–30.

———. 1994. *Houses in the Rainforest: Ethnicity and Inequality among Farmers and Foragers in Central Africa*. University of California Press.

Hart, J. A. 1978. "From Subsistence to Market: A Case Study of the Mbuti Net Hunters." *Human Ecology* 6(3):325–53.

———. 1979. *Nomadic Hunters and Village Cultivators: A Study of Subsistence Interdependence in the Ituri Forest of Zaire*. Michigan State University.

Hattori, S. 2012. *Challenge for the Coexistence of the Forests and People: Conservation of Tropical Rainforests and the Culture of Hunter-Gatherers in Cameroon*. Kyoto: Syoukoudou-shoten (in Japanese).

Hewlett, B. S. 1991. *Intimate Fathers: The Nature and Context of Aka Pygmy Paternal Infant Care*. Ann Arbor: University of Michigan Press.

———. "Cultural Diversity among African Pygmies." In *Cultural Diversity among Twentieth-century Foragers: An African Perspective*, edited by S. Kent. Cambridge: Cambridge University Press, 215–44.

Ichikawa, M. 1978. "The Residential Ggroups of the Mbuti Pygmies." *Senri Ethnological Studies* (1):131–88.

———. 1986. "Ecological Bases of Symbiosis, Territoriality and Intra-Band Cooperation of the Mbuti Pygmies." *Sprache und Geschichte in Afrika* 7 (1):161–88.

———. 1991. "The Impact of Commoditisation on the Mbuti of Eastern Zaire." *Senri ethnological studies* (30):135–62.

Joiris, D. V. 1998. *La chasse, la chance, le chant: Aspects du système rituel des Baka du Cameroun*. Thèse de doctorat, Université Libre de Bruxelles.

———. 2003. "The Framework of Central African Hunter-gatherers and Neighbouring Societies. *African Study Monographs*, supplement 28:57–79.

Kazadi, M. 1981. "Méprisés et admirés: l'anvivalence des relations entre les Bacwa (Pygmées) et les Bahemba (Bantu)." *Africa* 51 (4):837–47.

Kenrick, J. 2005. "Equalising Processes, Processes of Discrimination and the Forest People of Central Africa." In *Property and equality* 2, edited by T. Widlok and W. G. Tadesse. Oxford: Berghahn Books, 104–28.

Kimura, D. 2001. "Utterance Overlap and Long Silence among the Baka Pygmies: Comparison with Bantu Farmers and Japanese University Students. *African Study Monographs*, supplement 26:103–21.

Kitanishi, K. 2003. "Cultivation by the Baka Hunter-gatherers in the Tropical Rain Forest of Ccentral Africa." *African Study Monographs*, supplement 28:143–57.

———. 2006. "The Impact of Cash and Commoditization on the Baka Hunter-gatherer Society in Southeastern Cameroon. *African Study Monographs*, supplement 33:121–42.

Klieman, K. A. 2003. *"The Pygmies Were Our Compass": Bantu and Batwa in the History of West Central Africa, Early Times to c. 1900 C.E, Social History of Africa*. Portsmouth, NH: Heinemann.

Köhler, A. 2005. "Of Apes and Men: Baka and Bantu Attitudes to Wildlife and the Making of Eco-goodies and Baddies." *Conservation and Society* 3(2):407–435.
———. 2005. "Money Makes the World Go Round? Commodity Sharing, Gifting and Exchange in the Baka (Pygmy) Economy." In *Property and equality* 2, edited by T. Widlok and W. G. Tadesse. Oxford: Berghahn Books, 32–55.
Köhler, A. and J. Lewis. 2002. "Putting Hunter-Gatherer and Farmer Relations in Perspective. A Commentary from Central Africa." In *Ethnicity, Hunter-Gatherers, and the "Other": Association or Assimilation in Africa*, edited by S. Kent, Washington, DC: Smithsonian Instituion Press, 276–305.
Lewis, J. 2002. *Forest Hunter-Gatherers and Their World. A Study of the Mbendjele Yaka Pygmies of Congo-Brazzaville and Their Secular and Religious Activities and Representations*. Ph.D. Thesis, University of London.
———. 2005. "Whose Forest Is It Anyway? Mbendjele Yaka Pygmies, the Ndoki Forest and the Wider World." In *Property and equality* 2, edited by T. Widlok and W. G. Tadesse. Oxford: Berghahn Books, 56–78.
Matsuura, N. 2006. "Sedentary Lifestyle and Social Relationship among Babongo in Southern Gabon." *African Study Monographs*, supplement 33:71–93.
———. 2012. *The Modern "Forest People": An Ethnography on Babongo Pygmies in Central Africa*. Kyoto: Syowado (in Japanese).
Noss, A. J. 1997. "The Economic Importance of Communal Net Hunting among the Baaka of the Central African Republic." *Human Ecology* 25(1):71–89.
Oishi, T. 2012. "Cash Crop Cultivation and Interethnic Relations of the Baka Hunter-Gathrers in Southeastern Cameroon." *African Study Monographs*, supplement 43:115–36.
———. 2012. "Gorilla-man and Man-gorilla: Human-animal Boundaries and Interethnic Relationships in Central African Rainforest." In *Anthropology of Man and Animal*, editied by K. Okuno, M. Yamaguchi and S. Kondo. Yokohama: Syunpusha (in Japanese), 93–129.
Pagezy, H. 1988. "Coping with Uncertainty in Food Supply among the Oto and the Twa Living in the Equatorial Flooded Forest Near Lake Tumba, Zaire." In *Coping with Uncertainty in Food Supply*, edited by I. De Garine and G. A. Harrison. Oxford: Clarendon, 175–209.
Patin, E., G. Laval, L. B. Barreiro, A. Salas, O. Semino, et al. 2009. "Inferring the Demographic History of African Farmers and Pygmy Hunter-gatherers Using a Multilocus Resequencing Data Set." *PLOS Genetics* 5(4):e1000448. doi:10.1371/journal.pgen.1000448.
Pottier, J. 2007. "Rights Violations, Rumour, and Rhetoric: Making Sense of Cannibalism in Mambasa, Ituri (Democratic Republic of Congo)." *Journal of the Royal Anthropological Institute* 13(4):825–843.
———. 2008. "Displacement and Ethnic Reintegration in Ituri, DR Congo: Challenges Ahead." *The Journal of Modern African Studies* 46(03): 427–450.
Quintana-Murci, L., H. Quach, C. Harmant, F. Luca, B. Massonnet, et al. 2008. "Maternal Traces of Deep Common Ancestry and Asymmetric Gene Flow between Pygmy Hunter-gatherers and Bantu-speaking Farmers." *Proceedings of the National Academy of Sciences of the United States of America* 105(5):1596–1601.

Robillard, M. 2010. *Pygmées Baka et voisins dans la tourmente des politiques environnementales en Afrique centrale.* Thèse. Paris: Muséum National d'Histoire Naturelle.
Rupp, S. 2003. "Interethnic Relations in Southeastern Cameroon: Challenging the 'Hunter-Gatherer'–'Farmer' Dichotomy." *African Study Monographs*, supplement 28:37–56.
———. 2011. *Forests of Belonging: Identities, Ethnicities, and Stereotypes in the Congo River Basin (Culture, Place, and Nature).* Seattle: University of Washington Press.
Sakanashi, K. 2010. "Utilization of Labor in Cacao Production in Central African Rain Forest: The Case of Fang Bantu Farmers Living in the Southern Cameroon." In *Socilal Aspects of Forest Living: Human, Nature and Histori of African Tropical Forests*, edited by D. Kimura and K. Kitanishi. Kyoto: Kyoto University Press (in Japanese), 129–49.
Schadeberg, T. 1999. "Batwa: The Bantu name for the Invisible People." In *Challenging Elusiveness: Central African Hunter-gatherers in a Multidisciplinary Perspective*, edited by K. Biesbrouck, S. Elders, and G. Rossel. Leiden: Research School for Asian, African, and Amerindian Studies (CNWS), Universiteit Leiden, 21–39.
Schebesta, P. 1936. *My Pygmy and Negro Hosts.* London: Hutchinson & Co.
Schultz, M. 1986. "Economic Relations between the Batua and Baoto of Bibelo Village, Bikoro Zone, Zaïre: A Preliminary Report on New Fieldwork." *Sprache und Geschichte in Afrika* 7(2):339–59.
Sinang, J. 2011. "Servitude and Domination of Baka Pygmies in South-eastern Cameroon.: A Case of Ignored Slavery." Unpublished paper.
Soengas, B. 2009. "Preliminary Ethnographic Research on the Bakoya in Gabon." *African Study Monographs* 30(4):187–208.
Sporcq, J. 1975. The Bira of the Savanna and the Bira of the Rain Forest: A Comparative Study of Two Populations of the Democratic Republic of the Congo." *Journal of Human Evolution* 4(6):505–16.
Takeuchi, K. 1995. "Subsistence Hunting in African Tropical Forest: Hunting Technigues and Activities among the Aka Hunter-gatherers in Northastern Congo." *Zooarchaeology* 4:27–52 (in Japanese).
———. 2005. The Ambivalent Symbiosis between the Aka Hunter-gatherers and Neighboring Farmers." In *Culture Conservation and Development in African Rain Forest (Research Report of MEXT Grant-in-aid for Scientific Research)*, edited by K. Takeuchi, 11–28. http://www.academia.edu/4651515/The_Ambivalent_Symbiosis_between_the_Aka_Hunter-Gatherers_and_Neighboring_Farmers
Terashima, H. 1986. "Economic Exchange and the Symbiotic Relationship between the Mbuti (Efe) Pygmies and the Neighboring Farmers." *Sprache und Geschichte in Afrika* 7(1):391–405.
———. 1987. Why Efe Girls Marry Farmers?: Socio-ecological Backgrounds of Inter-ethnic Marriage in the Ituri Forest of Central Africa. *African Study Monographs*, supplement 6:65–83.
Turnbull, C. M. 1961. *The Forest People. A Study of the Pygmies of the Congo.* London: Jonathan Cape Ltd.
———. 1965. *Wayward Servants.* New York: Natural History Press.

———. 1983. *The Mbuti Pygmies: Change and Adaptation*. New York: Holt Rinehart and Winston.

Van de Sandt, J. 1999. "Struggle for Control over Natural Resources in Bagyeli-Fang Relations: Five Ways of Coping with Changing Relations." *Challenging Elusiveness: Central African Hunter-gatherers in a Multidisciplinary Perspective,* edited by K. Biesbrouck, S. Elders and G. RosselLeiden. Research School for Asian, African, and Amerindian Studies (CNWS), Universiteit Leiden, 221–39.

Vansina, J. 1990. *Paths in the Rainforests: Toward a History of Political Tradition in Equatorial Africa*. Madison: University of Wisconsin Press.

Verdu, P., F. Austerlitz, A. Estoup, R. Vitalis, M. Georges, et al. 2009. "Origins and Genetic Diversity of Pygmy Hunter-Gatherers from Western Central Africa." *Current Biology* 19(4):312–8.

Wæhle, E. 1986. "Efe (Mbuti Pygmy) Relations to Lese Dese Villagers in the Ituri Forest, Zaire: Historical Changes during the Last 150 Years." *Sprache und Geschichte in Afrika* 7(2):375–411.

Wilkie, D. S, and B. Curran. 1993. "Historical Trends in Forager and Farmer Exchange in the Ituri Rain Forest of Northeastern Zaire." *Human ecology* 21(4):389–417.

Yasuoka, H. 2006. "The Sustainability of Duiker (Cephalophus Spp.) Hunting for the Baka Hunter-gatherers in Southeastern Cameroon." *African Study Monographs*, supplement 33:95–120.

12

Forest Conservation and Indigenous Peoples in The Congo Basin: New Trends toward Reconciliation between Global Issues and Local Interest

Mitsuo Ichikawa

Introduction

Contemporary hunter–gatherers do not live in an identical situation to Paleolithic humans. Although some aspects of their lifestyle can provide clues to understanding past human societies (Ichikawa 2004), the foraging peoples of the Congo Basin are modern humans facing modern human problems. Poverty, sickness, and environmental destruction are having direct and substantial impacts on their livelihood and culture. The effects, both regional and global, of environmental destruction in particular gained world notoriety at the 1992 Rio Summit in Brazil. Following the summit, the tropical rainforests, and the immense biodiversity they are home to, became a central focus of conservationist movements.

In central Africa, conservationist efforts have led to the creation of numerous reserves to protect the region's floral and faunal diversity. Conservation organizations such as the World Wildlife Fund (WWF), the Wildlife Conservation Society (WCS), and other large international nongovernmental organizations (NGOs) have been actively promoting such projects. In Cameroon, the Lobeke area in the Eastern Province was declared a national park in 1999, following a recommendation from WWF. In the same year, a protected area of tri-national parks was formed, comprising Lobeke in Cameroon, the Dzanga-Sangha Forest Reserve in the Central African Republic, and the

Nouabalé-Ndoki National Park in the Republic of Congo. To the west of Lobeke National Park, Cameroon and Gabon are working on the Tridom (Tri-National Dja-Odzala-Minkebe) project to create a tri-national "interzone" bordered by the Minkebe, Boumba-Bek, Nki, and Odzala National Parks and the Dja Wildlife Reserve. Projects such as these have created a network of protected areas over the last fifteen years. Areas protected for wildlife conservation in Cameroon now cover more than seven million hectares, 15 percent of the nation's territory (Topa et al. 2009).

However, areas with rich biological diversity often overlap with the areas inhabited by indigenous forest people. The expansion of protected areas may eventually deprive indigenous forest people of their traditional means of livelihood. Many inhabitants of the central African forests now face restricted access to forest resources that they have been routinely using for many centuries. This poses a serious dilemma for indigenous hunter–gatherers, who rely almost exclusively on forest resources for their livelihood. There are now questions regarding how the global issue of forest conservation can be harmonized with the needs of local people, that is, how to secure a means of livelihood for indigenous peoples while maintaining the rich biological diversity of forest ecosystems. In this paper, I will discuss the possibilities for reconciliation between these seemingly opposing interests.

Covering an area of 150–170 million hectares, the Congo Basin is the second-largest stand of tropical forest in the world, after the Amazon. The forests in this region attract global attention for a number of reasons. First, timber is a major export product and is expected to generate substantial national revenue. Second, the forests are valued for their biological diversity, which is the greatest in Africa in terms of mammal and bird species. The large floral diversity is also highly valued (Debroux et al. 2007). The forests contain rare and endangered species, including lowland and mountain gorillas, chimpanzees and bonobos, okapis, leopards, forest elephants, and others. In consideration of such rich biological diversity, conservationists have set aside eleven areas in the region for nature conservation and sustainable management of the natural environment (OFAC 2008). Third, there is increasing interest in global environmental services, and particularly in the forest as a means of carbon sequestration for the purpose of mitigating air pollution and climate change. Tropical forests in the Congo Basin are estimated to store 30–40 billion tons of carbon, or 200–300 tons per hectare of forest area (Debroux et al. 2007). Therefore, there is a global incentive to conserve the forests in this region.

International support has gained momentum for both conservationist and indigenous rights movements. Conservationists have benefited from the influence of *The Economics of Climate Change,* a 2006 publication by the British government also known as *The Stern Review,* which reports on climate change and its effects. According to the *review*, "the overall costs and risks of climate change will be . . . losing 5% global GDP . . . [and] the damage . . . could rise to 20%. . . . In contrast, the costs of . . . reducing greenhouse gas emissions can be limited to around 1% of global GDP each year." Therefore, action to reduce CO_2 emissions is both necessary and economically viable. There are, however, almost sixty million Africans who depend daily on the forest as a source for fuels, food, medicine, construction materials, tools and ornaments, and other cultural materials (Ichikawa 1992, 2009). The access of indigenous peoples, in particular, to forest resources has garnered support from the *UN Declaration for the Rights of Indigenous Peoples,* issued in 2007. According to the declaration, "indigenous peoples have the right to the full enjoyment . . . of all human rights . . . free from any kind of discrimination . . . to freely pursue their economic, social and cultural development." Although this declaration is not legally binding, it has had a large impact on recent indigenous social and political movements in central Africa. The very difficult and pressing question we now face is whether and how it may be possible reconcile forest conservation with the welfare of indigenous peoples.

Global Forest Programs and Local Issues: Emissions

The Stern Review (2006) advised that to reduce global carbon dioxide (CO_2) emissions cheaply and quickly, further deforestation must be prevented. According to the United Nations Environmental Program (UNEP), tropical rainforests store 283 gigatons of carbon in their biomass (cited by Debroux 2007). The combined effects of deforestation and forest degradation are estimated by the World Bank to constitute the second-largest human-induced contribution to global warming and are responsible for approximately 20 percent of global CO_2 emissions, around 1.6 billion tons (FAO 2007). Deforestation is the main source of CO_2 emissions in many developing countries, with large amounts generated in Africa, where more than half of the annual destruction of the world's forests (over 4 million ha out of a global total of 7.3 million ha) has occurred over the 2000–2005 period (FAO 2007). Actions to make it more economically viable for these countries to preserve their forests rather than destroying them would therefore reduce CO_2 emissions.

For this reason, the World Bank and related agencies have been eager to initiate a program that can assist countries to reduce CO_2 emissions by protecting forests and the people who benefit from them. The Forest Carbon Partnership Facility (FCPF) has thus been initiated with the aim of reducing deforestation and forest degradation by compensating developing countries for the reduction of CO_2 emissions. The FCPF, approved by the World Bank's board of executive directors in September 2007, became functionally operational on June 25, 2008. The initial fourteen participating tropical and subtropical countries, which included the Democratic Republic of Congo (DRC), Central African Republic (CAR), and Republic of Congo, are expected to receive grant support as they build their capacity for reduced emission from deforestation and degradation (REDD). By April 2009, the number of countries participating in the FCPF had increased to thirty-seven, including most of the central African countries (Kitanishi 2010).

Under this arrangement, it becomes economically viable for the developing countries, as well as for the global community, to preserve the forests. It is estimated that, on a per ton of carbon basis, these timber-producing countries can earn only \$1–\$5 (all dollar amounts in US currency) from harvesting the trees, whereas they can earn around \$20 (the market price of carbon credits in Europe) for protecting them. *The Stern Review* estimated that each ton of CO_2 emitted could cause \$85 of damage to the global economy (Olivier 2008).

For example, the DRC (formerly Zaire), one of the pilot countries participating in the FCPF, will receive grant support for REDD. Annual deforestation in the DRC is estimated to be 580,000 ha (Debroux et al. 2007), and each hectare of rain forest contains 200–300 tons of carbon. If 20 percent of the deforestation is prevented, and the carbon credit price is paid at \$4/ton for forest preservation, then the potential remuneration from 'avoided deforestation' would be \$85–100 million per year.

However, the arrangement has some problems. The conservation of forests may bring enormous economic benefits, but researchers Ernsting and Rughani have warned that (1) without adequate consultation or prior strengthening of community land tenure rights and development of a forest law enforcement capacity within participating countries, FCPF-REDD may not provide equitable benefits for poor residents (particularly forest-dependent indigenous peoples); and (2) without proper planning and implementation, FCPF-REDD risks exacerbating conflicts between conservation groups and forest-dependent populations. FCPF-REDD could eventually lead to the expropriation of indigenous people and violation of their customary rights (2007).

There are strong reasons for such concerns. Logging was once expected to contribute to the alleviation of poverty in rural forest areas. However, the industry has developed a bad reputation in central Africa and has been criticized for contributing only marginally to the welfare of local people (Burnham 2000; World Bank 2005). Attempting to improve the situation, central African countries reformed the forestry sector and adopted the policy that the central government is to distribute a part of the revenue from logging concessions to local decentralized entities. The government of Cameroon, for example, distributes 40 percent of the logging concession fee to the provincial government and 10 percent to the local community (government of Cameroon 1994). However, most of the money distributed in this way has reportedly been spent with no concern for the welfare of the local people and with no account of the expenditures (Debroux et al. 2007).

Because such outcomes were to some extent expected, some countries have requested that the logging companies provide local communities with direct, in-kind contributions. The logging companies are obliged to provide the local people with welfare services (which would otherwise be provided by the state or provincial government). When a logging concession is granted, the logging company must pay a concession fee and submit a sustainable management plan (the company has five years to produce the plan). Additionally, the company must make an agreement with the local people, known as a *Cahier des Charges* (note of obligations—a social contract agreement with local communities), in which in-kind contributions to the local community, such as building a dispensary, school, bridge, or other infrastructure, are promised. However, these promises are often not kept for various reasons. For example, workers may not be available for the required construction work because of low wages being offered by the logging company. Bribes may also be given to village chiefs in order to avoid having to follow through on project commitments. In other cases, the promise may simply be ignored, as the local people have no means to protest directly against the logging company (Inspection Panel 2007).

The situation is even worse in areas where indigenous Pygmy people live alongside the Bantu-speaking farmers. In the DRC, the new 2002 Forest Code and 2005 Presidential Decree call for consultation with the local community prior to granting a new logging concession, but often only the Bantu farmers are consulted, and no meaningful participation by the indigenous Pygmy people is invited. The *Cahier des Charges* is likewise only negotiated between the logging company and the Bantu farmers (Inspection Panel 2007). The logging companies have been reported to build roads penetrating Pygmy settlements and to cut trees even in burial sites and parts of the forest considered sacred without consulting with Pygmy groups or compensating them for the loss of forest resources (Inspection Panel 2007). When the Pygmy people have complained to the logging companies about these acts, the companies have advised them to travel to Kinshasa and appeal to the administration (as stipulated in the Forest Code of 2002); however, they have no means to do so.

Integrating Global and Local Concerns: Nontimber Forest Products

The use of nontimber forest products (NTFPs) provides an alternative method for harmonizing forest conservation with the welfare of local people. The use of NTFPs, which are extracted from the living trees and animals that depend on the plants, is compatible with carbon sequestration and other environmental services, insofar as it does not result in deterioration of the forest ecosystem. This is a method of using the "interest" (i.e., fruit) without damaging the "capital" (i.e., trees). It will enable the integration of global interests (REDD, reducing emission from deforestation and degradation) with local welfare (exploiting the economic value of NTFPs).

World Bank experts have estimated the economic value of NTFPs produced by the forests of the DRC (see table 12.1). According to their report, the market value from timber production, formal and informal combined, is estimated at US $160 million per annum. By contrast, the total economic value of the resources used by local people, such as firewood, bushmeat, honey,

Table 12.1. DRC's estimated economic value of forest goods and services in US dollars.*

Formal timber	Approximately 60 million
Informal timber	Approximately 100 million
Fuel wood	Tentatively estimated over 1 billion
Bushmeat	Tentatively estimated over 1 billion
Other foods	No figures available
Medicines	No figures available
Materials, implements	No figures available
Watershed protection	Tentatively estimated 0.1–1 billion
Ecotourism	Marginal
Carbon	Zero
Option, existence values	Approximately 18 million
Cultural and political	No figures available

*Market value, replacement value, or willingness to pay.
Source: Data from Debroux et al., 2007.

medicines, and other nontimber forest products, is more important by far, amounting to well over US $2 billion per annum. The value of other social and cultural aspects of the forest is practically immeasurable. Ecotourism and carbon sequestration may further add substantial value to these forests in the near future.

In Cameroon, following the economic crisis of the late 1980s and early 1990s, large-scale logging activities by foreign companies have been accelerating (Ichikawa 2006). The nation has become one of the world's largest timber-exporting countries, after New Guinea, Gabon, and Malaysia. Timber exports from Cameroon account for as much as one-tenth of annual global production. After a revision of the Forest Law (1994), the logging area (both actual and planned) expanded to 76 percent of the country's total forest area by 1999, as compared with a mere 8 percent in 1959 (Bikie et al. 2000). Although slightly reduced in recent years, logging still comprises one of the major export industries of Cameroon. The total value of Cameroon's forest products as estimated from the available data is about $580 million. Of this, however, $120 million derived from products other than timber. The value of fuel wood is estimated at $130 million. If bushmeat, fruit, medicinal plants and the value of biodiversity conservation and environmental services are added, the total value of nontimber forest products will well exceed that of timber products (Topa et al. 2009). In the South and Central Provinces, the annual revenue from *Irvingia* nuts (wild nuts used for cooking) for individual producers is estimated at 213,206 CFA (about 325 €), or approximately 5,000 tons of the nuts estimated at 4 billion CFA for the Provinces (Awono et al. 2009).

The World Bank, the Food and Agricultural Organization (FAO), and other international development organizations had until recently been promoting sustainable logging operations. The improvement of infrastructure and forest-related legal reforms (establishing or revising the forest law) for the development of the logging industry were believed to lead toward poverty reduction in rural areas. It has become apparent, however, that logging is likely to bring wealth only to a small number of people and will not aid in the alleviation of poverty in local communities (World Bank 2005). It may even risk further impoverishing the local people by depriving them of their means of livelihood. Against such a background, the World Bank has changed its forest policy to pay more attention to small-scale forest industries in addition to the continued promotion of large-scale logging (Debroux et al. 2007). Small-scale forest industries focus on nontimber forest products, which are now attracting increasing attention as a means of integrating the global issue of forest conservation with the welfare of local communities.

To summarize the issues:

1) Globally, deforestation accounts for one-fifth of CO_2 emissions; it is the second largest source of emission after the energy sector.
2) Developing countries currently receive US $1–5/ton per carbon base, whereas each ton emitted could cause damage worth US $85 to the global economy
3) Therefore, forest conservation is an effective means for reducing CO_2 emissions.
4) Paying countries for forest conservation may eventually lead to the expropriation of local people and violations of their customary rights. Underprivileged indigenous peoples are particularly at risk.
5) NTFPs may produce more economic value than logged timber. Therefore, the exploitation of NTFPs may be more effective for poverty alleviation (tourism and environmental services are expected to add even more value in the future).
6) An effective solution for the two problems (forest conservation and securing the livelihood of local people) together may be to combine forest conservation with the commercialization of NTFPs.

Indigenous Peoples in the Congo Basin

The term "indigenous people" has been used without a clear definition in this article. For a more meaningful discussion of indigenous access to forest resources to take place such a definition is of great significance. This is particularly true in the Congo Basin where the distinction between those who are indigenous and those who are not is not at all times obvious.

For implementation of their projects, the operational policy (OP) and operational directives (OD) of the World Bank require special safeguard procedures for indigenous peoples. According to OD4.20 (later revised as OP 4.10), indigenous peoples have:

- A close attachment to ancestral territories and to the natural resources in these areas.
- Self-identification and identification by others as members of a distinct cultural group.
- An indigenous language, which is often different from the national language.
- The presence of customary social and political institutions.
- A primarily subsistence-oriented production system.

In the New World, these criteria suffice in most cases because it is usually clear who are the indigenous people within a particular country. However, these criteria do not fit the African situation, in which many dominant ethnic groups also meet these criteria. In the African context, the African Commission of Human and Peoples' Rights stated that

> almost all African states host a rich variety of different ethnic groups.... Basically all of these groups are indigenous to Africa. However, some are in a structurally subordinate position to the dominating groups and the state leading to marginalization and discrimination. It is this situation, which the indigenous concept in its modern analytic form, and the international legal framework attached to it, addresses. (ACHPR and IWGIA 2006)

In this context, the Pygmy people discussed in this article, are indigenous people in the central African forests because they are marginalized, discriminated against, and placed in a structurally subordinate position.

The indigenous forest people in central Africa (known collectively as Pygmies) are known by different names in different areas, such as Mbuti and Efe in the Ituri Forest of the DRC, Batwa in the Great Lake region, Batswa in the equatorial region of DRC, Baka in Cameroon, and Aka in the Central African Republic and Congo-Brazzaville. Their number is estimated to be between 200,000 and 1,000,000 in total, although no reliable census data are available. Most of them live in small residential groups of 20–60 people and lead at least a seminomadic life in the forest. They have no traditional chief or other central authority and enjoy an egalitarian social life characterized by extensive sharing. They subsist mainly by hunting, especially for forest duikers, and collecting a variety of forest plants for food, medicine, and material culture.

Most of these Pygmy groups maintain an exchange relationship with neighboring agricultural groups. The relationship probably dates back centuries, or even millennia (Bahuchet and Guillaum 1983). Through these long-standing relationships, Pygmy people have specialized as hunter–gatherers and as such are loosely connected with other parts of a wider society. However, these relationships have not developed on equal terms. As central African people have been gradually incorporated into the modern state system during the course of colonization and independence, only a limited number of opportunities have been available for the indigenous Pygmy people to participate in the activities of the broader world. Few Pygmy people are employed as civil servants, and political representation in the wider society has been largely monopolized by the dominant agriculturalist villagers. Moreover, Pygmies have had no legal rights to the forest land in which they live.

Under the laws of central African countries, forest land belongs officially to the state, but there are also customary rights by which local people regulate the use of forest land. Customary rights are acknowledged on the principle of first come, first served: "under almost all African customary systems, occupancy is generally the key to 'ownership' and land is allocated by those claiming prior occupancy through lineages and clans" (Colchester 2001). The areas where Pygmy people coexist with other groups today are also used by agricultural villagers who use the forest for hunting, fishing, and other procurement activities. In these areas, customary rights to the forests usually overlap. The perspective of the agricultural villagers is that the clan or lineage of the "patron" villagers "owns" the forest and its inhabitants. Under such patronage, the Pygmy people are "allowed" to use the forest. While this interpretation may not be accepted by the Pygmy people, the opinion of the more powerful villagers is likely to be respected in the wider societal context, and the rights of the Pygmy people are often neglected. Even if they are acknowledged as the first inhabitants of the area, "throughout the Congo Basin, governments have ignored politically weak hunter-gatherer groups because they do not make investments in land generally recognized by authorities, such as clearing, farming, or mining" (IRIN 2006). Hunting and gathering is not considered a legitimate or sustainable use of land compared with the agricultural practices of the Bantu people. This commonly results in discrimination against nomadic Pygmy peoples. While the constitutions of central African countries guarantee fundamental rights and equality for all citizens, including cultural minority peoples (see, for example, the Constitution of DRC), the reality is quite different. It is clear that cultural minorities are actually underprivileged (CAMV et al. 2006).

The first International Forum of Indigenous People in Central Africa (FIPAC), which considered the plight of the Pygmy peoples in the Congo Basin, was held at Impfondo, on the right bank of the Ubangi River, Congo-Brazzaville, on May 12, 2007. It was the first attempt on the part of the Pygmy people to promote their common identity and appeal for rights to the forests through a large conference. The delegates came from the DRC, Cameroon, Gabon, and other countries in central Africa. Some delegates traveled from South Kivu in the DRC, almost 2,000 km from Impfondo. They appealed to the world regarding the difficulties they faced and claimed that indigenous people from the forests of central Africa represented "the third world of the third world" (James 2007). Following this conference, the networking of Pygmy associations has accelerated considerably, and many NGOs have been formed to represent Pygmy people in various parts of central Africa. The African network is also linked to global networks of indigenous peoples, within which they have begun to develop their own identity. The second FIPAC took place from March 15–18, 2011, again at Impfondo, with more than 300 participants representing various indigenous peoples in central Africa. The delegates discussed the rights of indigenous people and the dynamics of biodiversity conservation in the Congo Basin (CBFP 2011). These movements are backed by the UN Declaration on the Rights of Indigenous Peoples. Whether such a movement will eventually achieve success depends on both state governments and international support.

Establishing Customary Rights to Forest Resources

The Need for Customary Rights

World Bank experts stated in their report on the DRC that "[70 percent] of 60 million people in DRC . . . derive their meager income from small . . . non-timber forest-based businesses" (cited in Debroux et al. 2007). To maintain their livelihood and sources of income, it is paramount that indigenous forest peoples be able to secure the customary use of NTFPs. However, "the current level of harvesting for several forest products is probably not sustainable. Wildlife surveys seem to confirm that the current scale and intensity of hunting outpace renewal for many species" (Debroux et al. 2007). This constitutes the primary reason why the new Forest Code of DRC restricts commercial exploitation of forest resources primarily for this reason. The commercial exploitation of forest resources, and hunting in particular, may lead to overexploitation, deterioration of the resource base, and depletion of rare and endangered animals such as great apes, forest elephants, okapis, and leopards.

Overexploitation could potentially be avoided through effective resource management for sustainable use by the forest people. If indigenous Pygmy groups were to be given exclusive rights to the forest resources, they could initially manage the forests with external assistance. The indigenous forest people have rich knowledge of forest wildlife, as has been shown in previous studies (Bahuchet 1985; Ichikawa 1992; Terashima and Ichikawa 2003). Their expertise in forest ecology dynamics and the availability of forest resources would be invaluable in creating and maintaining a sustainable management system.

Until recently, NTFPs were used based on *de facto* rights of the forest-dwelling Pygmy people, but there is no legal basis for preventing outsiders from encroaching and exploiting forest areas in which indigenous people have been living for many centuries. Because of such deficiencies in the law, there has often been an influx of migrant hunters and traders into the forest. In some areas of the Congo Basin, this has resulted in "empty forest syndrome" (Redford 1992), where few animals are found even if an area's forest landscape remains intact due to overexploitation. Encroachment by outsiders has had a severe and negative impact on the indigenous Pygmy people's ability to manage forest resources on a long-term sustainable basis. The most important task for policy makers, therefore, is to establish firm, customary rights to the forests on which Pygmy people.

Customary rights are the "rights which are acquired by custom. They differ from prescriptive rights in that the former are local usages, belonging to all the inhabitants of a particular place . . . . [whereas] the latter are rights of individuals, independent of the place of their residence" (The Free Dictionary by FAREX n.d.). Thus, such rights to the forests derive from the indigenous people's centuries-long presence in—and use of—the area. Their lifestyle is sustained through their interactive and interdependent relationship with the forest environment, which their daily activities reproduce and reinforce. This interdependence should be considered the basis for granting customary rights to the indigenous Pygmy people of the Congo Basin. It is important to determine through further study exactly how the Pygmy people depend on the forest and its resources and how their activities influence the forest environment.

Forest Culture

Observations of a typical Pygmy forest camp reveal that forest products play an important role to daily life, serving as food, medicine, construction materials, decoration, and other materials (figure 12.1). The small huts are made entirely of wild plant materials, and the bulk of food consumed at the forest camp is obtained from the forest. A variety of cultural materials are made from forest plants, such as hunting and gathering tools, household utensils, ornaments, and so on. During the initial stages of our studies in the Ituri Forest, we documented impressive use of forest resources and collected many plant specimens along with their vernacular names, information on their uses by Pygmy people, and other ethnobotanical information (Ichikawa 1992; Terashima and Ichikawa 2003).

The tropical rain forest is often described as a cornucopia of food (Meyers 1992), the horn of abundance in the Greek mythology. The Mbuti people in the Ituri Rainforest use more than 100 species for food, including the oil-rich nuts of *Irvingia* spp., which are widely used and traded in the region. Energy-rich starchy food, including *Canarium* fruit as well as wild yams, sweet and sour fruits of *Landolphia*, *Annonidium*, and *Aframomum*, are frequently eaten by local inhabitants and are sold at local markets (Ichikawa 1992).

The tropical rainforest can be considered analogous to a chemical factory (Meyers 1992). Many plants in tropical rainforests accumulate various toxic secondary compounds as an adaptation against predation by herbivores. Many of these, if administered properly, can be used as medicine for curing diseases or as poisons for hunting and fishing. We recorded more than 200 species of such medicinal plants in the Ituri Forest (Terashima and Ichikawa 2003).

Over 300 species of plants in the Ituri Forest have documented use in indigenous cultural materials. Traditional cultural materials are simple, consisting of fewer than 100 items in total, including materials for construction, transportation, hunting, gathering, cooking, and decoration. Almost 80 percent of cultural materials are made at least partially from plant materials that are abundant in the forest (Tanno 1981).

In addition to the direct use of plants, hundreds of plants are used in indirect ways, for example, as sources of nectar and bait for use in hunting, fishing, and collecting animals. Many of the taller trees have no direct use, but they are home to the beehives that provide honey, one of the most highly sought after delicacies in the rainforest. The vegetation also feeds insects and wild animals, which the forest people depend on for much of their meat and protein.

The Mbuti believe that the forest is imbued with a supernatural being called *Apakumandura* ("Father of Forest"), who controls all the life in the forest. Certain tree species, such as *Uvariopsis congolana* (Annonaceae), called *akobisi* by the Mbuti, are ritually protected because they are considered a visible agent of this spiritual being. It is strictly forbidden to cut or damage these trees (Ichikawa 1996).

Some forest products are important sources of cash income in the Congo Basin. These include food plants such as *Irvingia* nuts and *Canarium* fruit and plant materials such as palm lianas and shoots for making baskets and mats. In addition to these, the most important forest product, bushmeat, is sold at local markets and exchanged with traders for agricultural products, clothes, and other manufactured goods. These NTFPs have the potential for commercialization and hence for increasing the income of the forest people.

While many other forest plants have little or no commercial value, they are indispensable to the maintenance of daily life and traditional culture. For example, the consumption of some wild animals is thought to cause bad diseases in certain people. Parents with small children avoid eating these "bad animals," which have "evil spirits," and may cause diseases in their dependent children (Ichikawa 1987; Lewis 2008). While these animals have a negative value, they also provide the forest people with culturally constructed reasons for diseases and other misfortunes (Ichikawa 2007).

In these ways, the forest peoples utilize the diverse potential of the forest. The forest plants and animals also provide the people with the basis of their cultural identity. Their life and culture cannot be maintained without the

Figure 12.1. An Mbuti forest camp.

forest in its entirety, and thus the conservation of the forest ecosystem and the culture of forest people are inseparable. The destruction of the forest would result in the deterioration of a culture that is heavily dependent on, and in very significant ways a part of, the Congo Basin rainforest.

The Contribution of Forest People to Biological Diversity

Although humans have a variety of impacts on the forest environment, here we discuss two examples of positive impacts that are considered to enhance the forest's natural biodiversity. One is the soil-nutrients concentration effect and the other is the food-plant concentration effect, each of which is a product of the indigenous people's habitation of the rainforest.

Through ethnobotanical research in the Ituri Forest, we observed the importance to the Mbuti of "sun trees," those that favor open or disturbed habitats with abundant sunlight (Ichikawa 1996, 2001). An example is *Canarium scwenfurthii,* which produces flavorful, olive-like fruit and is often found in locations deep in the forest said to have been inhabited at one time by humans as a village or camp site (Ichikawa 2001). These trees could only have grown in gaps or clearings where sufficient sunlight once reached the ground. Although scattered gaps are often created by storms and other natural causes, some gaps in the forest are the result of human activity. When forest people open a new camp, they clear the small to medium-sized trees, allowing more sunlight to reach the ground. Hence, human habitation improves the light conditions in the certain areas of the forest. In these areas, seedlings from discarded food plants also germinate under the favorable light conditions, resulting in high food concentration.

The soil improvement effect results from the large quantities of minerals and organic matter that are concentrated in forest peoples' camps. Food, fuels, and other resources collected from the forest and consumed at the camp are subsequently deposited in the soil. Although not all of these are absorbed by the soil, the nitrogen supplied from the food alone in a camp of fifty people is estimated to be equivalent to 200 kg of ammonium sulfate (Ichikawa 2001). Therefore, forest people concentrate the otherwise thinly distributed soil nutrients in localized patches around a camp site.

Tropical rainforests are characterized by poor soils, and the soil nutrients are quickly absorbed by plants and stored in the plant body. When plants are eaten by humans and animals, the nutrients are eventually returned to the soils in a concentrated form. In this way, human activity contributes to the circulation of materials in the forest, as long as the forest remains a closed system. However, once this circulating system ceases to function, deterioration of the forest ecosystem results.

Another example of positive human intervention in the forest environment, the food-plant concentration effect, is more closely related to the legitimacy of indigenous peoples' customary rights to the forest. By examining the long history of human habitation and coexistence with the forests, a legitimate basis for the rights of people to the use of the forests can be established. This history has been addressed by the "Wild Yam Question," a hypothesis proposed by Bailey et al. (1989) and Headland (1987) in the late 1980s. According to this hypothesis, there are not enough wild food resources in the forest to support human subsistence throughout the year, particularly in the lean period of the dry season. Various researchers have considered this problem, including Bahuchet at al. (1991) and Sato (2001), but few have ever examined the problem based on quantitative data obtained from observations of actual foraging in the forest.

Yasuoka (2006) collected quantitative data on the diet and subsistence activities of the Baka Pygmies. He showed that during the *molongo* (a long-term foraging expedition), the Baka depended exclusively on wild food. Their food consumption was adequate during this period and even exceeded amounts consumed in the village camp where agricultural food is available. The most important foods during the *molongo* were two species of wild annual yams (yams with annual stems), which provided more than 60 percent of the total energy intake.

Yasuoka's (2009a and b) surveys of the vegetation and distribution of yam species in the forest led to the consideration of the forest landscapes within a historical context. He showed that the two annual yam species are found in forest locations with an open canopy (gaps), and that while gaps are omnipresent in the forest, annual yams were found only in limited gaps. Where they were found, they were concentrated and copious, providing a large quantity of food, while in other similarly open areas yam stems were virtually absent (Yasuoka 2009b; Ichikawa 2012).

The meaning of this uneven distribution of yam patches has been considered in our study. An old map published a hundred years ago by the German cartographer Max Moisel (1910) provided us with a clue to understanding the uneven distribution. The old map shows several old village sites, which are now located in the Nki National Park. Yasuoka also located other old village sites through interviews with the Baka elders. Interestingly, locations with high abundance of yams were located close to the old village sites (Yasuoka 2009a). Presumably, there is some link between the distribution of annual yams and past human habitation. The formation of large clearings near settlements is likely to have created a favorable environment for the light-demanding annual yams to grow. A kind of para-cultivation, which, as described by Dounias (2001), involves replanting the yam stems in the same places; or more advanced proto-cultivation, which involves more elaborate care and shifting of planting sites, may have also have contributed to the creation of locations where yams are abundant. The evidence for human intervention in the formation of yam patches in the central African forests is compelling.

We have investigated the impacts of human activity on the forest environment, on the growing conditions of edible plants, and on various types of forest vegetation. We were particularly interested in the distribution of edible plants and the impact of past human activity on vegetation. Studies of the history of shifting cultivators have revealed patches of old secondary forest derived from past human settlements, which are located on land now designated as national parks (Shikata 2006). If the positive impacts of human activity on the forest environment are supported by further studies, we will be able to confirm that indigenous forest people do not merely depend on the forest and its products, they also contribute to forest regeneration (see also Bailey et al. 1992).

Furthermore, we suspect that past human activity had a role in the formation of forest vegetation in this region, as suggested by Van Gemerden et al. (2003). These researchers concluded that the "present-day tree species composition of . . . [the] Central African rain forest still echoes historical large-scale disturbances, most probably caused by human land use . . . which is relevant for biodiversity conservation planning." Recognition of human involvement in the formation of such spaces (e.g., clearing forest land for agricultural fields, villages, and camp sites) has the potential to change perspectives on the conservation and use of the forests in this region considerably. This has an important implication for the legitimacy of the customary rights to the forests because it suggests that human intervention has a hand in forest formation processes.

To summarize, even as human life and culture depend heavily on diverse forest resources, the forest ecosystem and its resource base are, in turn, maintained by human habitation and associated activities. In other words, human activity and the forest are interdependent. The present forest landscape can be, at least in part, regarded as a historical product, created through long-term interactions with indigenous forest people. This perspective is indispensable to legitimizing forest people's customary rights to the forests.

The Impact of Market Forces: Bushmeat and Logging

A system of interdependence between humans and forests has been maintained on the condition that forest resources circulate in a closed system. But this is no longer the case. When forest resources are transported to the outside world for commercial purposes (in an open system), this circulation within the forest ecosystem can no longer be maintained. The resource base will deteriorate if no sustainable management system can be established.

One of the problems caused by the incorporation of local subsistence economies into the outside market system is the bushmeat crisis, which is the overexploitation of animal resources. Wild animals have long been hunted for subsistence purposes by the forest people in central Africa. As long as animals are hunted only for subsistence purposes, there is no problem (Yasuoka 2006). Animals are exploited only for meeting the limited demand of a small number of local residents. The materials in the animals are digested, excreted, and absorbed again by the soil in the vicinity of their settlements. However, once commercial hunting is introduced, large amounts of bushmeat are traded and transported outside the forest to distant towns. Bushmeat is important in these towns as a cheap protein source. It is also valued culturally by urban residents who have lost contact with the wilderness of the forest as a source of wild power, which they believe cannot be obtained from the meat of domesticated animals. Commercial hunting has been on the rise since the introduction of meat trading. It has now become a threat to certain animals and particularly to the rare and endangered species in the region. It may ultimately lead to the "empty forest syndrome" (Redford 1992), in which animals are few or absent, even if the forest itself remains intact. Furthermore, the loss of animal populations will eventually lead to the deterioration of the forest ecosystem because animals play an important role in seed dispersal and forest regeneration (Wilkie et al. 2011).

It is now necessary to establish a means for the conservation and sustainable management of animal resources. Customary rights to forest resources will provide the forest people with such a means. If the forest people have exclusive rights to the forest resources, it would prevent outsiders from encroaching. From a long-term perspective, this would make resource management much easier.

Another cause of environmental deterioration is large-scale logging. Particularly since the late 1980s and early 1990s, when central African countries were suffering from an economic crisis and conceded to the structural adjustment policy recommended by the IMF, logging operations have been rapidly accelerating. Large quantities of minerals and organic matter that are contained in wood have been transported out of the forest. The capacity of forests to capture water has also been considerably reduced; with neither the protection of the canopy nor root systems for anchorage, the soil in logged areas is quickly eroded into the rivers. The circulation system in the forest can no longer be maintained. These extensive logging operations have now reached the interior forest, even to the villages of the indigenous Pygmy people in the remotest parts of the forest. Loss of forest is, therefore, endangering the life of forest peoples as well as causing global problems such as greenhouse gas emissions and loss of biodiversity.

Concluding Remarks

The conservation of tropical rainforests involves both global and local issues. The global issues relate to the environmental services provided by forests, particularly the role of carbon sequestration, for which payment to forested countries within the REDD framework is now being discussed. The local issue is provision of security for the lifestyle and culture of indigenous forest peoples. Such security could potentially be made possible by income generated through the commercialization of NTFPs. In addition, the proceeds would pay for indigenous people's increased access to medical care, education, and other modernized services. To create a mutually satisfying solution, a social system in which the indigenous people themselves are responsible for the conservation and sustainable use of the forest is essential.

A number of steps will be required for the establishment of such a system. First, a fair and equitable share of the benefits from REDD must be conveyed to the local people. The appropriate use of the revenue generated from the forests is integral to the successful implementation of REDD. Second, it is necessary to secure customary rights to the forests on which an effective management system can be established for conservation and sustainable use of the resources. As forecast in the "tragedy of the commons" theory, an open-access system where outsiders freely encroach on the forests will rapidly bring about the depletion of forest resources.

Various trials of such systems have been attempted in central Africa. Forest-living people have attempted to manage forest resources by themselves through the establishment of "community forest," with the approval of the government authorities. In Cameroon, following the 1994 forestry reforms, the first two community forests were established in 1997. By 2008, agreements to create 135 community forests on 621,246 hectares were finalized, and an additional 700,000 hectares were under consideration for future community forest status (Topa et al. 2009). The indigenous BaGyeli people in southern Cameroon and the Baka communities in the East and South Provinces have also obtained their own community forests with the support of international NGOs.

Although these community forests in Cameroon are not primarily designated for the extraction of NTFPs, community forestry for NTFPs has been successful elsewhere, especially in South America, where the extraction of NTFPs has functioned for more than twenty years (see articles on "extractivism" in Hladik et al., 1993). A similar system is possible in Africa and may be a better option because community forestry for timber production has not been successful in attaining the intended goals of providing welfare to local communities. This lack of success has been attributed to interference on the part of the local elites who are protecting their own vested interests (Topa et al. 2009).

To strengthen the management capacity of forest people, participatory mapping has been undertaken by indigenous Pygmy people with the support of international NGOs (for examples, see Lewis 2012; Tchoumba and Nelson 2006). These projects have been cited in appeals to government authorities requesting recognition of the longstanding forest use by the indigenous people and the instatement of their customary rights to the forest resources.

To date, these mapping exercises have mainly been used as a basis for claiming the rights of local people to the forests or for locating illegal logging activities (Keane 2008). But mapping of customary forest use has potential beyond just appealing to the government and other external forces. It will provide the people with a useful means to regulate forest use between individual indigenous groups and enable them to allocate resource uses from a long-term perspective. Through visualizing the spatial and temporal patterns of land and resource use in the form of maps, forest people can obtain clearer images of how they and other groups in the vicinity use the space and resources. They can then regulate their activities so that their actions do not lead to excessive competition and overexploitation.

The forest-living indigenous people are engaged in collecting GPS data as part of this participatory mapping project and have begun to use the results for their own benefit (Lewis and Nkuintchua 2012). In this sense, the exercise has the potential to become a new type of citizen's science by combining the collective knowledge of the forest people with modern technology and related expertise. As Rowland (2012) points out, "the next generation of citizen science attempts to make communities active stakeholders in research that affects them, and use their work to push forward policy changes." I hope the time may soon come when the indigenous Pygmy people become expert GPS users and active advocates for their own rights.

References

ACHPR and IWGIA. 2006. *Indigenous Peoples in Africa: The Forgotten Peoples?* African Commission on Human and Peoples' Rights (ACHPR) and International Work Group for Indigenous Affairs (IWGIA), Copenhagen.

Awono, A., D. Manirakiza, and V. Ingram. 2009. "Etude de Base de la Filiere Ndo'o (*Irvingia* spp.)." Region du Centre, Sud et Littoral, Cameroun. l'Union Européenne.

Bahuchet, S. and H. Guillaum. 1983. "Aka-farmer Relationship in Central African Rain Forest." In *Politics and History in Band Societies*, edited by R B. Lee and E. Leacock. Cambridge: Cambridge University Press, 189–211.

Bahuchet, S., D. Mckey, and I. De Garine. 1991. "Wild Yams Revisited: Is Independence from Agriculture Possible for Rain Forest Hunter-gatherers?" *Human Ecology* 19(2):213–43.

Bailey, R. C., G. Head, M. Jenike, B. Owen, R. Rechtman, and E. Zechenter. 1989. "Hunting and Gathering in Tropical Forest: Is It Possible?" *American Anthropologist* 91(1):59–82.

Bailey, R. C., B. Hewlett, and S. Bahuchet. 1992. "Development in the Central African Rainforest: Concern for Forest People." In *Conservation of West and Central African Rainforests*, edited by K. Cleaver, M. Munasinghe, M. Dyson, N. Egli, A. Peuker, and F. Wencélius. The World Bank.

Bauchet, S. 1985. *Les Pygmées Aka et la forêt centrafricaine: ethnologie écologique*, Paris, SELAF.

Bikie, H., J-G. Collombo, L. Djombo, S. Minnemeyer, R. Ngoufo, and S. Nguiffo. 2000. *An Overview of Logging in Cameroon.* Washington DC: World Resources Institute.

Burnham, P., 2000 Whose Forest? Whose Myth? Conceptualisation of Community Forests in Cameroon. In: Abramson. A. & D. Theodossopoulos (eds.), *Land Law and Environment.* Pluto Press, London, 31–58.

CAMV (Centre d'Accompagnement des Autochtones Pygmées et Minoritaires Vulnérables), ARAP (Association Pour le Regroupement et l'Autopromotion des Pygmées), CPAKI/RDC (Collectif pour les Peuples Autochtones au Kivu), PDMAC (Action Pour la Promotion des Droits des Minorités Autochtones en Afrique Centrale), SIPA (Solidarité pour les Initiatives des Peuples Autochtones), UEFA (Union Pour l'Emancipation de la Femme Autochtones), FPP (Forest Peoples Programme). 2006. *Persistent and Pervasive Racial discrimination against Indigenous peoples in the Democratic Republic of Congo.* www.forestpeoples.org/sites/fpp/files/publication/2010/08/drccerdjun06eng.pdf. Accessed July 20, 2012.

CBFP, 2011 *Impfondo hosts the 2nd edition of the International Forum of Indigenous People Central African (FIPAC): Significant progress in the recognition of indigenous people in Central Africa.* www.cbfp.org/news_en/items/FIPAC_2_-_Impfondo_-institutionalisation_FIPAC-E.html. Accessed on July 15, 2012

Colchester, M., ed., 2001. *A Survey of Indigenous Land Tenure: A Report for the Land Tenure Service of the Food and Agriculture Organization.*

Debroux, L., T. Hart, D. Kaimowitz, A. Karsenty, and G. Topa, eds. 2007. *Forests in Post-Conflict Democratic Republic of Congo: Analysis of a Priority Agenda.* A Joint Report by Teams of the World Bank, CIFOR, CIRAD, AWF, CNONGD, CI, GTF, LINAPYCO, SNV, WCS, ICRAF. 1–82.

Dounias, E., 1993 Perception and use of wild yams by the Baka hunter-gatherers in south Cameroon. In (Hladik, C. M., Hladik, A., Linares, O. F., Pagezy, H., Semple, A. and Hadley, M., eds.) *Tropical Forests, People and Food: Biocultural Interactions and Applications to Development*, UNESCO, Paris, 621–32.

———. 2001. The management of wild yam tubers by the Baka Pygmies in southern Cameroon. *African Study Monographs Supplementary Issue* no.26: 135–56.

Ernsting, A. and D. Rughani. 2007. *Reduced Emissions From Deforestation: Can Carbon Trading Save Our Ecosystems?* www.biofuelwatch.org.uk.

FAO. 2007. *Forestry Country Profiles–Legislation. Global Forest Resources Assesment* 2005 (updated) www.fao.org/forestry/site/30816/en/cod/page.jsp.
Government of Cameroon 1994 *Law No. 94–01 of 20 January 1994.*
Government of DRC, 2002 *Loi N°011/2002 du 29 Aout 2002 Portant Code Forestier.*
Headland, T. N. 1987. The wild yam question: How well could independent hunter-gatherers live in a tropical rain forest ecosystem? Human Ecology 15(4):463–91.
———. 1997. Revisionism in ecological anthropology. Current Anthropology, 38(4):605–30.
Hladik, C. M., Hladik, A., Linares, O. F., Pagezy, H., Semple, A. and Hadley, M., eds., 1993 *Tropical Forests, People and Food: Biocultural Interactions and Applications to Development,* UNESCO, Paris.
Ichikawa, M., 1987 Food Restrictions of the Mbuti Pygmies, Eastern Zaire. *African Study Monographs, Supplementary Issue,* no. 6: 97–121
———. 1992 Diversity and Multiplicity in the Forest Use by the African Hunter-gatherers. *Tropics,* vol. 2 (2):107–21.
———. 1996 Co-existence of Man and Nature in the African Rain Forest. Fukui and Ellen (eds.), *Redefining Nature: Ecology. Culture and Domestication,* Berg Publishers: 467–92.
———. 2001. The forest world as circulation system: The impact of Mbuti habitation and subsistence activities on the forest environment," *African Study Monographs, supplementary issue,* no. 26: 157–68.
———. 2004. Benefit of Foresight: From evolutionary interest to global environmental problems. *Before Farming (On-Line version),* 2004(4): Article 7.
———. 2006 Problems in the Conservation of Rainforests in Cameroon. *African Study Monographs, supplementary issue,* no. 33:3–20.
———. 2007 L'évitement alimentaire des viandes d'animaux sauvages chez les chasseurs-cueilleurs d'Afrique centrale. Dounias, E., Motte-Florac, E. et Mesnil, M. (eds.) *Le symbolisme des animaux—L'animal "clé de voûte" dans la tradition orale et les interactions homme-nature. (Colloques et Séminaires-IRD),* Paris.
———. 2009 Forests and Indigenous People in Post-conflict Democratic Republic of Congo. In : Kato, G. A. Uyar (eds.), *Question of Poverty and Development in Conflict and Conflict Resolution,* Afrasian Studies, Ryukoku University, 211–26.
———. 2012 Historical Ecology and Contemporary Problems in Central Afircan Forests.
African Study Monographs, supplementary issue, no. 43:3–14.
Inspection Panel (World Bank), 2007. *Investigation Report on DRC: Transitional Support for Economic Recovery Grant (TSERO) and Emergency Economic and Social Reunification Support Project (EESRSP),* http://siteresources.worldbank.org/EXTINSPECTIONPANEL /Resources /EligibilityReportFinal.pdf
IRIN, 2006 *Minorities under Siege: Pygmies Today in Africa.*
James, J., 2007 A Pygmy Conference in the Rainforest. http://news.bbc.co.uk/2/hi/programmes/from_our_own_correspondent/6646115.stm. Accessed November 15, 2007.
Keane, F., 2008 *GPS Helps Pygmies Defend Forest.* BBC News: 2008/01/30 19:51:32 GMT. http://news.bbc.co.uk/2/hi/africa/7218078.stm
Kitanishi, K., 2010. African Tropical Rain Forests and Globalization. In, Kimura D., and K. Kitanishi, *Ecology of Forest-living Peoples,* Kyoto University Press, 59–76 (in Japanese).

Lewis, J. 2008 Ekila: Blood, Bodies, and Egalitarian Societies. *Journal of the Royal Anthropological Institute (N.S.)* 14, 297–315.

———. 2012 Technological Leap-frogging in the Congo Basin, Pygmies and Global Positioning Systems in Central Africa: What has happened and Where Is It going? *African Study Monographs, Supplementary Issue* No.43:45–59.

Lewis, J. and T. Nkuintchua, 2012 Accessible technologies and FPIC: independent monitoring with forest communities in Cameroon, 151–65. http://pubs.iied.org/pdfs/G03408.pdf. Accessed July 18, 2012.

Meyers, N. 1992 (1985) *The Primary Source: Tropical Forests and Our Future.* W. W. Norton & Company, New York.

OFAC (Observatory for the Forests of Central Africa), 2008. *The Forests of Congo Basin: State of the Forest.* Downloaded from the Website of OFAC. www.observatoire-comifac.net. Accessed in April 2011.

Olivier, R., 2008. *All About Forests and Carbon Trading Story Highlight.*
(www.cnn.com/2008/TECH/02/10/eco.carbon/#cnnSTCText, February 11, 2008).

Redford, K. H., 1992 The Empty Forest. *BioScience*, vol. 42(6):412–22

Rowland, K., 2012 Citizen Science Goes 'Extreme': Researchers Push for Wider Use of Community-generated Data in Science and Policy-making. *NATURE / NEWS*, February 17, 2012. www.nature.com/news/citizen-science-goes-extreme-1.10054.

Sato, H., 2001. The Potential of Edible Wild Yams and Yam-like Plants as a staple Food Resource in the African Tropical Rain Forest. *African Study Monographs, supplementary Issue*, No. 26:123–34.

Shikata, K., 2006. What the forest tells us: The relationship between the shifting cultivators' life and Forest vegetation in the tropical rain forest of southeastern Cameroon. Newsletter of the Japanese Society of Tropical Ecology, no. 62:1–7

Stern, N., 2006 *The Economics of Climate Change: The Stern Review.* Cambridge University Press.

Tanno, T., 1981 Plant Utilization of the Mbuti Pygmies: With Special Reference to Their Material Culture and Use of Wild Vegetable Foods. *African Study Monographs*, vol. 1:1–53.

Tchoumba, B., and J. Nelson, 2006 *Protecting and Encouraging Customary use of Biological Resources by the Baka in the West of the Dja Biosphere Reserve: Contribution to the implementation of Article 10(c) of the Convention on Biological Diversity*. Forest Peoples Program. www.forestpeoples.org/sites/fpp/files/publication/2010/08/cameroon10cjun06eng.pdf. Accessed July 12, 2012).

Terashima, H. and M. Ichikawa, 2003. A Comparative Ethnobotany of the Mbuti and Efe Hunter-gatherers in the Ituri forest, Democratic Republic of Congo. *African Study Monographs* 24(1, 2):1–168

The Free Dictionary by FAREX, N.D. http://legal-dictionary.thefreedictionary.com. Accessed September 12, 2012.

Topa, G., C. Megevand, A. Karsenty, and L. Debroux, 2009 *The Rainforest of Cameroon*, pp. 1–198, Washington, World Bank United Nations, 2007 *United Nations Declaration on the Rights of Indigenous Peoples.* www.un.org/esa/socdev/unpfii/documents/DRIPS_en.pdfN.

Van Gemerden, B. S., M. P. E. Parren, and F. Bongers, 2003 The Pristine Rain Forest? Remnants of Historical Human Impacts on Current Tree Species Composition and Diversity. *Journal of Biogeography* 30, 1381–90.

Wilkie, D., S. Wilkie1, E. L. Bennett, C. A. Peres and A. A. Cunningham, 2011 The Empty Forest Revisited. *Annals of the New York Academy of Sciences,* vol. 1223:120–8.

World Bank, 1991. OD 4.20 (later modified as OP4.10 in 2005) on Indigenous peoples, July 2005.

———. 2005. Bank Management Response to Request for Inspection Panel Review of the Democratic Republic of Congo: Transitional Support for Economic and Social Recovery Operation and Emergency Economic and Social Reunification Support Project, http://siteresources.worldbank.org/EXTINSPECTIONPANEL/Resources/DRCfinalJan15.pdf.

Yasuoka, H., 2006a. The Sustainability of Duiker (*Cephalophus* spp.) Hunting for the Baka Hunter-gatherers in Southeastern Cameroon. *African Study Monographs, supplementary Issue,* no. 33:95–120.

———. 2006b. Long-term Foraging Expedition (*molongo*) among the Baka Hunter-Gatherers in the Northwestern Congo Basin, with Special Reference to the "Wild Yam Question." *Human Ecology* 34(2):275–96.

———. 2009a. Concentrated distribution of wild yam patches: Historical ecology and the subsistence of African rainforest hunter-gatherers. Human Ecology (online version).

———. 2009b. The variety of forest vegetations in southeastern Cameroon, with special reference to the availability of wild yams for the forest hunter-gatherers. *African Study Monographs* 30(2):89–119.

List of Contributors

Dr. Serge Bahuchet
Musée National d'Histoire Naturelle
Paris, France

Dr. Luigi Luca Cavalli-Sforza
Stanford University
Stanford, CA, USA

Dr. Alain Froment
Institut de Recherche Pour le Développement
Musée de l'Homme
Paris, France

Dr. Susanne Fürniss
Centre National de la Recherche Scientifique
Paris, France

Dr. Barry S. Hewlett
Washington State University
Vancouver, WA, USA

Dr. Mitsuo Ichikawa
Kyoto University
Kyoto, Japan

Mr. Christopher Kiahtipes
Southern Methodist University
Dallas, TX, USA

Dr. Jerome Lewis
University College London
London, UK

Dr. Karen Lupo
Southern Methodist University
Dallas, TX, USA

Dr. Robert E. Moïse
Independent scholar
Felton, CA, USA

Mr. Alfred Jean-Paul Ndanga
Director, Centre Universitaire de Recherche et de Documentation en Histoire et Archeologie Centrafricaines
Bangui, Central African Republic

Dr. Stephanie Rupp
City University of New York
New York, NY, USA

Dr. Hiroaki Sato
Hamamatsu University School of Medicine
Hamamatsu, Japan

Dr. Kiyoshi Takeuchi
Independent scholar
Kyoto, Japan

Dr. Paul Verdu
Centre National de la Recherche Scientifique
Paris, France

Index